Building Energy Performance Modelling and Simulation

Building Energy Performance Modelling and Simulation

Guest Editors

Joanna Ferdyn-Grygierek
Krzysztof Grygierek
Agnes Psikuta

MDPI

Basel • Beijing • Wuhan • Barcelona • Belgrade • Novi Sad • Cluj • Manchester

Guest Editors
Joanna Ferdyn-Grygierek
Department of Heating, Ventilation and Dust Removal Technology
Silesian University of Technology
Gliwice
Poland

Krzysztof Grygierek
Department of Mechanics and Bridges
Silesian University of Technology
Gliwice
Poland

Agnes Psikuta
Laboratory for Biomimetic Membranes and Textiles
Empa-Swiss Federal Laboratories for Materials Science and Technology
St. Gallen
Switzerland

Editorial Office
MDPI AG
Grosspeteranlage 5
4052 Basel, Switzerland

This is a reprint of the Special Issue, published open access by the journal *Energies* (ISSN 1996-1073), freely accessible at: https://www.mdpi.com/journal/energies/special_issues/R4BBI2EJV3.

For citation purposes, cite each article independently as indicated on the article page online and as indicated below:

Lastname, A.A.; Lastname, B.B. Article Title. *Journal Name* **Year**, *Volume Number*, Page Range.

ISBN 978-3-7258-5999-3 (Hbk)
ISBN 978-3-7258-6000-5 (PDF)
https://doi.org/10.3390/books978-3-7258-6000-5

Contents

About the Editors

Joanna Ferdyn-Grygierek

Joanna Ferdyn-Grygierek is an associate professor in the Department of Heating, Ventilation, and Dust Removal Technology at the Faculty of Energy and Environmental Engineering, Silesian University of Technology, Poland. Her research focuses on building performance simulation, indoor environmental quality, natural ventilation, thermal comfort, and energy modelling. She works on optimising building design and HVAC systems with respect to energy use and indoor environmental conditions. She has authored or co-authored numerous scientific publications and has extensive experience in both experimental studies and advanced numerical modelling, including the use of EnergyPlus, ESP-r, and multizone airflow modelling techniques. Her research also involves in situ environmental measurements in real buildings. Professor Ferdyn-Grygierek has participated in numerous research projects, serving as both principal investigator and team member, and has prepared multiple expert studies for industry partners. She collaborates with design offices and is a co-author of several dozen HVAC and heating system designs for buildings. She is a certified energy auditor and energy performance assessor for buildings. In addition, she is a member of IBPSA (International Building Performance Simulation Association), contributing to the development and promotion of building simulation methods.

Krzysztof Grygierek

Krzysztof Grygierek is an associate professor at the Department of Mechanics and Bridges, Faculty of Civil Engineering, Silesian University of Technology, Poland. His research focuses on the application of artificial intelligence to the optimal design of buildings with respect to indoor environmental quality, energy use, and life-cycle assessment. His scientific interests include metaheuristic optimisation, fuzzy logic, and machine learning methods, particularly artificial neural networks (ANN). He has extensive experience in advanced numerical modelling, including building performance simulation using EnergyPlus and multizone airflow modelling in CONTAM. His work also covers CFD simulations of indoor airflows as well as finite element methods (FEM) analysis of building structures. He has participated in numerous research projects and has published widely in the fields of building physics, environmental engineering, and computational modelling.

Agnes Psikuta

Agnes Psikuta is a senior researcher in the Materials-Body Interaction group of the Laboratory for Biomimetic Membranes and Textiles at Empa, Switzerland. Having a background in building systems and architecture, she specialises in exploring and modelling the complex thermal interactions between the human body, clothing systems, and the environment through continuous development and improvement of a variety of physical and physiological models. Her research focuses on expanding the application of human thermal modelling in research related to built environment measurements and simulations, and on applications demanding detailed simulation of human thermal response, such as personal environmental control systems and their efficiency at the human level, occupational exposure, and interaction with conditioning systems.

Preface

This Reprint brings together a collection of studies that reflect the current scientific advances in the modelling and simulation of building energy performance and indoor environmental quality. The growing impact of climate change, the need to reduce greenhouse gas emissions, and the rising expectations of building occupants have created new challenges for researchers and practitioners working in the field of building engineering. These challenges motivate the need for improved tools capable of predicting, assessing, and optimising the performance of buildings under real and increasingly dynamic conditions.

The Reprint aims to provide a comprehensive overview of modern approaches used to analyse energy demand, thermal comfort, indoor air quality, and the complex interactions between building systems and occupant behaviour. It addresses the needs of scientists, engineers, consultants, and students who seek a deeper understanding of advanced simulation methods, their practical applications, and the current research trends shaping the future of building performance analysis.

By presenting recent theoretical developments, case studies, model validation efforts, and innovative numerical techniques, this Reprint offers a valuable resource for anyone interested in simulation-based strategies for improving building energy efficiency and indoor environmental quality. It highlights both methodological progress and practical insights that can assist in designing, operating, and evaluating energy-efficient and comfortable buildings.

Joanna Ferdyn-Grygierek, Krzysztof Grygierek, and Agnes Psikuta
Guest Editors

energies

MDPI

Editorial

Building Energy Performance Modelling and Simulation

Joanna Ferdyn-Grygierek [1,*], Krzysztof Grygierek [2] and Agnes Psikuta [3]

[1] Department of Heating, Ventilation and Dust Removal Technology, Faculty of Energy and Environmental Engineering, Silesian University of Technology, Konarskiego 20, 44-100 Gliwice, Poland
[2] Department of Mechanics and Bridges, Faculty of Civil Engineering, Silesian University of Technology, Akademicka 5, 44-100 Gliwice, Poland; krzysztof.grygierek@polsl.pl
[3] Empa—Swiss Federal Laboratories for Material Technology and Science, 9014 St. Gallen, Switzerland; agnes.psikuta@empa.ch
* Correspondence: joanna.ferdyn-grygierek@polsl.pl

The building sector is currently facing two conflicting challenges—the urgent need to reduce energy consumption and greenhouse gas emissions, and the continuous growth of expectations regarding thermal comfort, indoor air quality, and health [1]. Addressing such a dual challenge requires robust analytical tools. Building Energy Modelling (BEM) and Building Performance Simulation (BPS) have therefore become indispensable tools to support energy-efficient design, optimize building operations, and guide retrofit strategies [2–4].

The recent literature highlights five major directions for BEM application [5]:

- Performance-driven design of new and retrofitted buildings;
- Operational optimization of HVAC and energy systems;
- Integration with real-time data and the development of digital twins;
- Urban Building Energy Modelling (UBEM) to support city-level energy planning;
- Building-to-grid interaction for demand response and resilience.

Simulation models in this field can be broadly divided into physical (forward) models—based on energy and mass balance equations—and data-driven (inverse) models, which rely on statistical or machine learning techniques. While the former offer strong physical interpretability, the latter provide efficiency and adaptability, particularly when combined in hybrid approaches [6].

The application of simulation methods has a long history [6]. Early overviews highlighted the growing potential of building simulation for design and operation [7], while more recent developments have demonstrated its use in co-simulation frameworks [8] and as a methodological basis for retrofitting strategies in existing buildings [9]. A wide range of software tools, such as EnergyPlus, ESP-r, and TRNSYS, now support both research and industry practice, each with their own strengths and limitations [10].

Contemporary research highlights the role of simulation in enhancing performance-driven design [5], as well as in operational optimization of building systems through advanced control strategies [8]. Furthermore, simulation has extended beyond the building scale towards UBEM, which provides valuable insights for city-scale planning and integration with renewable energy systems [11–13].

Optimization remains a key application of BPS, with multi-objective algorithms such as NSGA-II enabling simultaneous improvement of energy efficiency and thermal comfort [14]. Deterministic and probabilistic approaches to optimization have been applied for their ability to handle complex, non-smooth performance functions [15]. Case studies further demonstrate the value of optimization in tailoring retrofit and comfort strategies to local contexts [16].

Finally, current advances increasingly integrate data-driven methods into traditional simulation workflows. Hybrid frameworks combining physical modelling with machine learning enable predictive maintenance, advanced control, and digital twin applications [5]. The rapid expansion of sensor networks and IoT technologies is accelerating these developments, pointing towards a future where BPS will be tightly integrated with real-time building management and urban energy systems [17,18].

BPS methods based on macroscale models are complemented by computational fluid dynamics (CFD) methods for a detailed analysis of the distribution of air parameters in the room, which allows, for example, for the assessment of local human thermal sensations or the distribution of pollutants in the zone.

Against this background, this Special Issue of Energies, "Building Energy Performance Modelling and Simulation", brings together a diverse set of research contributions. The published papers demonstrate the breadth of applications of simulation and modelling techniques—from building retrofits and new design concepts, through data-driven optimization, to advanced computational fluid dynamics (CFD) analysis. Together, they form a comprehensive picture of current advances in building energy performance research.

Retrofitting and Climate-Resilient Design

The issue features three contributions that focus on the retrofit of existing housing stock. Ferdyn-Grygierek and Grygierek present a comparative analysis of passive and active retrofit solutions in ageing multifamily housing in Central Europe. Their co-simulation approach, combining EnergyPlus and CONTAM, reveals the high potential of hybrid passive measures such as solar protective glazing and reflective roofs to reduce summer overheating while maintaining winter energy efficiency. The work highlights the importance of tailoring retrofit strategies to future climate conditions.

A complementary perspective is offered by Menconi et al., who analyze retrofit strategies for traditional listed dwellings in the UK. Their study underlines the tension between heritage preservation and energy efficiency, proposing systemic approaches to passive retrofitting that respect the thermo-hygrometric balance of historic constructions. This contribution demonstrates that responsible retrofit can simultaneously safeguard cultural value and contribute to climate targets.

Ciuman et al. provide insights into newly designed single-family houses within the framework of the "4E Idea"—energy-saving, ecological, ergonomic, and economic. Their simulations in IDA ICE highlight the role of integrated design combining efficient HVAC systems, renewable energy sources, and cost optimization. The results not only serve as a proof-of-concept but also provide benchmarks for future sustainable residential designs.

Data-Driven Modelling and Machine Learning

Two papers explore the integration of artificial intelligence and digital methods into building energy assessment. Nassif et al. apply machine learning techniques—ranging from linear regression to neural networks and ensemble methods—to predict heating coil performance in HVAC systems. Their results point to bagging and neural networks as highly promising approaches, capable of improving system efficiency and reducing operational uncertainties.

Tsikas et al. advance this line of research by proposing a BIM-based machine learning application for a parametric assessment of residential building energy performance. By training statistical and AI models on a dataset of 337 BIM-derived instances, they demonstrate that artificial neural networks can serve as accurate surrogate models for complex energy simulations. Importantly, they also deliver a user-friendly interface tool, making such predictive modelling accessible for practical design applications.

Advanced Thermal and Airflow Simulation

The Special Issue also contains valuable contributions addressing the detailed physical modelling of building components and airflows. Urzędowski et al. use CFD and statistical analysis to investigate how material parameters, such as surface emissivity and roughness, affect the thermal resistance of ETICS wall systems. Their results show the potential of reflective coatings and microstructural optimization to significantly enhance wall performance while reducing thickness, with implications for prefabricated elements.

Hurnik et al. conduct a comprehensive validation of different turbulence models in predicting sidewall jet airflow in rooms. Their study confirms the reliability of the standard k-ε and EVTM models while highlighting the limitations of other approaches. This research provides practical guidance for selecting CFD models in HVAC design, particularly in the early stages where accurate airflow prediction is crucial for comfort assessment.

Urban and Occupancy-Related Factors

Beyond the scale of individual components and systems, the issue also includes research on urban and human-related aspects of energy performance. Sadłowska-Sałęga and Wąs evaluate the impact of shading from neighbouring buildings in the historic centre of Kraków. Their simulations show that shading can increase heating demand while substantially reducing cooling loads, emphasizing that urban morphology must be accounted for in retrofit strategies and HVAC system sizing.

Nam and Kim explore energy performance improvements in Korean senior centres. Their combination of simulation and economic feasibility analysis reveals that although multiple measures improve efficiency, boiler replacement is the only intervention with a positive long-term economic return without government subsidies. This study is a reminder of the importance of coupling technical performance with financial viability in real-world decision-making.

Norouziasl et al. investigate occupancy factors in a US small office buildings using agent-based modelling combined with EnergyPlus simulations. They identify occupant density as the most influential parameter across climate zones. The study highlights the stochastic nature of occupancy and demonstrates the value of dynamic schedules for more realistic energy modelling.

Conclusions and Perspectives

Taken together, the contributions in this Special Issue reflect both the maturity and the evolving frontiers of building energy performance simulation. On one hand, the works confirm the central role of simulation in assessing retrofit strategies, system performance, and comfort conditions. On the other, they showcase emerging opportunities—from AI-based surrogate modelling to hybrid physical-data approaches—that promise to accelerate and enhance decision-making in both design and operation.

Looking forward, several trends are likely to shape the field:

- the integration of physics-based and machine learning models into hybrid frameworks;
- increased emphasis on resilience to climate change, particularly in retrofitting existing and historic stock;
- the extension of simulation from individual buildings to urban contexts, incorporating morphology and microclimate;
- and the systematic inclusion of user behaviour and economic feasibility in performance assessments.

We hope that the articles presented here will serve as inspiration and a valuable reference for researchers, practitioners, and policymakers engaged in creating energy-efficient, resilient, and comfortable built environments.

Funding: This research received no external funding.

Acknowledgments: We sincerely thank the authors for their contributions, the reviewers for their constructive feedback, and the editorial team of Energies for their support in preparing this Special Issue.

Conflicts of Interest: The authors declare no conflicts of interest.

List of Contributions:

1. Ferdyn-Grygierek, J.; Grygierek, K. Towards Climate-Resilient Dwellings: A Comparative Analysis of Passive and Active Retrofit Solutions in Aging Central European Housing Stock. *Energies* **2025**, *18*, 4386. https://doi.org/10.3390/en18164386.
2. Menconi, M.; Painting, N.; Piroozfar, P. Simulated Results of a Passive Energy Retrofit Approach for Traditional Listed Dwellings in the UK. *Energies* **2025**, *18*, 850. https://doi.org/10.3390/en18040850.
3. Ciuman, P.; Kaczmarczyk, J.; Winnicka-Jasłowska, D. Investigation of Energy-Efficient Solutions for a Single-Family House Based on the 4E Idea in Poland. *Energies* **2025**, *18*, 449. https://doi.org/10.3390/en18020449.
4. Nassif, A.; Dharmasena, P.; Nassif, N. Application of Machine Learning Techniques for Predicting Heating Coil Performance in Building Heating Ventilation and Air Conditioning Systems. *Energies* **2025**, *18*, 2314. https://doi.org/10.3390/en18092314.
5. Tsikas, P.; Chassiakos, A.; Papadimitropoulos, V.; Papamanolis, A. BIM-Based Machine Learning Application for Parametric Assessment of Building Energy Performance. *Energies* **2025**, *18*, 201. https://doi.org/10.3390/en18010201.
6. Urzędowski, A.; Sachajdak, A.; Syta, A.; Zaburko, J. CFD and Statistical Analysis of the Impact of Surface Physical Parameters on the Thermal Resistance of Layered Partitions in ETICS Systems. *Energies* **2025**, *18*, 107. https://doi.org/10.3390/en18010107.
7. Hurnik, M.; Ciuman, P.; Popiolek, Z. Eddy–Viscosity Reynolds-Averaged Navier–Stokes Modeling of Air Distribution in a Sidewall Jet Supplied into a Room. *Energies* **2024**, *17*, 1261. https://doi.org/10.3390/en17051261.
8. Sadłowska-Sałęga, A.; Wąs, K. Impact of Shading Effect from Nearby Buildings on Energy Demand and Load Calculations for Historic City Centres in Central Europe. *Energies* **2024**, *17*, 6400. https://doi.org/10.3390/en17246400.
9. Nam, A.; Kim, Y.I. Prioritizing Energy Performance Improvement Factors for Senior Centers Based on Building Energy Simulation and Economic Feasibility. *Energies* **2024**, *17*, 5576. https://doi.org/10.3390/en17225576.
10. Norouziasl, S.; Vosoughkhosravi, S.; Jafari, A.; Pang, Z. Assessing the Influence of Occupancy Factors on Energy Performance in US Small Office Buildings. *Energies* **2024**, *17*, 5277. https://doi.org/10.3390/en17215277.

References

1. Buildings—Energy System. Available online: https://www.iea.org/energy-system/buildings (accessed on 21 September 2025).
2. Harish, V.S.K.V.; Kumar, A. A Review on Modeling and Simulation of Building Energy Systems. *Renew. Sustain. Energy Rev.* **2016**, *56*, 1272–1292. [CrossRef]
3. Negendahl, K. Building Performance Simulation in the Early Design Stage: An Introduction to Integrated Dynamic Models. *Autom. Constr.* **2015**, *54*, 39–53. [CrossRef]
4. Østergård, T.; Jensen, R.L.; Maagaard, S.E. Building Simulations Supporting Decision Making in Early Design—A Review. *Renew. Sustain. Energy Rev.* **2016**, *61*, 187–201. [CrossRef]

5. Pan, Y.; Zhu, M.; Lv, Y.; Yang, Y.; Liang, Y.; Yin, R.; Yang, Y.; Jia, X.; Wang, X.; Zeng, F.; et al. Building Energy Simulation and Its Application for Building Performance Optimization: A Review of Methods, Tools, and Case Studies. *Adv. Appl. Energy* **2023**, *10*, 100135. [CrossRef]
6. Wang, H.; Zhai, Z. Advances in Building Simulation and Computational Techniques: A Review between 1987 and 2014. *Energy Build.* **2016**, *128*, 319–335. [CrossRef]
7. Hong, T.; Chou, S.K.; Bong, T.Y. Building Simulation: An Overview of Developments and Information Sources. *Build. Environ.* **2000**, *35*, 347–361. [CrossRef]
8. Wetter, M. Co-Simulation of Building Energy and Control Systems with the Building Controls Virtual Test Bed. *J. Build. Perform. Simul.* **2011**, *4*, 185–203. [CrossRef]
9. Ma, Z.; Cooper, P.; Daly, D.; Ledo, L. Existing Building Retrofits: Methodology and State-of-the-Art. *Energy Build.* **2012**, *55*, 889–902. [CrossRef]
10. Crawley, D.B.; Hand, J.W.; Kummert, M.; Griffith, B.T. Contrasting the Capabilities of Building Energy Performance Simulation Programs. *Build. Environ.* **2008**, *43*, 661–673. [CrossRef]
11. Reinhart, C.F.; Cerezo Davila, C. Urban Building Energy Modeling—A Review of a Nascent Field. *Build. Environ.* **2016**, *97*, 196–202. [CrossRef]
12. Kontokosta, C.E.; Tull, C. A Data-Driven Predictive Model of City-Scale Energy Use in Buildings. *Appl. Energy* **2017**, *197*, 303–317. [CrossRef]
13. Na, W.; Wang, M. A Bayesian Approach with Urban-Scale Energy Model to Calibrate Building Energy Consumption for Space Heating: A Case Study of Application in Beijing. *Energy* **2022**, *247*, 123341. [CrossRef]
14. Bre, F.; Fachinotti, V.D. A Computational Multi-Objective Optimization Method to Improve Energy Efficiency and Thermal Comfort in Dwellings. *Energy Build.* **2017**, *154*, 283–294. [CrossRef]
15. Wetter, M.; Wright, J. A Comparison of Deterministic and Probabilistic Optimization Algorithms for Nonsmooth Simulation-Based Optimization. *Build. Environ.* **2004**, *39*, 989–999. [CrossRef]
16. Griego, D.; Krarti, M.; Hernández-Guerrero, A. Optimization of Energy Efficiency and Thermal Comfort Measures for Residential Buildings in Salamanca, Mexico. *Energy Build.* **2012**, *54*, 540–549. [CrossRef]
17. Agouzoul, A.; Tabaa, M.; Chegari, B.; Simeu, E.; Dandache, A.; Alami, K. Towards a Digital Twin Model for Building Energy Management: Case of Morocco. *Procedia Comput. Sci.* **2021**, *184*, 404–410. [CrossRef]
18. Wang, W.; Guo, H.; Li, X.; Tang, S.; Xia, J.; Lv, Z. Deep Learning for Assessment of Environmental Satisfaction Using BIM Big Data in Energy Efficient Building Digital Twins. *Sustain. Energy Technol. Assess.* **2022**, *50*, 101897. [CrossRef]

energies

MDPI

Article

Towards Climate-Resilient Dwellings: A Comparative Analysis of Passive and Active Retrofit Solutions in Aging Central European Housing Stock

Joanna Ferdyn-Grygierek [1,*] and Krzysztof Grygierek [2]

1 Department of Heating, Ventilation and Dust Removal Technology, Faculty of Energy and Environmental Engineering, Silesian University of Technology, Konarskiego 20, 44-100 Gliwice, Poland

2 Department of Mechanics and Bridges, Faculty of Civil Engineering, Silesian University of Technology, Akademicka 5, 44-100 Gliwice, Poland; krzysztof.grygierek@polsl.pl

* Correspondence: joanna.ferdyn-grygierek@polsl.pl

Abstract

This article evaluates the effectiveness of various energy retrofit solutions—both passive and active—for reducing energy demand and improving indoor thermal conditions in apartments of typical multifamily buildings in Central Europe, considering current and future climate conditions. This study combines computer-based co-simulations (EnergyPlus and CONTAM) with in situ thermal measurements to identify challenges in maintaining indoor thermal conditions and to support model validation. Key indicators include the number of thermal discomfort hours and heating and cooling demand. The evaluated strategies include passive measures (wall insulation, green or reflective roofs, roller blinds, solar protective glazing) and active solutions such as mechanical cooling. The comfort operative temperature range of the adaptive model is adopted as a measure of thermal comfort and the energy demand in individual apartments as a measure of energy efficiency. The simulation results showed that solar protective glazing combined with a reflective roof reduced thermal discomfort hours by up to 95%, while modern windows alone decreased them by 90% and lowered heating demand by 18%. In contrast, typical passive solutions such as internal blinds or balconies were significantly less effective, reducing discomfort hours by only 11–42%. These findings highlight that, while no single retrofit measure is universally optimal, well-selected passive or hybrid strategies can substantially improve summer comfort, maintain winter efficiency, and reduce long-term reliance on energy-intensive cooling systems in aging multifamily housing.

Keywords: thermal comfort; heating and cooling demand; multifamily buildings; energy simulation; climate change

1. Introduction

In the face of ongoing climate change, the energy retrofit of residential buildings is becoming not only a matter of energy efficiency and cost savings, but also one of residents' comfort and health. With rising average temperatures and the increasing frequency of extreme weather events (such as heatwaves), the challenges associated with building modernization are taking on a new dimension. Traditional energy retrofits have primarily focused on reducing heat loss and improving energy efficiency during the winter season. While many buildings now have insulated facades, reducing heating use, global warming has led to apartment overheating during summers [1–5]. Furthermore, insufficient ventilation [6,7] and the behavior of residents, such as the improper use of windows or blinds [8,9], can exacerbate the risk of overheating. Many older residential buildings, particularly in Central and Eastern Europe, were not designed with high temperatures and increased heat loads in mind. Transforming these structures into more heat-resilient infrastructures can be both costly and technically challenging, due to structural or urban planning limitations, for example. This necessitates a thorough analysis of how new solutions can be implemented in existing buildings, such as appropriate solar insulation, shading systems, or passive ventilative cooling, which were not always considered in older designs.

The problem of the overheating of dwellings covers various climatic areas, not only the warmest ones [10]. The report of the Ministry of Housing, Communities and Local Government [11] lists, among others, Germany and France, where air-conditioned buildings are not standard. The literature clearly indicates the benefits of having buildings with a more efficient thermal envelope; but, relying solely on retrofit strategies based on building insulation can yield unexpected results if the risk of overheating is not taken into account [12,13]. Unfortunately, heatwaves are expected to become more frequent and with higher temperatures as global warming continues [14]. To address this problem, building retrofits in the temperate climatic conditions of Europe should balance winter heat loss reduction with passive cooling measures in summer [15–17].

1.1. Literature Review

Numerous studies have demonstrated the potential of passive strategies in reducing energy demand and mitigating summer overheating in existing buildings. These strategies include external insulation, shading devices, reflective surfaces, night ventilation, and improvements in window design [10,16,18–26]. In their conclusions, the authors highlight the crucial importance of properly selecting window glazing, given its substantial impact on energy efficiency. Several simulation-based analyses confirm that combining passive solutions—such as insulation and shading—with hybrid systems that integrate passive techniques and active air conditioning can significantly reduce energy use and improve indoor human comfort in both hot and temperate climates [23–35]. The authors acknowledge that ventilative cooling, particularly night-time ventilation when integrated with other passive strategies, can result in a near-complete reduction in thermal discomfort. The benefits of ventilative cooling have also been widely recognized and highlighted within the framework of the International Energy Agency's Annex 62 [36]. Recent reports from international bodies such as UNEP (United Nations Environment Programme) [37] and GlobalABC [38] also emphasize the importance and feasibility of implementing passive cooling during retrofit interventions, particularly in urban housing.

Despite the availability of passive technologies, one of the most critical and often overlooked aspects is user behavior. Studies show that occupant actions—especially window operation—can dramatically influence both cooling potential and overall building performance [39–42]. Behaviorally driven ventilation effectiveness varies widely, and probabilistic models have been proposed to account for this variability [43–45]. So far, no universal model for modeling the behavior of building users has been developed. This may be particularly difficult for residential buildings, where the users are people of different ages who have very different preferences regarding indoor conditions and, consequently, very different settings related to heating, ventilation, and air conditioning systems. Although the ASHRAE Global Occupant Behavior Database [46,47] provides a useful international reference, its coverage of window-opening behavior—especially for passive cooling purposes in residential buildings—is limited, both geographically and seasonally. As a result, tailored or localized modeling approaches are still needed to realistically capture human interaction with natural ventilation systems.

This variability in occupant behavior becomes even more significant when considered in the context of climate change, which is projected to exacerbate the risk of summer overheating and thermal discomfort—even in buildings that have undergone energy retrofits [48–53]. For example Escandón et al. [51] predicted an increase of 35% in the thermal discomfort hours in summer for multifamily buildings constructed in Spain. Similar conclusions were reached by Moazami et al. [52], who calculated the energy use of 16 standard ASHRAE buildings under different climate scenarios for Geneva. Recent multi-city projections of cooling degree days in seven European countries indicate that this indicator may roughly double by 2050—with the largest relative increases occurring in northern locations—highlighting the need to anticipate changes to the building stock and shifts in occupant behavior when planning retrofit strategies for European climates [54]. While passive methods such as night ventilation and solar shading can delay or mitigate the need for mechanical cooling, their effectiveness under future climate scenarios—and in relation to user behavior—remains insufficiently studied [55].

1.2. Research Gap, Aim, and Scientific Novelty

Buildings in Central and Eastern Europe, particularly older multifamily housing, face significant challenges in maintaining thermal comfort with low energy use under changing climatic conditions. In Poland, residential buildings account for a substantial share of the country's total energy use, reaching 20.2% in 2021 [56], which is similar to the EU average. Of the 6.8 million residential buildings [57] (14.6 million apartments), 68% are in multifamily structures, mostly built in the 1960s–1970s with prefabricated concrete [58]. Thermally upgrading this aging building stock has become a strategic priority, not only for energy savings but also for ensuring acceptable indoor thermal environments.

Despite growing interest in methods of reducing room overheating, many areas are still not sufficiently researched. There is a lack of systematic studies on the synergistic effect of combining passive and active approaches. There is a gap in the analysis of how the effectiveness of these methods changes with predicted climate changes, such as increasing average temperatures or intensifying heatwaves, especially for naturally ventilated buildings. Most simulation studies assume constant air exchange, which greatly simplifies the problem and does not take into account the variability of the infiltrating air flow, especially in the case of the probabilistic behavior of people in terms of opening windows. The influence of wind and the stack effect are practically omitted in building performance simulation due to the difficulty of modeling gravity ducts in buildings, which affects the accuracy of the assessment of instantaneous heat loads, especially in multi-story buildings. A significant part of the research is focused on the design of new buildings that are resistant to climate change. Meanwhile, the issues of energy-efficient passive renovation methods of existing buildings are relatively poorly developed. This is important because most of the global building stock consists of older structures that require adaptation. Solving these problems can significantly contribute to sustainable building development and enhance energy efficiency in the context of climate change.

Apartments on the top floor of multifamily buildings are more exposed to the influence of the external environment than other apartments due to contact with the roof, especially when the roof is not insulated. Therefore, this research aims to assess renovation solutions that improve thermal conditions in such apartments, taking into account the probabilistic behavior of residents in operating windows in Polish buildings with natural ventilation.

This study answers two key questions: (1) How effective are different passive cooling energy retrofit solutions—individually, in combination, and together with mechanical cooling—in controlling the indoor thermal climate and saving energy in the current and future conditions of central Europe? (2) Will passive solutions provide thermal comfort conditions in dwellings in the future? The effectiveness of several energy retrofit measures—including shading, envelope insulation, and solar reflectance—is assessed using the building performance simulation method. First, the temperature measurements are taken in existing buildings to determine the problem of maintaining the desired thermal conditions to ensure thermal comfort for residents. Then, a multivariant thermal assessment of apartments for the entire year is performed on a building model covering the top and penultimate floors based on co-simulations using EnergyPlus and CONTAM programs.

The main contributions and innovative aspects of this study are summarized as follows:

- Dynamic thermal assessment with ventilation modeling: A novel method is developed to integrate mass and energy transfer calculations, enabling the determination of variable ventilation airflow at each simulation step. This allows for precise evaluation of instantaneous human thermal comfort and air cooling potential throughout the year. The approach addresses a significant research gap in assessing naturally ventilated buildings, especially those with gravity ventilation ducts, under real-time internal and external thermal loads.
- Behavior-integrated energy modeling: By incorporating realistic occupant behavior, particularly window opening patterns, this study reveals how human actions influence the performance of HVAC systems and passive cooling techniques. This behavioral aspect enhances the realism and applicability of simulation results in residential settings.
- Comparative evaluation of passive and active cooling strategies: This study compares passive, active, and hybrid cooling methods, both individually and in combination, identifying the most effective solutions for maintaining indoor thermal comfort in the face of climate warming. This comparative framework is especially relevant for retrofitting older buildings in Central and Eastern Europe.

Together, these innovations provide a comprehensive and practical framework for improving the thermal performance of aging residential infrastructures in a changing climate.

2. Methods

2.1. Research Object

One segment of a multifamily building with five residential floors, built in the 1960s and located in southern Poland, was selected for this study. Each floor contains three apartments with areas ranging from 43 to 63 m^2. Each apartment consists of two main rooms, kitchen and bathroom with toilet. Six apartments on the two upper floors were taken into account (Figure 1).

The apartments have windows facing north-east and south-west. In its current form, the building has a reinforced concrete structure (prefabricated walls and nonventilated reinforced concrete roof) and has been partially renovated (only insulation of external walls). The building has typical double-glazed windows (argon gas 90% in the chamber) with a solar heat gain coefficient of 0.64 and a heat transfer coefficient U = 1.1 W/m^2K. The building is equipped with a central heating system with radiators. Fresh air enters the building through leaks in windows and doors; after mixing with air in the rooms, it is removed through ventilation grilles connected to gravity chimneys. Each apartment is equipped with two gravity chimneys; one is located in the kitchen, the other in the bathroom. This is a typical ventilation system solution in residential buildings in Central and Eastern Europe.

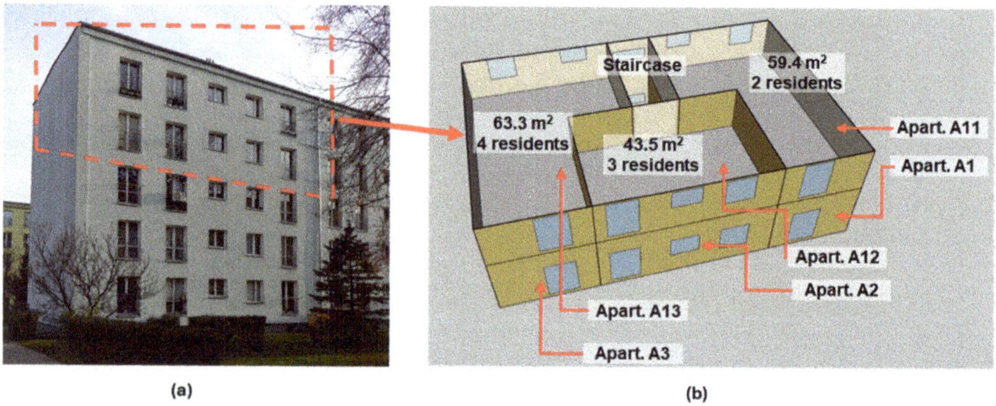

Figure 1. Building under consideration: (**a**) building view and (**b**) view of thermal model geometry.

2.2. Locations and Climate Scenarios

The Central Europe location selected for analysis, which corresponds to one of the major cities in southern Poland, was Katowice. This location is characterized by a temperate transitional climate (Dfb class according to the Köppen–Geiger classification [59]). For the simulations, the typical meteorological year (TMY) [60] and future climate were calculated based on global warming forecasts for 2050 (Figure 2). In the TMY, the temperature changed from −18.7 °C to 31.0 °C during the year, with an average value of 8.1 °C. The predicted climate data were calculated from the A2 emissions scenario (one of the most popular scenarios in the literature [61]). For the climate, the temperature ranged from −13.9 °C to 37.6 °C, with an average annual temperature of 11 °C.

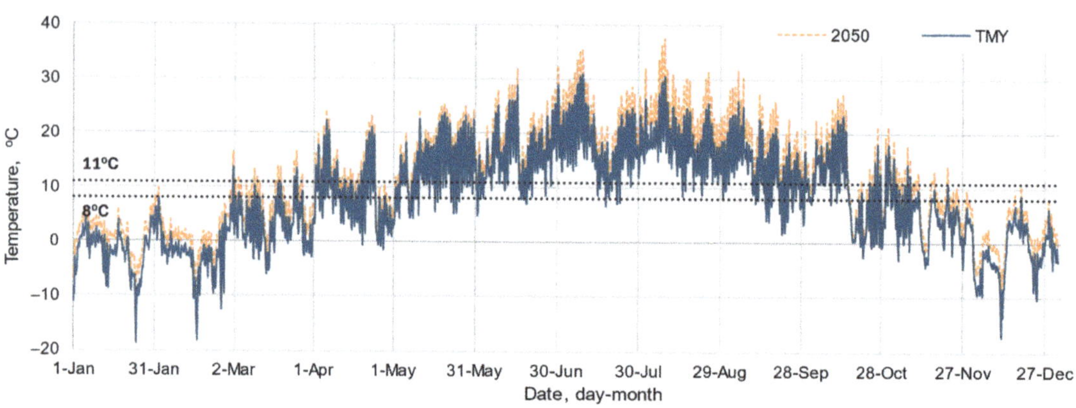

Figure 2. Variability of hourly external air temperature for this study.

2.3. Measurements In Situ

The measurements had two aims: to present the problem of overheating in apartments and to validate the thermal model. For a period of 8 months in 2021, the indoor air temperature and relative humidity were recorded (with a 5 min time step) in two apartments

(A11 and A12). Data loggers AR236.B (APAR Control, Raszyn, Poland) with an accuracy of ±0.3 °C for temperature and ±2%RH for humidity were used. In each apartment, there were two recorders from January to August to check the thermal conditions in different periods of the year. For the analysis, the average value of temperature in each apartment (at each recording step) was calculated. During the measurements, the residents used the rooms freely, including manual operation of windows and blinds. Historical climate data were imported from the nearest weather station of the Institute of Meteorology and Water Management–State Research Institute [62].

2.4. Building Thermal Model and Simulation

The thermal calculation using the multizone model of the selected building fragment was performed using EnergyPlus 9.4 (US Department of Energy, Washington, DC, USA) [63] connected with CONTAM 3.4 (National Institute of Standards and Technology, Gaithersburg, MD, USA) [64] simulation programs (Figure 3). In this research, due to the lack of a gravity chimney model in the air flow network (AFN) module of EnergyPlus and available in CONTAM and the simplicity of modeling in the CONTAM program, it was decided to model the natural ventilation in this program. This connection was made using the functional mock-up unit (FMI) standard [65].

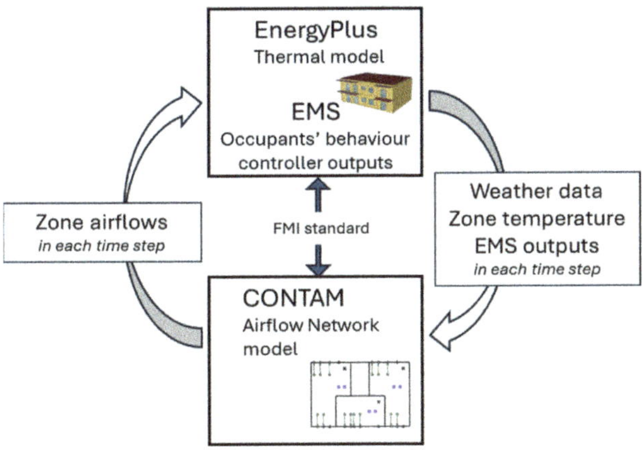

Figure 3. Co-simulation scheme of CONTAM and EnergyPlus.

The apartments on the top two floors were modeled (Figure 1). Each apartment was treated as a separate thermal zone, which was used as an open space due to its small size. This simplification had little impact on the results because, in reality, internal doors in apartments are usually open, allowing air to flow freely between rooms. Internal walls were considered as additional thermal mass inside the apartments. This model took into account uncontrolled airflows, penetrating through openings and leaks in the envelope of the building. The following airflows of natural (gravitational) ventilation were balanced: flow through gaps in windows, flow through an open or tilted window, and flow through gravitational ventilation ducts. It was assumed that the large balcony window with dimensions of 62 by 192 cm could be opened or tilted, and the small window with dimensions of 62 by 97 cm could be tilted. The model was described in detail in ref. [7]. The input data for the thermal model used in the case studies are listed in Table 1.

The key part of this research was to properly model people's behavior regarding window opening. In reality, this activity is uncontrolled and depends on the individual preferences of the residents. In this study, a stochastic window opening control model was created, taking into account the behavior of the residents, which depended on the outdoor temperature, indoor comfort temperature, wind speed, and air change rates. It proposed a probabilistic approach in which the window settings could change during the day (with probabilities of 0.5) and at night (0.25), with specific time and degree restrictions for opening and closing windows. The key assumptions included the following: (i) initial constraints: window opening depended on the wind speed, the difference in indoor and outdoor temperatures, and the comfort temperature (windows could only be opened if the wind speed was low, the ambient temperature was lower than the indoor temperature, and the indoor operative temperature was above the comfort level); (ii) window opening levels: tilting the balcony window, tilting the balcony window and a small window, fully opening the balcony window, and tilting the small window, (iii) time constraints: changing the window opening level was possible during the day at hourly intervals; (iv) additional rules: constraints were introduced to avoid draughts and excessive cooling of the rooms. The limit values of input parameters (e.g., wind speed and temperature differences) at which the window opening was changed were optimized using genetic algorithms. The parameters were selected to minimize the number of hours of thermal discomfort with limited air change rates. The window opening controller was integrated with the EMS module of EnergyPlus. The window setting was changed, taking into account the stochastic nature of human behavior in controlling the windows. Each retrofitting scenario was simulated ten times, with variations in results arising from the applied stochastic model. The final results used in the analysis represent the average across all simulations for each case. A detailed description of the model and optimization can be found in our previous paper [7].

Table 1. Input data for the thermal model.

Model Component	Input Value	Operating Time	Comment
Heating	Indoor temperature setpoints: apartments: 21 °C; staircase: 16 °C	From September to May	According to EN 16798-1:2019 standard [66]
Cooling	Indoor temperature setpoints: 26 °C (only apartments)	All year; only in cases with a cooling system	According to EN 16798-1:2019 standard [66]
Occupants	126 W per person during the day, 73 W per person during the night (sleeping)	According to the individual schedule in each apartment (Figure A1), there was always at least one person in the apartment	According to ASHRAE-55 standard [67]
Lighting	Apartments 3.5 W/m^2	Day: turned on when the lighting intensity is lower than 250 lm/m^2 Night: turned off	LED lamps and typical lighting intensity for residential premises were adopted
Equipment	Electric hob: 500 W; fridge: 150 W or 250 W; computer: 100 W; and TV: 175 W	According to the individual schedule in each apartment (Figure A2)	According to the typical home equipment power
Infiltration	Airflow calculated in each time step for each zone	All year	One-way flow using POWERLAW model
Opening windows	Variable airflow calculated in each time step for each zone	According to schedule	Two-way flow model (single opening)
Mechanical ventilation	Constant airflow in each time step for each zone	All year; only in cases with a cooling system	According to EN 16798-1:2019 standard [66]
Window blinds	Internal blinds with a solar transmittance of 0.4 and a solar reflectance of 0.4	ON-OFF operating. ON mode: the operative indoor temperature exceeds the comfort temperature by 1.5 K and the perpendicular to the window solar radiation exceeds 150 W/m^2	Probability of blinds being closed and opened: 0.5

2.5. Thermal Model Validation

Validation of simulation results was carried out based on indoor temperature measurements in two apartments, A11 and A12, in the summer period (June–August). During this period, the heating system was not working and the indoor temperature resulted only from the heat balance in the apartments. The real use of the rooms was not recorded. Therefore, for validation, the indoor heat gains and the window opening schedule were calibrated to match the temperature variation courses. The model also included historical climate data.

The two indicators given in the ASHRAE guide [68] were used to assess the compliance of the simulation model: the normalized mean bias error (NMBE) and the coefficient of variation of root mean squared error CV(RMSE), which are calculated from

$$\text{NMBE} = \frac{\sum_{i=1}^{n}(M_i - S_i)}{(n-p)\cdot\overline{M_i}} \times 100, \tag{1}$$

$$\text{CV(RMSE)} = \frac{1}{\overline{M_i}}\sqrt{\frac{\sum_{i=1}^{n}(M_i - S_i)^2}{n-p}} \times 100. \tag{2}$$

where M_i is the measured value, S_i is the simulated value, n is the number of compared values, $\overline{M_i}$ is the mean of the measurement values, and p is the number of adjustable model parameters. The ASHRAE guide suggests these as maximum errors for energy use in buildings, but some researchers also take them into account when comparing indoor temperatures.

Measured indoor temperature data, originally recorded with a 5 min time step, were averaged over three consecutive intervals to match the 15 min time step used in the simulation results for validation purposes (Figure 4). Empirical verification regarding indoor temperature makes the results of the simulation calculations reliable, accurate, and valid for use in this study. The values of the NMBE and CV(RMSE) indicators are small: NMBE is 0.5% and CV(RMSE) is 2% in apartment A11 and 3% in apartment A12. Correlation coefficients R are very high and range from 0.92 in A12 to 0.97 in A11.

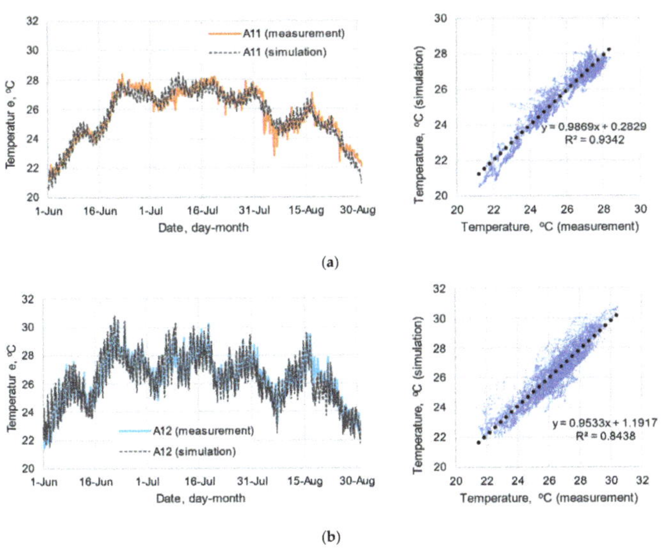

Figure 4. Variation of 15 min value and correlation of measured and calculated indoor air temperature from June to August for (**a**) apartment A11 and (**b**) apartment A12.

2.6. Energy Retrofit Improvements and Evaluation Criteria

The analysis was carried out for three main groups of improvements—building insulation, solar radiation control, and mechanical cooling (Figure 5)—which were divided into nineteen detailed variants of solutions affecting thermal conditions and energy demand in the building (Table 2). In the first part of the analysis, the influence of the envelope insulation was checked. Next, simulations were performed using passive solar control, i.e., balconies, green roofs, and a roof covered with a membrane reflecting solar radiation, alongside a reflective roof and windows with better parameters from the point of view of solar protection. Finally, active thermal condition control, i.e., a mechanical cooling system, was simulated. Some combinations of improvements were also calculated, such as solar protective glazing and a cool roof, mechanical cooling and solar protective glazing, and mechanical cooling and a cool roof.

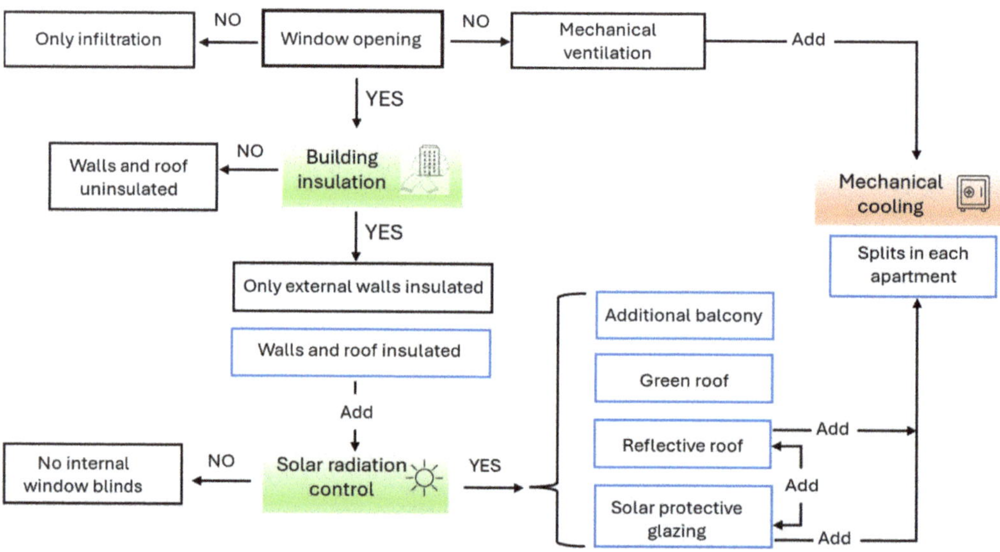

Figure 5. Flow chart illustrating the process of selecting and narrowing down simulation scenarios.

Six cases (in blue line in Figure 5) were simulated using both current climate and future predicted climate data. In all cases (except for two), the rooms were aired by opening windows and internal roller blinds were turned on. In order to demonstrate the validity of using ventilation by opening windows and blinds on windows, additional simulations were carried out without these elements. All the simulations were performed with a 15 min time step for the whole year.

For cases without mechanical cooling, the number of thermal discomfort hours throughout the year and the indicators of the heating demand per square meter of the floor were evaluated. For cases with mechanical cooling, it was assumed that the comfort conditions were met, but this criterion was not assessed while the heating and cooling energy were evaluated. The annual energy demand in individual apartments was adopted as an energy efficiency index, expressed per square meter of floor area. In turn, an adaptive model based on the recommended ranges of indoor operative temperature according to the EN 16798-1:2019 standard [66] was adopted as a measure of human thermal comfort. In this study, the limits of the comfort temperature were calculated for category II of the thermal environment, i.e.,

$$To_{comfort} = 0.33 \times Tout + 18.8 \pm 3.0, \tag{3}$$

where *Tout* is the weighted mean of the previous 7-day daily mean outdoor air temperature (°C); the limits only apply when the running mean outdoor temperature is greater than 10 °C. Below this threshold—which applies during the colder seasons of the year—the adaptive comfort model assumes a constant of the operative comfort temperature of 20 °C for category II indoor environments. This condition was maintained by the heating system. In EnergyPlus, the operative temperature is calculated as the average of the indoor air temperature and the mean radiant temperature of the zone. The "zone-average" method was used to calculate the mean radiant temperature as a weighted average. This method assumes that the person is at the center of the zone. This model is commonly used in the thermal simulation of residential buildings where the location of the occupants changes dynamically.

Table 2. Considered cases and solutions to improve thermal conditions in apartments.

Improvement		Case																		
		1	2	3 *	4	5	6 **	7	8	9	10	11	12	13	14 *	15 **	16	17	18	19
Building insulation	Walls insulated		✓	✓	✓	✓	✓	✓	✓	✓	✓	✓	✓	✓	✓	✓	✓	✓	✓	✓
	Roof insulated			✓	✓	✓	✓	✓	✓	✓	✓	✓	✓	✓	✓	✓	✓	✓	✓	✓
Natural ventilation	All windows closed					✓														
	Opening windows	✓	✓	✓	✓		✓	✓	✓	✓		✓			✓	✓	✓	✓	✓	
Solar radiation control	Internal blinds	✓	✓	✓		✓	✓	✓	✓	✓	✓	✓	✓	✓	✓	✓	✓	✓	✓	✓
	Balcony						✓									✓				
	Green roof							✓									✓			
	Reflective roof								✓			✓	✓					✓		
	Solar protective glazing									✓		✓		✓					✓	
Mechanical cooling	Air conditioners, mechanical ventilation										✓		✓	✓						✓
Climate	Current (TMY)	✓	✓	✓	✓	✓	✓	✓	✓	✓	✓	✓	✓	✓						
	Future 2050														✓	✓	✓	✓	✓	✓

* Base models for current and future climates. ** Includes two sub-cases a and b with different balcony depths.

2.6.1. Building Insulation

Three cases were considered as follows:

- Completely insulated building (used as a base model for comparison): insulation of the flat roof with 20 cm of mineral wool ($U = 0.146$ W/m^{2}K) and the existing condition of the external walls insulated by 15 cm of mineral wool ($U = 0.196$ W/m^{2}K);
- Partly insulated building (the building in its current state): external wall insulated with 15 cm of mineral wool ($U = 0.196$ W/m^{2}K) and the flat roof uninsulated ($U = 0.632$ W/m^{2}K);

- Uninsulated building (original construction condition from the 1960s): the external wall has only old insulation made with 5 cm of mineral wool between the structural layers of the wall (U = 0.750 W/m^{2}K) and flat roof uninsulated (U = 0.632 W/m^{2}K).

2.6.2. Solar Radiation Control

Two variants with balconies of 120 and 160 cm depth for balcony doors were considered (Figure 6). The width of the balcony for apartments A1, A3, A11, and A13 was the width of the external walls, i.e., 4.30 m. For apartments A2 and A12, the width of the balcony was 2.60 m, which was approximately 1/3 of the width of the external wall. For apartments A11–A13, a slab acting as a roof was also planned.

Figure 6. Proposition of balcony construction and model geometry view.

When designing the green roof structure, the main assumption was to select parameters that would best reduce indoor temperature in the summer. The main parameter considered is the leaf area index (LAI), which means the ratio of leaf area to ground area. The higher the parameter value, the greater the plant density in relation to the ground area in which it is located. The LAI = 5 was assumed aligning with Mahmoodzadeh [69]. The model took into account the height of the plants (400 mm) and the thickness of the substrate (150 mm).

Another analyzed solution that affected the thermal conditions in apartments was the use of a cool roof. Such a solution is used in buildings to reduce energy demand by using reflective coatings (with a high solar reflectance coefficient). In this study, a layer with a solar energy absorption coefficient of 0.22 was assumed.

As one of the variants, Pilkington solar control windows were tested, consisting of Pilkington Suncool 70/35 glass on the outside, Argon gas (90%), Pilkington Optifloat Clear glass on the inside, Argon gas (90%), and Pilkington Optitherm SI3 glass on the inside (solar heat gain coefficient of 0.35) [70].

2.6.3. Mechanical Cooling and Ventilation

Split air conditioners with a cooling air temperature setpoint of 26 °C were used for mechanical cooling. Air conditioners are one of the most commonly used solutions to improve thermal conditions in residential buildings in the summer. Due to the mechanical cooling system, the windows were not opened by residents. The calculation model took into account the use of constant air volume ventilation with heat recovery with an efficiency of 70% to provide the required amount of fresh air. The system provided constant airflow throughout the year, which was 126 m^3/h for each apartment, according to the EN 16798-1:2019 standard [66] (design air flow rates by room and building type—two main rooms in the dwelling, kitchen and bathroom).

3. Results

3.1. Assessment of Indoor Thermal Conditions Based on In Situ Measurements

The recorded values of temperature and relative humidity are presented in Figure 7. The apartment A12 was characterized by greater load variability, which resulted in greater instantaneous fluctuations in the indoor temperature. The same annual trend can be seen in both apartments. In the winter–spring period, the indoor temperature was in the range of 20–23 °C. During this period, the thermal conditions of the apartments were shaped by the heating system, which in Poland generally operates from September to May. From June, the indoor temperature increased above 26 °C, with values reaching as high as 30 °C. During the recording, residents could open windows and use sunshades according to their preferences, and these activities were not recorded. Despite this, the temperature above 26 °C occurred for 49% and 62% of the summertime in apartments A12 and A11, respectively. Relatively high humidity was recorded in apartment A12 in winter, which was the result of using a humidifier. In summer, relative humidity in both apartments was in the range of 40 to 60% (with minor exceedances up to a maximum of 70%), thus meeting the recommended requirements of the EN 16798-1:2019 standard [66].

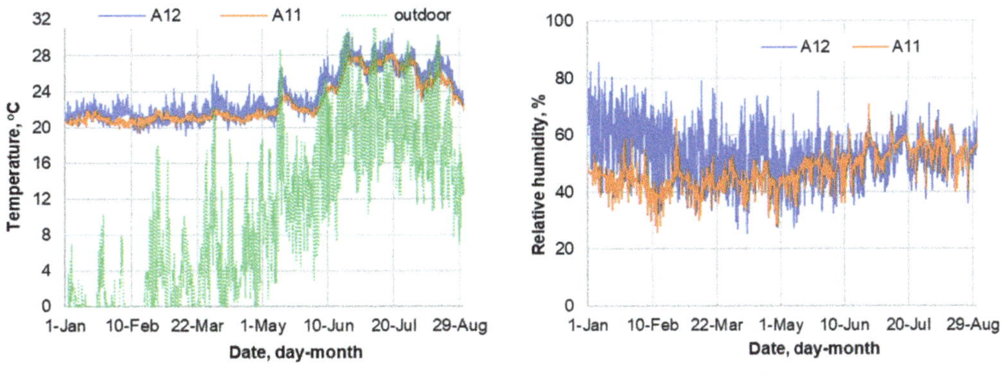

(a) (b)

Figure 7. Variability of measured (**a**) indoor and outdoor air temperature and (**b**) indoor relative humidity in apartments.

It should be noted that the actual use of the rooms was not systematically recorded during the monitoring period. The absence of detailed information on occupant behavior (e.g., window opening patterns, shading use, occupancy schedules) introduces some uncertainty in interpreting the measured data and limits the ability to fully correlate observed thermal conditions with specific user actions.

3.2. Numerical Analysis

The following symbols were adopted for apartments in the analyses: on the penultimate floor, A1, A2, and A3; on the top floor, A11, A12, and A13 (Figure 1).

3.2.1. Impact of External Envelope Insulation (Cases 1 to 3)

Insulating the external walls reduced heating demand by an average of 42%, with the largest drop (75%) in corner apartment A3 without direct roof exposure (Figure 8a).

Adding roof insulation brought further reductions, particularly for top-floor units (e.g., −26 kWh/m^2 in A13); this was a 10-fold reduction. The reduction in heat demand was also visible on the lower floor, which does not touch the roof. This was caused by the increase in the indoor temperature on the top floor in the transitional period (i.e., fall and spring) and, therefore, additional heat gains in the apartments below were present. It can also be seen that, after the full insulation of the building envelope, the annual heat demand was small and amounted to a maximum of 5.5 kWh/m^2 in apartment A12.

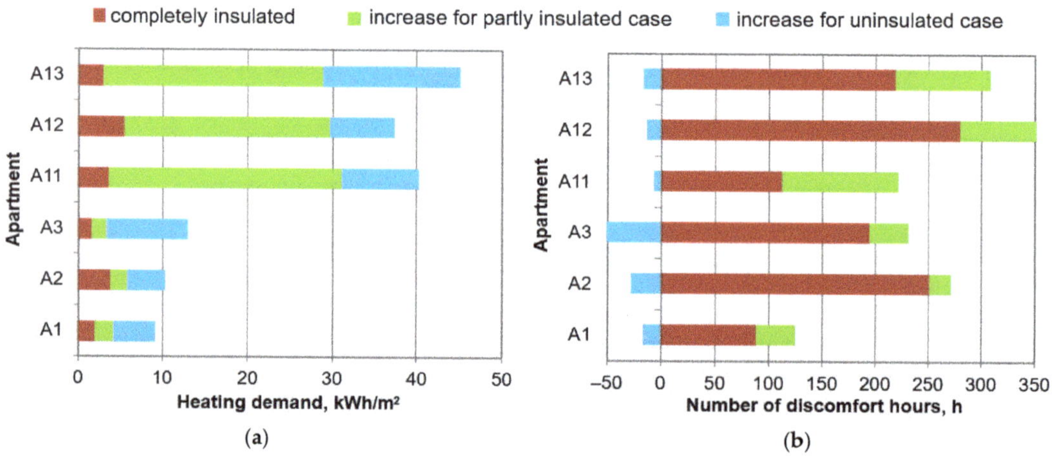

Figure 8. (**a**) Heating demand and (**b**) number of discomfort hours before insulation (case 1), after insulation of the walls (case 2), and after insulation of the walls and flat roof (case 3).

However, wall insulation alone slightly worsened the summer comfort of the residents (due to higher heat accumulation) by 11–29% in apartments on the penultimate floor and by 3–6% on the upper floor (Figure 8b). On the other hand, the reduction in the heat transfer coefficient of the flat roof (after insulation) had a positive effect on the thermal condition in the building, increasing the period of thermal comfort compared with the building before the renovation. For example, for apartment A11, the number of hours of thermal discomfort was reduced by about 110 h per year, which is half of these hours in the same apartment but with an uninsulated roof. The beneficial effect was not influenced by the insulation of the partition itself (because this procedure worsens the thermal conditions) but by the ventilation of the building by opening windows, which was more frequent and intensive due to higher indoor temperatures. The effect was also visible in the apartments on the lower floor, but it was, on average, only 18% (in the range of 8–30%).

3.2.2. Impact of Interior Blinds on Windows (Case 4)

Internal blinds had negligible influence on heating demand (<2%) but reduced annual discomfort hours by 11–19% (Figure 9). The effect was more pronounced on the penultimate floor (over 17% reduction) than the top floor (13%), reflecting differences in solar exposure. The heating demand effect was poor because, in winter, the blinds were very rarely used.

Internal blinds may be a temporary solution and, as climate changes, their impact on improving indoor conditions will decrease [7]. Due to the construction of the blinds on the inside of the window, they provide little insulation against solar radiation and external gains, i.e., heat penetrates the interior of the room.

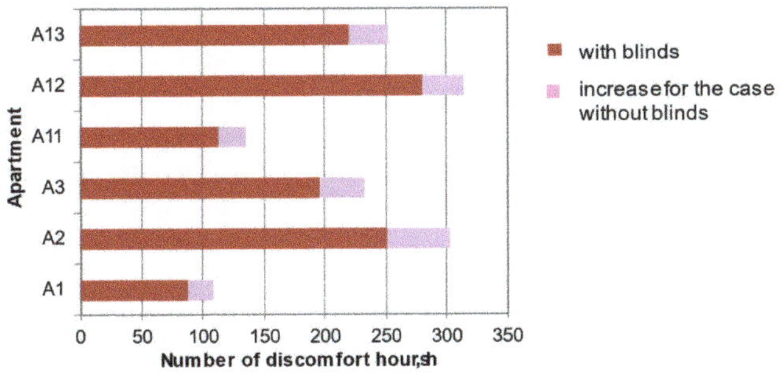

Figure 9. Number of discomfort hours for cases with (case 3) and without (case 4) internal blinds on windows.

3.2.3. Impact of Additional Ventilation by Open Windows (Cases 3 and 5)

Keeping the windows closed reduced the heating demand by 27% in apartment A13 to 37% in apartment A02, but caused extreme summer overheating, with the number of discomfort hours reaching 4000 h, which is almost half of the year. Window ventilation increased air change rates nine-fold and reduced discomfort by over 90% (Figure 10). With closed windows, the air change rate oscillated around the value of approximately 0.25 h^{-1}, which, for rooms with a volume of 130 to 190 m^3, provided a fresh airflow in the amount of 33 to 48 m^3/h. This airflow is too small and does not meet hygiene standards [66].

The effect of occupant behavior on window operation was also examined with respect to air exchange and thermal conditions. The maximum and minimum results from 10 simulations of case 3 were compared. Since the daily probability of opening or closing windows was set at 0.5, any user "errors" (e.g., not opening or closing a window when needed) could occur at different times of the year in each simulation. This variability significantly impacted indoor thermal conditions. While the average annual air change rate remained virtually unchanged, the number of discomfort hours varied by 34 to 66 h between simulations (Figure 10). In the case of apartment A11, this difference amounted to nearly half of the average discomfort hours across all simulations, highlighting the substantial influence of occupant behavior on indoor environmental quality. The impact of the ventilation system on thermal conditions and the heating demand for apartments is described in detail our previous study [7].

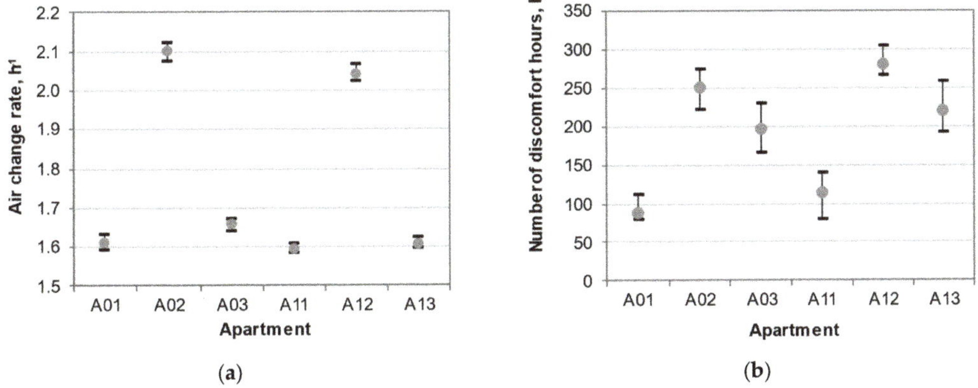

Figure 10. Minimum, maximum, and average (**a**) air change rate and (**b**) number of discomfort hours for apartments in simulations of case 3.

3.2.4. Impact of Additional Solar Radiation Control (Cases 6 to 9)

Solar radiation control strategies had varied effects (Figure 11). Balconies negligibly increased annual heating demand (1–2% due to lower solar gains in winter) but significantly reduced summer discomfort. For example, in apartment A12, a 160 cm deep balcony reduced discomfort hours by 86 h (31% reduction), while a 120 cm balcony reduced them by 67 h. The highest percentage reduction in discomfort hours—up to 42%—was observed in apartments A1–A11 with the deeper balconies. An increase in balcony depth by 40 cm resulted in an average improvement of 21 h in thermal comfort.

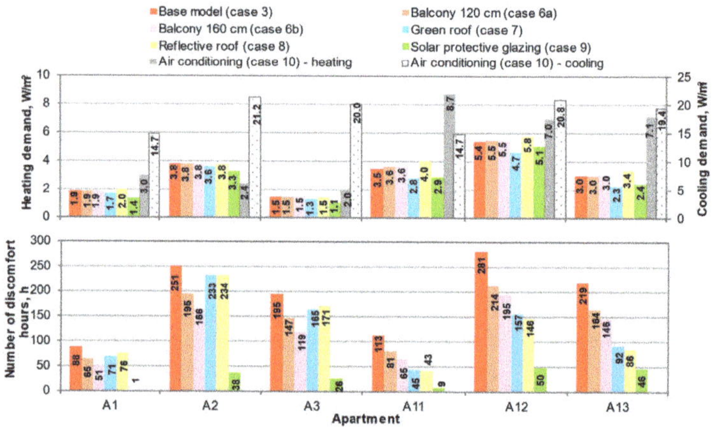

Figure 11. Energy demand for heating and cooling (only case 10) and number of discomfort hours for standard climate.

The green roof yielded a 20% reduction in heating demand on the top floor and 7% on the penultimate floor due to inter-floor thermal interaction. The top-floor apartments also experienced the largest reduction in discomfort hours—up to 60%—while the average improvement on the floor below reached 14%. These results indicate that cooler top-floor temperatures contribute to reduced thermal loads in the apartments below, supporting the use of green roofs as a sustainable alternative to mechanical cooling. Similarly, reflective roofs provided up to 57% discomfort reduction on the top floor but increased heating demand in winter (up to +13%).

Installing solar-protective glazing with low U-values reduced heating demand by 18% on average, with greater savings (by 7 percentage points) on the penultimate floor due to a higher window-to-surface ratio. The smallest savings were observed in apartments A2 (12%) and A12 (7%) due to their central location and limited glazing exposure. Nevertheless, this solution provided the greatest thermal comfort improvement: an average 90% reduction in discomfort hours on the penultimate floor and 85% on the top floor. In apartment A1, discomfort hours were eliminated entirely; in A11, they were reduced to 8 h. These apartments with fewer occupants (only one to two persons, see Figure A1), and thus lower internal gains, experienced the largest proportional benefits from solar gain reduction.

3.2.5. Mechanical Ventilation and Cooling (Case 10)

The model that includes mechanical cooling and mechanical ventilation is characterized by increased heating demand compared with a building without such systems (Figure 11). The increase in heat demand was a consequence of ensuring thermal comfort conditions and providing a larger amount of fresh air required by hygiene standards. The heat demand increased in most apartments by about 35–60% for apartments on the penultimate floor and almost 1.5 times for apartments on the top floor. In this case, the heat for ventilation increased despite the use of heat recovery. In apartment A12, the increase was small, only 28% and, in apartment A2, a decrease in heat demand was even observed. In the case of the middle apartments, which are most exposed to overheating, the windows were opened most often in the base model, resulting in the highest heat demand index in the base variant. Therefore, the increase in the heat demand for ventilation was no longer significant in this case and, in apartment A2, the heat demand for ventilation was even lower in the case of mechanical ventilation. More on this topic is provided elsewhere [7].

3.2.6. Combination of Passive and Active Solutions (Cases 11 to 13)

Typically, during renovation, several major works are carried out with the aim of modernizing the building. To make the best use of the solutions presented in this study, several variants were combined in different ways, thus creating their combinations. Figure 12 compares heating and cooling demands in buildings with such solutions as sun protection windows with a reflective roof, mechanical cooling with a reflective roof, and mechanical cooling with new windows. The solar protective glazing was the most advantageous variant so far in terms of ensuring thermal comfort to residents. Combining it with a reflective roof gave an even better effect, especially on the top floor, where the maximum number of hours of discomfort was only 19 h in apartment A12 (without the reflective roof, it was 50 h). Heating demand increased, on average, in apartments by 9% compared with the variant with only a sun protection window.

In apartments with a mechanical cooling system, replacing the window with a sun-proof version was a good solution. It caused a decrease in the demand for cooling by 27% on average in apartments, but also a decrease in the heating demand by 8%. In the case of combining mechanical cooling and a reflective roof, a decrease in cooling demand was achieved by 13% (with the main effect on the top floor), but the demand for heat increased by 7% on average in apartments. However, this is a much cheaper solution.

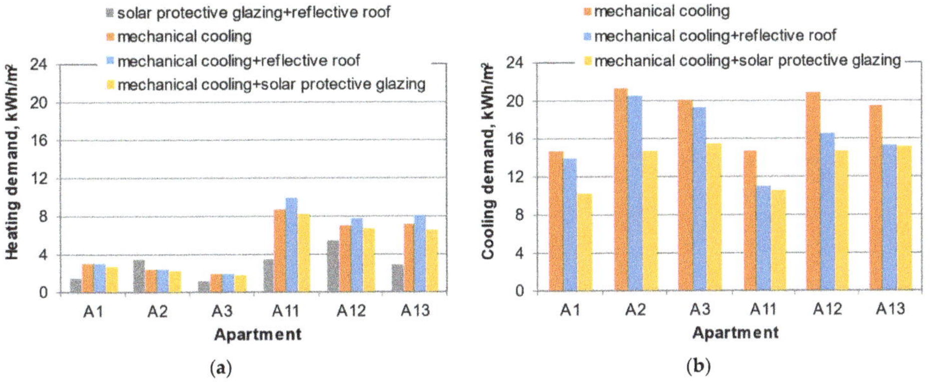

Figure 12. (a) Heating and (b) cooling demand for apartments with various combinations of improvements.

3.2.7. Future Climate Conditions (Cases 14 to 19)

In the projected warmer climate, the heating demand in the apartments of the base building drastically decreased in some apartments by as much as 6 times (Figure 13). The average heat demand index was very low and did not exceed 1 kWh/m^2. It can be concluded that, in the future, in well-insulated buildings, the heating cost will be negligible. Unfortunately, increasing outdoor temperature will deepen the problem of thermal discomfort for residents. For example, for apartment A11, the number of thermal discomfort hours increased by as much as 780 h per year and, for apartment A12, by 900 h/year, i.e., an eight- and four-fold increase, respectively. The worst conditions (similar to the standard climate) were recorded in apartments A2 and A12. The unfavorable conditions lasted there for more than 13% of the year. Considering that this was mainly the summer period from May to September, this was more than 30% of this period.

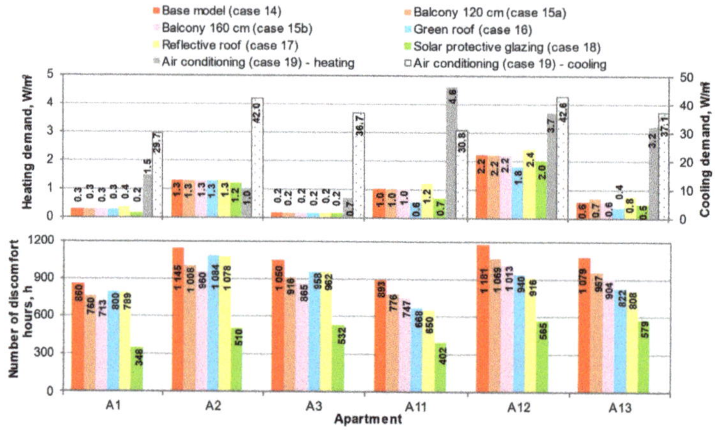

Figure 13. Energy demand for heating and cooling (only case 19) and number of discomfort hours for future climate.

The use of shading in the form of balconies provided a similar insignificant effect on the heat demand, as in the case of the standard climate. However, comparing the effect of the balcony in the standard and the warm climate on the thermal conditions, the results showed a decrease in the effect with the increase in the average global temperature. The improvement in thermal conditions was more than half as small in the climate of 2050.

For the warm climate, a greater effect of using green and reflective roofs was observed to reduce heat demand. For the upper floor, using a green roof resulted in a 28% decrease in heat demand. Similar to the standard climate, the smallest effect of green and reflective roofs on reduced heat demand was observed for apartments A2 and A12. The effect of using a green roof and a reflective roof on reducing the number of thermal discomfort hours was twice that of 2050.

The use of new windows with solar protective glazing reduced the heat demand for the apartments. The heat demand decreased by 20% on average, which was very similar to the standard climate. The smallest decrease in heating demand was achieved for apartments A2 and A12, located in the middle of the segment. The use of new windows also increased the thermal conditions in the warmer climate. However, this effect was much smaller than in the standard climate and improved thermal conditions by 53% on average. The results on the individual trends in the number of thermal discomfort hours were the same as for the standard climate, and the least significant for middle apartments with windows facing the southwest.

The cooling demand for the climate of 2050 was twice as high as in the standard climate, and the increase was very even across all apartments. The values of heat demand in the warmer climate were low and accounted for 2% to 15% of the cooling demand for individual rooms. On average, heat demand represented 7% of cooling demand in the warmer climate. The presented analysis reflects the expected energy use of air conditioning in apartments required to ensure thermal comfort, without implementing additional solutions to reduce cooling demand.

4. Discussion

Thermal comfort was not assessed in the field measurements of this study (this analysis was carried out based on numerical calculations); however, measurements showed that overheating of dwellings in summer is a problem not only in warm climates—for example, as shown by Murtyasa et al. [71] in a terraced house in Malaysia—but even in temperate climates. Similar indoor temperature ranges (from 23 °C to 28 °C) were recorded in living spaces in the UK climate during the heatwave in June 2018 [1]. Even higher indoor temperatures (exceeding 30 °C) were reported in the same climate by Gupta et al. [28] in 2019 and by Zahiri and Gupta in 2022 [72]. Overheating was also observed during measurement campaigns in Spain [73] and Finland in summer of 2020 [74] and 2021 [75]. Overheating may be defined differently depending on the region, but there is a general agreement that exceeding 26–27 °C is problematic [31].

The study reaffirms previous findings [71,76] that highly insulated envelopes—while beneficial in winter—can trap heat in summer and worsen thermal conditions in buildings. Similar conclusions were drawn by Fallah et al. [77], who emphasized that insulation retrofitting, especially in hot climates, can increase cooling energy demand. D'Agostino et al. [78] also advised caution with excessive insulation in buildings with high internal heat loads. A systematic review by Hu et al. [79] showed that passive techniques can reduce indoor temperature by 2 °C on average, cut cooling loads by 30%, and increase thermal comfort hours by 23% in warm climates. The results of this study confirmed that such effects can also be relevant in temperate climates.

In the context of thermal retrofitting, improvements should target the weakest thermal elements, e.g., walls, windows, and especially roofs, which accumulate heat due to solar exposure [20]. Green roofs and solar protective windows significantly reduce thermal loads. However, their applicability may be constrained by structural limitations or winter performance trade-offs, e.g., reflective roofs increase heating demand due to reduced solar gains in winter compared with conventional or green roofs [20].

In the projected 2050 warmer climate, the relative effectiveness of individual retrofitting measures remains similar but the magnitude of their impact changes. The average energy savings from selected passive measures (green roofs and solar windows) increases from 16% in standard climate to 18% in warmer conditions. As ambient temperatures rise, annual heating demand drops—by up to 72% in some simulated scenarios—thus extending the payback periods of insulation-focused retrofits.

Importantly, warmer climates significantly increase thermal discomfort. Simulations predict a four-fold rise in discomfort hours in multifamily apartments. In extreme cases, some studies predict that acceptable thermal conditions may be met less than 5% of the time annually [71]. Among evaluated passive strategies, solar protective glazing has the strongest effect, followed by balconies and canopies. However, their effectiveness varies by apartment layout and location. For top-floor apartments, green or cool roofs can reduce discomfort hours by over 50%, consistent with findings from Spain [20] and sub-Saharan Africa [27].

Given the time horizon of this study and the declining relevance of historical climate baselines, future-oriented scenarios were adopted. Notably, 2023 was the hottest year on record [80], underlining the urgency of adaptation strategies, especially for durable prefabricated concrete buildings that still have long operational lifespans [81].

However, some limitations must be acknowledged. First, climate variability and uncertainty present inherent challenges in long-term simulations. Although representative climate data for 2050 were used, such projections are based on specific scenarios and assumptions. Real-world climate evolution—especially the frequency, intensity, and duration of heatwaves—may differ significantly from modeled scenarios. Therefore, future studies should consider multiple representative concentration pathways (RCPs) to capture a range of possible outcomes. Second, occupant behavior modeling, particularly window operation, remains a major source of uncertainty. In this study, a stochastic behavioral model was employed to reflect probable user interaction with windows. Nevertheless, behavioral responses can vary greatly across demographics, cultures, and daily routines. While the use of probabilistic patterns improves realism compared with deterministic schedules, the complexity of human decision making cannot be fully captured by simulations. Additionally, the calibration of the model relies on the available literature data, which may not fully reflect local behavioral patterns. Future research should therefore include field studies to validate and refine behavioral assumptions. Moreover, no indoor comfort survey was conducted in parallel with measurements, which limited the direct linkage between subjective comfort and simulated metrics. Incorporating qualitative occupant feedback in future studies would enrich the interpretability of results. Finally, while this study presents a comparative analysis of passive and active strategies, economic aspects (e.g., cost–benefit analysis) and real-world implementation feasibility were not covered in depth. From an economic perspective, the feasibility of implementing passive and hybrid strategies can vary significantly across solutions. Mechanical cooling systems, especially when paired with decentralized mechanical ventilation, tend to incur the highest installation costs. While they ensure thermal comfort, their long-term energy use and maintenance requirements raise concerns about affordability and sustainability. Similarly, the addition of balconies, though architecturally beneficial, typically involves higher construction costs and offers only moderate improvements in indoor thermal conditions. By contrast, solar protective glazing can provide the greatest reduction in discomfort hours at a similar investment cost to balconies, suggesting a more favorable cost-to-effectiveness ratio. The most economically advantageous strategies may be cool or green roofs, which are up to four times cheaper than solar protective windows. However, their effectiveness diminishes on lower floors, limiting their overall impact in multi-story buildings. These considerations suggest that passive strategies should be prioritized not only for their environmental benefits but also for their potential economic viability, especially in large-scale retrofitting programs targeting aging multifamily housing stock.

Despite these limitations, the findings contribute to a better understanding of the potential and constraints of passive retrofitting strategies under changing climate conditions. Given the risks associated with increasing air-conditioning reliance [82,83], prioritizing passive and hybrid solutions [28,71] remains essential for sustainable adaptation in the building sector.

5. Conclusions

The passive methods, suitable for the wide applications presented in this study, seem to be able to significantly reduce the demand for mechanical cooling on a large scale of retrofitted multifamily buildings in the context of climate change.

The main conclusions and recommendations are as follows:

- Insulation of the walls and roof significantly reduces heat demand, for example, a decrease of 42–75% depending on the apartment. Unfortunately, at the same time, thermal conditions worsen in summer due to the increase in heat accumulation in the partitions of the building.

- Internal blinds reduce the thermal discomfort number hours by only 11–19%, and their effect on reducing heat demand is negligible.
- The balconies reduce the overheating of the apartments in the summer, reducing the number of discomfort hours by 27–42%; they slightly increase the demand for heat in the winter (by 1 to 2%).
- A green roof reduces heat demand by 20% on the top floor and improves thermal conditions.
- A reflective roof effectively reduces the number of thermal discomfort hours in apartments on the top floor but increases the heat demand in the winter.
- Modern windows with solar protective glazing reduce the period of human thermal discomfort by an average of 90% and reduce the heat demand by 18%.
- Mechanical cooling effectively eliminates thermal discomfort but significantly increases energy demand, especially for mechanical ventilation.
- In future climate conditions (in 2050), the heat demand will decrease by 72%, but the number of thermal discomfort hours will increase by up to four times.
- Solutions such as balconies, green roofs, or modern windows lose some of their efficiency in warmer climates.
- The most effective combination is solar protective glazing and reflective roof, which reduces the number of thermal discomfort hours by 95%.
- Mechanical cooling should be avoided as a primary solution; instead, more passive measures are recommended to improve thermal comfort.

This study covered only one building; however, the building selected is a typical multifamily residential structure. Such buildings were constructed on a large scale between the 1960s and 1980s to address housing shortages caused by urbanization and population growth in Central and Eastern European countries, including Poland. Apartment blocks from this period were characterized by standardized design, construction technology, and functional layout. As a result, they serve as typical and easily comparable research subjects for energy retrofit analyses. A significant portion of the population in Poland and the region still resides in these dwellings. Therefore, energy retrofitting of this type of housing can bring substantial benefits to a large number of residents. In summary, the multifamily building from the 1960s is not only a symbol of the era of urban development but also a typical example of construction that requires adaptation to contemporary climatic and energy challenges. The results presented have practical implications for energy auditors, designers, urban planners, and decision makers, supporting the development of energy-efficient strategies for improving thermal conditions in aging residential buildings. These recommendations can be summarized as follows:

- Revise building regulations and energy retrofit programs to address both winter heat retention and summer overheating risks, especially in the context of a warming climate.
- Promote passive and hybrid solutions—such as solar protective glazing, green roofs, external shading, and reflective surfaces—as preferable alternatives to mechanical cooling, which increases energy demand and peak loads.
- Include overheating risk and indoor human thermal comfort as standard criteria in energy audits and public funding schemes for building retrofits.
- Prioritize retrofit support for post-war multifamily apartment blocks, which represent a large portion of the housing stock in Central and Eastern Europe and are particularly vulnerable to thermal discomfort.
- Encourage the use of dynamic simulations in design and evaluation processes to better reflect future climatic scenarios and adaptive responses.

Future Research

In this study, only the part of the building that includes the apartments most exposed to the impact of external conditions was analyzed. To expand this research, an analysis of the heat demand of the entire building is planned, which allows one to determine costs and the payback time of the investment. When a specific heat source is taken into account, it will also be possible to determine the positive or negative impact of the proposed solutions on the external environment. Additionally, future work may include a subjective assessment of occupants' thermal comfort and behavioral patterns through questionnaires. Such surveys would require the engagement of a large number of residents to provide reliable and representative results, and could serve as a valuable complement to the simulation-based findings presented in this study.

Author Contributions: Conceptualization, J.F.-G. and K.G.; methodology, J.F.-G. and K.G.; software, K.G.; formal analysis, J.F.-G.; data curation, K.G.; writing—original draft preparation, J.F.-G. and K.G. All authors have read and agreed to the published version of the manuscript.

Funding: This research received no external funding.

Data Availability Statement: Data are contained within the article.

Acknowledgments: The work was supported by the Polish Ministry of Science and Higher Education via a research subsidy. The authors would like to thank the residents for their consent to carry out measurements in their apartments.

Conflicts of Interest: The authors declare no conflicts of interest.

Appendix A

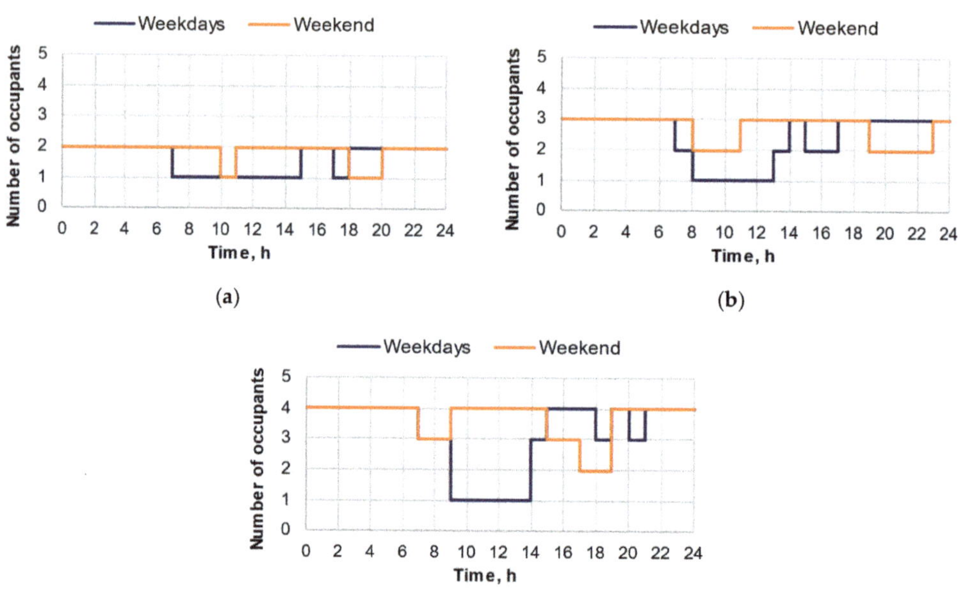

Figure A1. Occupancy schedule for apartments (**a**) A1 and A11, (**b**) A2 and A12, and (**c**) A3 and A13.

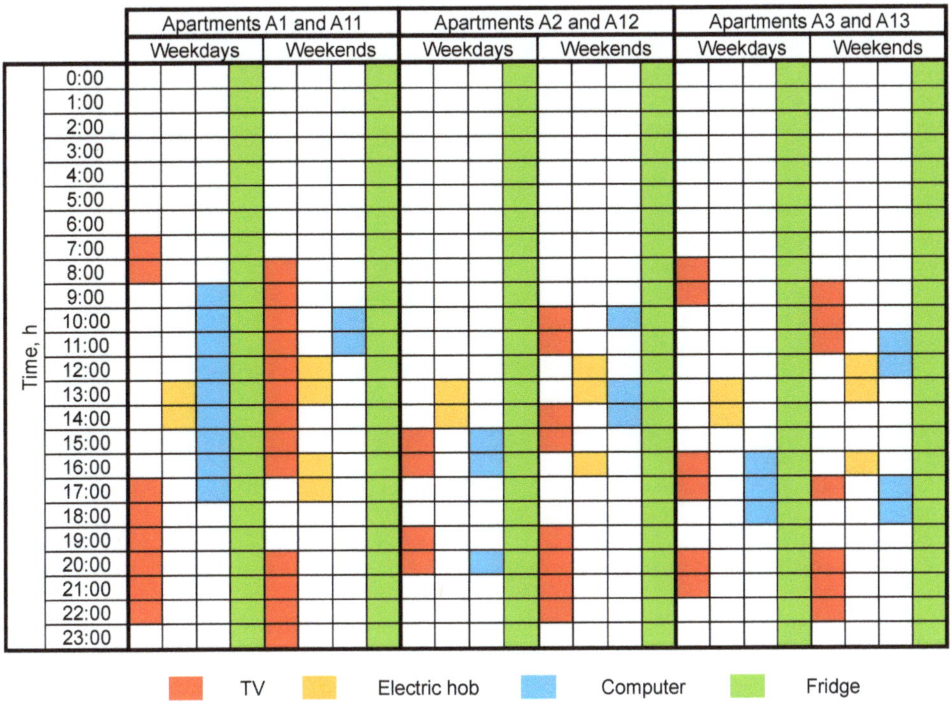

Figure A2. Schedule of use of equipment in apartments.

References

1. Ozarisoy, B. Energy Effectiveness of Passive Cooling Design Strategies to Reduce the Impact of Long-Term Heatwaves on Occupants' Thermal Comfort in Europe: Climate Change and Mitigation. *J. Clean. Prod.* **2022**, *330*, 129675. [CrossRef]
2. Laouadi, A.; Bartko, M.; Lacasse, M.A. A New Methodology of Evaluation of Overheating in Buildings. *Energy Build.* **2020**, *226*, 110360. [CrossRef]
3. Zhou, X.; Carmeliet, J.; Sulzer, M.; Derome, D. Energy-Efficient Mitigation Measures for Improving Indoor Thermal Comfort during Heat Waves. *Appl. Energy* **2020**, *278*, 115620. [CrossRef]
4. Mavrogianni, A.; Taylor, J.; Davies, M.; Thoua, C.; Kolm-Murray, J. Urban Social Housing Resilience to Excess Summer Heat. *Build. Res. Inf.* **2015**, *43*, 316–333. [CrossRef]
5. Lomas, K.J.; Watson, S.; Allinson, D.; Fateh, A.; Beaumont, A.; Allen, J.; Foster, H.; Garrett, H. Dwelling and Household Characteristics' Influence on Reported and Measured Summertime Overheating: A Glimpse of a Mild Climate in the 2050's. *Build. Environ.* **2021**, *201*, 107986. [CrossRef]
6. Tink, V.; Porritt, S.; Allinson, D.; Loveday, D. Measuring and Mitigating Overheating Risk in Solid Wall Dwellings Retrofitted with Internal Wall Insulation. *Build. Environ.* **2018**, *141*, 247–261. [CrossRef]
7. Ferdyn-Grygierek, J.; Grygierek, K.; Gumińska, A.; Krawiec, P.; Oćwieja, A.; Poloczek, R.; Szkarłat, J.; Zawartka, A.; Zobczyńska, D.; Żukowska-Tejsen, D. Passive Cooling Solutions to Improve Thermal Comfort in Polish Dwellings. *Energies* **2021**, *14*, 3648. [CrossRef]
8. Baborska-Narożny, M.; Stevenson, F.; Grudzińska, M. Overheating in Retrofitted Flats: Occupant Practices, Learning and Interventions. *Build. Res. Inf.* **2017**, *45*, 40–59. [CrossRef]
9. Dubrul, C. *Inhabitant Behaviour with Respect to Ventilation E A Summary Report of IEA Annex VIII*; Air Infiltration and Ventilation Centre: Montreal, QC, Canada, 1988.
10. He, Q.; Hossain, M.U.; Ng, S.T.; Augenbroe, G. Identifying Practical Sustainable Retrofit Measures for Existing High-Rise Residential Buildings in Various Climate Zones through an Integrated Energy-Cost Model. *Renew. Sustain. Energy Rev.* **2021**, *151*, 111578. [CrossRef]

11. Ministry of Housing, Communities and Local Government. *Research into Overheating in New Homes Phase 1 Report*; Ministry of Housing, Communities and Local Government: London, UK, 2019.
12. Domínguez-Amarillo, S.; Fernández-Agüera, J.; Sendra, J.J.; Roaf, S. The Performance of Mediterranean Low-Income Housing in Scenarios Involving Climate Change. *Energy Build.* **2019**, *202*, 109374. [CrossRef]
13. Rodrigues, E.; Fernandes, M.S. Overheating Risk in Mediterranean Residential Buildings: Comparison of Current and Future Climate Scenarios. *Appl. Energy* **2020**, *259*, 114110. [CrossRef]
14. Rahif, R.; Kazemi, M.; Attia, S. Overheating Analysis of Optimized Nearly Zero-Energy Dwelling during Current and Future Heatwaves Coincided with Cooling System Outage. *Energy Build.* **2023**, *287*, 112998. [CrossRef]
15. Pajek, L.; Košir, M. Exploring Climate-Change Impacts on Energy Efficiency and Overheating Vulnerability of Bioclimatic Residential Buildings under Central European Climate. *Sustainability* **2021**, *13*, 6791. [CrossRef]
16. Hao, L.; Herrera-Avellanosa, D.; Del Pero, C.; Troi, A. Overheating Risks and Adaptation Strategies of Energy Retrofitted Historic Buildings under the Impact of Climate Change: Case Studies in Alpine Region. *Appl. Sci.* **2022**, *12*, 7162. [CrossRef]
17. Liyanage, D.R.; Hewage, K.; Hussain, S.A.; Razi, F.; Sadiq, R. Climate Adaptation of Existing Buildings: A Critical Review on Planning Energy Retrofit Strategies for Future Climate. *Renew. Sustain. Energy Rev.* **2024**, *199*, 114476. [CrossRef]
18. Arriazu-Ramos, A.; Pons Izquierdo, J.J.; Ramos Ruiz, G.; Sánchez-Ostiz, A.; Monge-Barrio, A. Facing Climate Change in a Temperate European City: Urban-Scale Diagnosis of Indoor Overheating and Adaptation Strategies for Residential Buildings. *Buildings* **2024**, *14*, 1423. [CrossRef]
19. Chen, X.; Qu, K.; Calautit, J.; Ekambaram, A.; Lu, W.; Fox, C.; Gan, G.; Riffat, S. Multi-Criteria Assessment Approach for a Residential Building Retrofit in Norway. *Energy Build.* **2020**, *215*, 109668. [CrossRef]
20. Palma, R.M.; Medina, D.C.; Delgado, M.G.; Ramos, J.S.; Montero-Gutiérrez, P.; Domínguez, S.A. Enhancing the Building Resilience in a Changing Climate through a Passive Cooling Roof: A Case Study in Camas (Seville, Spain). *Energy Build.* **2024**, *321*, 114680. [CrossRef]
21. Gupta, S.K.; Chanda, P.R.; Biswas, A. A 2E, Energy and Environment Performance of an Optimized Vernacular House for Passive Cooling—Case of North-East India. *Build. Environ.* **2023**, *229*, 109909. [CrossRef]
22. Oropeza-Perez, I.; Østergaard, P.A. Active and Passive Cooling Methods for Dwellings: A Review. *Renew. Sustain. Energy Rev.* **2018**, *82*, 531–544. [CrossRef]
23. Rababa, W.; Asfour, O.S. Façade Retrofit Strategies for Energy Efficiency Improvement Considering the Hot Climatic Conditions of Saudi Arabia. *Appl. Sci.* **2024**, *14*, 10003. [CrossRef]
24. Ciardiello, A.; Dell'Olmo, J.; Ferrero, M.; Pastore, L.M.; Rosso, F.; Salata, F. Energy Retrofit Optimization by Means of Genetic Algorithms as an Answer to Fuel Poverty Mitigation in Social Housing Buildings. *Atmosphere* **2023**, *14*, 1. [CrossRef]
25. Xie, F.; Wu, Y.; Wang, X.; Zhou, X. Optimization Strategies for the Envelope of Student Dormitories in Hot Summer and Cold Winter Regions: Multi-Criteria Assessment Method. *Sustainability* **2024**, *16*, 6172. [CrossRef]
26. Ascione, F.; Bianco, N.; Mauro, G.M.; Napolitano, D.F.; Vanoli, G.P. A Multi-Criteria Approach to Achieve Constrained Cost-Optimal Energy Retrofits of Buildings by Mitigating Climate Change and Urban Overheating. *Climate* **2018**, *6*, 37. [CrossRef]
27. Rincón, L.; Carrobé, A.; Martorell, I.; Medrano, M. Improving Thermal Comfort of Earthen Dwellings in Sub-Saharan Africa with Passive Design. *J. Build. Eng.* **2019**, *24*, 100732. [CrossRef]
28. Gupta, R.; Howard, A.; Davies, M.; Mavrogianni, A.; Tsoulou, I.; Jain, N.; Oikonomou, E.; Wilkinson, P. Monitoring and Modelling the Risk of Summertime Overheating and Passive Solutions to Avoid Active Cooling in London Care Homes. *Energy Build.* **2021**, *252*, 111418. [CrossRef]
29. Porritt, S.M.; Cropper, P.C.; Shao, L.; Goodier, C.I. Ranking of Interventions to Reduce Dwelling Overheating during Heat Waves. *Energy Build.* **2012**, *55*, 16–27. [CrossRef]
30. Gupta, R.; Gregg, M. Care Provision Fit for a Warming Climate. *Archit. Sci. Rev.* **2017**, *60*, 275–285. [CrossRef]
31. Gupta, R.; Gregg, M. Assessing Energy Use and Overheating Risk in Net Zero Energy Dwellings in UK. *Energy Build.* **2018**, *158*, 897–905. [CrossRef]
32. Corcoran, L.; Saikia, P.; Ugalde-Loo, C.E.; Abeysekera, M. An Effective Methodology to Quantify Cooling Demand in the UK Housing Stock. *Appl. Energy* **2025**, *380*, 125002. [CrossRef]
33. Yannas, S.; Rodríguez-Álvarez, J. Domestic Overheating in a Temperate Climate: Feedback from London Residential Schemes. *Sustain. Cities Soc.* **2020**, *59*, 102189. [CrossRef]
34. Corcoran, L.; Ugalde-Loo, C.E.; King, L.; Demski, C.; Lowes, R. Analysing the Effects of Common Passive Cooling Strategies in UK Homes: 16th International Conference on Applied Energy, ICAE 2024. *Energy Proc.* **2024**, *55*, 11585. [CrossRef]
35. Zukowska-Tejsen, D.; Kolarik, J.; Sarey Khanie, M.; Nielsen, T. *Potential of Solar Control Solutions and Ventilation for Reducing Overheating Risk in Retrofitted Danish Apartment Buildings from the Period 1850–1970*; Technical University of Denmark: Lyngby, Denmark, 2019.

36. EA EBC Annex 62—Ventilative Cooling. Project Summary Report. 2019. Available online: https://venticool.eu/information-on-annex-62/annex-62-home/ (accessed on 13 June 2025).
37. UN Environment. Sustainable Buildings UNEP—UN Environment Programme. Available online: https://www.unep.org/topics/cities/buildings-and-construction/sustainable-buildings (accessed on 13 June 2025).
38. Passive Cooling Action Group. Available online: https://globalabc.org/passive-cooling-action-group (accessed on 13 June 2025).
39. Ascione, F.; Bianco, N.; De Masi, R.F.; Mastellone, M.; Mauro, G.M.; Vanoli, G.P. The Role of the Occupant Behavior in Affecting the Feasibility of Energy Refurbishment of Residential Buildings: Typical Effective Retrofits Compromised by Typical Wrong Habits. *Energy Build.* **2020**, *223*, 110217. [CrossRef]
40. He, Z.; Hong, T.; Chou, S.K. A Framework for Estimating the Energy-Saving Potential of Occupant Behaviour Improvement. *Appl. Energy* **2021**, *287*, 116591. [CrossRef]
41. Fabi, V.; Andersen, R.K.; Corgnati, S.P.; Venezia, F. Influence of User Behaviour on Indoor Environmental Quality and Heating Energy Consumptions in Danish Dwellings. In Proceedings of the 2nd International Conference on Building Energy and Environment, Boulder, CO, USA, 1–4 August 2012.
42. D'Oca, S.; Fabi, V.; Corgnati, S.P.; Andersen, R.K. Effect of Thermostat and Window Opening Occupant Behavior Models on Energy Use in Homes. *Build. Simul.* **2014**, *7*, 683–694. [CrossRef]
43. Fabi, V.; Andersen, R.V.; Corgnati, S.; Olesen, B.W. Occupants' Window Opening Behaviour: A Literature Review of Factors Influencing Occupant Behaviour and Models. *Build. Environ.* **2012**, *58*, 188–198. [CrossRef]
44. Delzendeh, E.; Wu, S.; Lee, A.; Zhou, Y. The Impact of Occupants' Behaviours on Building Energy Analysis: A Research Review. *Renew. Sustain. Energy Rev.* **2017**, *80*, 1061–1071. [CrossRef]
45. Ding, Y.; Han, S.; Tian, Z.; Yao, J.; Chen, W.; Zhang, Q. Review on Occupancy Detection and Prediction in Building Simulation. *Build. Simul.* **2022**, *15*, 333–356. [CrossRef]
46. Dong, B.; Liu, Y.; Mu, W.; Jiang, Z.; Pandey, P.; Hong, T.; Olesen, B.; Lawrence, T.; O'Neil, Z.; Andrews, C.; et al. A Global Building Occupant Behavior Database. *Sci Data* **2022**, *9*, 369. [CrossRef]
47. Liu, Y.; Dong, B.; Hong, T.; Olesen, B.; Lawrence, T.; O'Neill, Z. ASHRAE URP-1883: Development and Analysis of the ASHRAE Global Occupant Behavior Database. *Sci. Technol. Built Environ.* **2023**, *29*, 749–781. [CrossRef]
48. Andrić, I.; Pina, A.; Ferrão, P.; Fournier, J.; Lacarrière, B.; Le Corre, O. The Impact of Climate Change on Building Heat Demand in Different Climate Types. *Energy Build.* **2017**, *149*, 225–234. [CrossRef]
49. Carter, J.G.; Cavan, G.; Connelly, A.; Guy, S.; Handley, J.; Kazmierczak, A. Climate Change and the City: Building Capacity for Urban Adaptation. *Prog. Plan.* **2015**, *95*, 1–66. [CrossRef]
50. Camilleri, M.; Jaques, R.; Isaacs, N. Impacts of Climate Change on Building Performance in New Zealand. *Build. Res. Inf.* **2001**, *29*, 440–450. [CrossRef]
51. Escandón, R.; Suárez, R.; Sendra, J.J.; Ascione, F.; Bianco, N.; Mauro, G.M. Predicting the Impact of Climate Change on Thermal Comfort in A Building Category: The Case of Linear-Type Social Housing Stock in Southern Spain. *Energies* **2019**, *12*, 2238. [CrossRef]
52. Moazami, A.; Nik, V.M.; Carlucci, S.; Geving, S. Impacts of Future Weather Data Typology on Building Energy Performance—Investigating Long-Term Patterns of Climate Change and Extreme Weather Conditions. *Appl. Energy* **2019**, *238*, 696–720. [CrossRef]
53. Liu, S.; Kwok, Y.T.; Lau, K.K.-L.; Ouyang, W.; Ng, E. Effectiveness of Passive Design Strategies in Responding to Future Climate Change for Residential Buildings in Hot and Humid Hong Kong. *Energy Build.* **2020**, *228*, 110469. [CrossRef]
54. Castaño-Rosa, R.; Barrella, R.; Sánchez-Guevara, C.; Barbosa, R.; Kyprianou, I.; Paschalidou, E.; Thomaidis, N.S.; Dokupilova, D.; Gouveia, J.P.; Kádár, J.; et al. Cooling Degree Models and Future Energy Demand in the Residential Sector. A Seven-Country Case Study. *Sustainability* **2021**, *13*, 2987. [CrossRef]
55. Bjelland, D.; Brozovsky, J.; Hrynyszyn, B.D. Systematic Review: Upscaling Energy Retrofitting to the Multi-Building Level. *Renew. Sustain. Energy Rev.* **2024**, *198*, 114402. [CrossRef]
56. Statistics Poland: Energy. 2023. Available online: https://rzeszow.stat.gov.pl/publikacje-i-foldery/energia/energia-2023,4,1.html (accessed on 14 November 2024).
57. Statistics Poland: Housing Conditions in Poland According to the Results of the National Population and Housing Census. 2021. Available online: https://stat.gov.pl/en/national-census/national-population-and-housing-census-2021/final-results-of-the-national-population-and-housing-census-2021/housing-conditions-in-poland-according-to-the-resultsof-the-national-population-and-housing-census-2021,5,1.html (accessed on 14 November 2024).
58. National Energy Conservation Agency. *A Handbook of Typology of Residential Buildings with Examples of Measures to Reduce Their Energy Consumption*; NAPE: Warszawa, Poland, 2011.

59. Koppen Climate Classification Definition, System, & Map Britannica. Available online: https://www.britannica.com/science/Koppen-climate-classification (accessed on 4 January 2025).
60. Typical Meteorological Years and Statistical Climatic Data for the Area of Poland for Energy Calculations of Buildings. Available online: https://dane.gov.pl/pl/dataset/797,typowe-lata-meteorologiczne-i-statystyczne-dane-klimatyczne-dla-o (accessed on 4 January 2025).
61. Verichev, K.; Zamorano, M.; Carpio, M. Effects of Climate Change on Variations in Climatic Zones and Heating Energy Consumption of Residential Buildings in the Southern Chile. *Energy Build.* **2020**, *215*, 109874. [CrossRef]
62. Weather Data; Institute of Meteorology and Water Management—State Research Institute. Available online: https://meteo.imgw.pl/ (accessed on 4 January 2025).
63. Engineering Reference. *EnergyPlus TM*, Version 9.4.0. Documentation. US Department of Energy: Washington, DC, USA, 2020.
64. Dols, W.S.; Polidoro, B.J. *CONTAM User Guide and Program Documentation*, Version 3.4; National Institute of Standards and Technology: Gaithersburg, MD, USA, 2020.
65. Blochwitz, T.; Otter, M.; Arnold, M.; Bausch, C.; Clauss, C.; Elmqvist, H.; Junghanns, A.; Mauss, J.; Monteiro, M.; Neidhold, T.; et al. The Functional Mockup Interface for Tool independent Exchange of Simulation Models. In Proceedings of the 8th Modelica Conference, Dresden, Germany, 20–22 March 2011; pp. 105–114.
66. *EU Standard EN 16798-1:2019*; Energy Performance of Buildings-Ventilation for Buildings-Part 1: Indoor Environmental Input Parameters for Design and Assessment of Energy Performance of Buildings Addressing Indoor Air Quality, Thermal Environment, Lighting and Acoustics-Module M1-6. European Committee for Standardization: Brussels, Belgium, 2019.
67. *ANSI/ASHRAE Standard 55*; Thermal Environmental Conditions for Human Occupancy. ASHRAE: Atlanta, GA, USA, 2023.
68. ASHRAE. *Guideline 14-2014: Measurement of Energy and Demand Savings*; ASHRAE: Atlanta, GA, USA, 2014.
69. Mahmoodzadeh, M.; Mukhopadhyaya, P.; Valeo, C. Effects of Extensive Green Roofs on Energy Performance of School Buildings in Four North American Climates. *Water* **2020**, *12*, 6. [CrossRef]
70. Pilkington. Available online: https://www.pilkington.com/pl-pl/pl/produkty/funkcje-szkla/izolacja-cieplna/pilkington-insulight-therm (accessed on 22 November 2024).
71. Murtyas, S.; Qian, R.; Matsuo, T.; Tuck, N.W.; Zaki, S.A.; Hagishima, A. Thermal Comfort in a Two-Storey Malaysian Terrace House: Are Passive Cooling Methods Sufficient in Present and Future Climates? *J. Build. Eng.* **2024**, *96*, 110412. [CrossRef]
72. Zahiri, S.; Gupta, R. Examining the Risk of Summertime Overheating in UK Social Housing Dwellings Retrofitted with Heat Pumps. *Atmosphere* **2023**, *14*, 1617. [CrossRef]
73. Otaegi, J.; Hernández, R.J.; Oregi, X.; Martín-Garín, A.; Rodríguez-Vidal, I. Comparative Analysis of the Effect of the Evolution of Energy Saving Regulations on the Indoor Summer Comfort of Five Homes on the Coast of the Basque Country. *Buildings* **2022**, *12*, 1047. [CrossRef]
74. Velashjerdi Farahani, A.; Jokisalo, J.; Korhonen, N.; Jylhä, K.; Ruosteenoja, K.; Kosonen, R. Overheating Risk and Energy Demand of Nordic Old and New Apartment Buildings during Average and Extreme Weather Conditions under a Changing Climate. *Appl. Sci.* **2021**, *11*, 3972. [CrossRef]
75. Farahani, A.V.; Jokisalo, J.; Korhonen, N.; Jylhä, K.; Kosonen, R. Hot Summers in Nordic Apartments: Exploring the Correlation between Outdoor Weather Conditions and Indoor Temperature. *Buildings* **2024**, *14*, 1053. [CrossRef]
76. Du, L.; Leivo, V.; Prasauskas, T.; Turunen, M.; Kiviste, M.; Martuzevicius, D.; Haverinen-Shaughnessy, U. Energy Retrofits in Multi-Family Buildings in North-East Europe: The Impacts on Thermal Conditions. *Energy Procedia* **2015**, *78*, 860–864. [CrossRef]
77. Fallah, M.; Medghalchi, Z. Proposal of a New Approach for Avoiding Anti-Insulation in Residential Buildings by Considering Occupant's Comfort Condition. *Therm. Sci. Eng. Prog.* **2020**, *20*, 100721. [CrossRef]
78. D'Agostino, D.; de' Rossi, F.; Marigliano, M.; Marino, C.; Minichiello, F. Evaluation of the Optimal Thermal Insulation Thickness for an Office Building in Different Climates by Means of the Basic and Modified "Cost-Optimal" Methodology. *J. Build. Eng.* **2019**, *24*, 100743. [CrossRef]
79. Hu, M.; Zhang, K.; Nguyen, Q.; Tasdizen, T. The Effects of Passive Design on Indoor Thermal Comfort and Energy Savings for Residential Buildings in Hot Climates: A Systematic Review. *Urban Clim.* **2023**, *49*, 101466. [CrossRef]
80. NOAA National Centers for Environmental Information, Monthly Global Climate Report for Annual 2023. Available online: https://www.ncei.noaa.gov/access/monitoring/monthly-report/global/202313 (accessed on 4 January 2025).
81. Building Research Institute: Large-Panel Construction—Technical Condition Report. Available online: https://budowlaneabc.gov.pl/budownictwo-wielkoplytowe-raport-o-stanie-technicznym/ (accessed on 22 November 2024).

82. Wollschlaeger, S.; Sadhu, A.; Ebrahimi, G.; Woo, A. Investigation of Climate Change Impacts on Long-Term Care Facility Occupants. *City Environ. Interact.* **2022**, *13*, 100077. [CrossRef]
83. Stone, B., Jr.; Mallen, E.; Rajput, M.; Gronlund, C.J.; Broadbent, A.M.; Krayenhoff, E.S.; Augenbroe, G.; O'Neill, M.S.; Georgescu, M. Compound Climate and Infrastructure Events: How Electrical Grid Failure Alters Heat Wave Risk. *Environ. Sci. Technol.* **2021**, *55*, 6957–6964. [CrossRef] [PubMed]

energies

MDPI

Article

Simulated Results of a Passive Energy Retrofit Approach for Traditional Listed Dwellings in the UK

Michela Menconi *, Noel Painting and Poorang Piroozfar *

School of Architecture, Technology and Engineering, University of Brighton, Brighton BN2 4GJ, UK
* Correspondence: m.menconi4@brighton.ac.uk (M.M.); a.e.piroozfar@brighton.ac.uk (P.P.)

Abstract: Energy performance improvements in existing homes play a substantial role in the achievement of the UK's net-zero emissions target. However, retrofitting dwellings remains a particularly challenging task in the UK, where traditional dwellings make up a large part of the building repository. Traditional dwellings' contribution to decarbonization has not yet been fully realized due to the risks imposed to the thermo-hygrometric balance of their constructions and to their heritage value. These tend to hinder the "fabric-first" approach for the retrofit of such dwellings, where active measures are often prioritized. The aim of this research is to propose a systemic approach to intervene in Traditional Listed Dwellings (TLDs) to improve their energy performance by means of passive retrofit measures and to shape a more future-proof heritage. A mixed methodology was developed that utilizes 19th C TLD case studies (CSs) in South-East England and dynamic energy simulation (DES) to investigate their current energy performance and possible improvements using responsible, safe and effective energy retrofit scenarios. Providing an overview of the methodology adopted in this research, this paper presents the main results of this study. This paper highlights the savings associated with the best-performing combinations of retrofit measures and the areas of intervention where the highest energy and carbon savings can be achieved.

Keywords: building envelope; energy retrofit; heritage dwellings; listed buildings; traditional dwellings

1. Introduction

In the UK, buildings are responsible for 20% of total greenhouse gas emissions; most of such emissions are due to dwellings, from the combustion of natural gas for heating [1]. Therefore, energy efficiency improvements for this part of the stock have been newly emphasized as a key strategy to work towards the achievement of the zero emissions target [2]. Active retrofit measures, e.g., transitioning away from gas boilers, are relatively straightforward to implement, primarily subject to the establishment of appropriate regulatory frameworks. Passive measures, which focus on enhancing the building envelope to reduce energy demand, on the other hand, present significantly greater challenges. These challenges arise from a variety of factors, as outlined below.

Approximately one quarter of the UK's dwellings are considered traditional buildings, being built before 1919 [3]. Their envelope constructions are made of solid permeable masonry walls, single-glazed timber windows and uninsulated floors and roofs; therefore, these dwellings are generally poorly performing [4]. However, these are also "hard to treat" properties [5]; in fact, any retrofit intervention must consider their delicate thermo-hygrometric balance, or it might result in unintended consequences, detrimental for the health of the fabric and occupants [3,6]. Furthermore, most traditional buildings have some heritage value, contributing to the character and aesthetics of their area [7]. Consequently, approximately one quarter of the traditional housing stock in the UK is either listed as or within a conservation area [8]. This makes the energy upgrade of such dwellings even more challenging as the potential energy savings from retrofit measures need to be weighed against the damage they may pose to their heritage value. As a result, the contribution of traditional dwellings to carbon reduction efforts within the residential sector has remained relatively limited thus far. Nevertheless, in the view of multiple stakeholders, historic buildings can play a substantial part in achieving the net-zero target and should be included in decarbonization policies [9]. This stresses the need for a major planning reform to enable heritage buildings' full contribution to be realized. It is the government's intention [10] to review, update and simplify the planning framework for listed buildings and conservation areas to facilitate sympathetic interventions "that support their continued use and address climate change" [11]. This could be achieved by revising the National Planning Policy Framework (NPPF) [12] to allow for a closer relationship between heritage protection and environmental sustainability. This principle is already contained in the NPPF, which describes sustainable development as one that contributes to protecting the historic environment while prescribing "positive improvements" to it [12]. Retrofit interventions may ultimately contribute to the aim of listing by allowing these assets to be kept in use and fully enjoyed by future generations.

Given this agenda, the study described in this paper aimed to propose suitable passive retrofit interventions for Traditional Listed Dwellings (TLDs) in South-East England to reduce their heating energy consumption, and hence their carbon emissions. The geographical setting chosen for this study is the city of Brighton and Hove, where buildings of traditional construction account for nearly 40% of the total dwelling stock [13]. Most of these traditional buildings date back to the early 19th C, the Regency period, when Brighton flourished from a fishing town into a seaside resort [14]. Over time, the grand terraced houses built in this period (Figure 1) were split into flats, and many were rented out, leading to a high rate of deterioration [15]. This study selected representative case studies (CSs) amongst the 19th C TLD population of Brighton and utilized a mixed-method approach centered around dynamic energy simulation (DES) to assess their current energy performance and the energy and carbon savings achievable by means of a range of suitable passive retrofit measures.

Figure 1. Brunswick square, one of the earliest Regency developments in Brighton.

After an overview of the methodology adopted for this research, this paper presents the results of this study, highlighting the most effective combinations of responsible and safe interventions, as well as the areas of intervention where the highest energy and carbon savings could be attained.

2. Literature Review

2.1. Energy Retrofit for Buildings: The Body of Research

A critical literature review was carried out to aid in the decisions regarding the methodological approach for this study. The analysis of the literature cast light on a gap in research concerning the investigation of the current energy performance and thermal behavior of TLDs and highlighted the need for a holistic and comprehensive methodology to propose interventions and to balance environmental and cultural issues.

Table 1 reports a synthesis of the studies reviewed, their geographic setting (highlighted in grey those carried out in the UK), the main method(s) utilized and the retrofit interventions assessed (in bold the passive measures).

Table 1. Review of the literature concerning energy retrofit.

Reference	Setting	Case Studies	Main Research Method(s)	Retrofit Intervention(s)
[16]	Ireland	Traditional dwelling	Simulation	**GFI, IWI**
[17]	Italy	Heritage public building	Simulation	Be, Bo, **DGW, DP, IWI,** L
[18]	UK	Traditional dwellings	Field observations Secondary data collection Thermographic surveys	**EWI**
[19]	Italy	Dwellings	Simulation	**EWI**
[20]	UK	Traditional dwellings	Experiment Simulation	**DGW, DP, GFI, IWI, RI**
[21]	UK	Dwellings	Building performance evaluation methods	**CWI, DGW, DP, GFI, IWI, RI**
[22]	UK	Listed dwellings	Simulation	Be, Bo, **CWI, DP,** PI, **SGW**
[23]	Italy	Dwellings	Experiment Simulation	**RI**
[24]	Croatia	Heritage public building	Simulation	**DGW, GFI, IWI, RI, SGW**
[25]	UK	Social housing	Simulation	Bo, **EWI, GFI,** PV, **RI, TGW**
[26]	Sweden	Heritage dwelling	Simulation	Bo, **DGW, DP, EWI, IWI, RI**
[27]	Ireland	Dwelling	Monitoring Thermographic surveys	**CWI, RI**
[28]	UK	Traditional dwellings	Monitoring Experiment Thermographic surveys	**IWI**
[29]	Italy	Heritage dwelling	Experiment Simulation	**Bo, DGW, IWI,** PV, SP
[30]	Sweden	Heritage dwellings	Literature review	**EWI**
[31]	UK	Buildings	Building Performance Evaluation methods	**DGW, DP, GFI, RI, SGW**
[32]	Italy	Dwelling	Simulation	Bo, **DGW, IWI,** SP
[33]	Portugal	Traditional heritage dwellings	Simulation	Be, Bo, **DGW, DP, EWI, GFI, RI, SD**
[34]	Italy	House	Simulation	Bo, **DGW, EWI,** PV, SP
[35]	Italy	Heritage office building	Simulation	**DGW, IWI,** L, **RI, SD**
[36]	Italy	Heritage public building	Monitoring Simulation	**CWI**
[37]	Denmark	Heritage public building	Monitoring Simulation	**IWI**
[38]	UK	Dwellings	Simulation	**CWI, DGW, IWI, RI,**
[39]	Scotland	n.a.	Experiment	**sDGW**
[40]	Scotland	Traditional house	Simulation	Bo, **DGW, GFI, IWI, RI**
[41]	Scotland	Listed buildings	Experiment	**sDGW**
[42]	UK	Traditional dwellings	Thermographic surveys	**EWI**
[43]	Italy	Apartment buildings	Monitoring Simulation	Bo, **CWI, EWI, GFI,** L, PV, **RI**
[44]	Greece	Residential buildings	Simulation	Be, **EWI, IWI**

Table 1. *Cont.*

Reference	Setting	Case Studies	Main Research Method(s)	Retrofit Intervention(s)
[45]	Germany	Heritage buildings	Thermographic survey Experiment Simulation Life cycle assessment	**DGW, DP**
[46]	Scotland	Victorian tenement building	Experiment Simulation	**IWI**
[47]	Scotland	Traditional wall	Simulation	**IWI**
[48]	Australia	Residential building	Energy audit Simulation	**DGW, DP**, L, **RI**, SP
[49]	UK	Residential buildings	Simulation	Bo, **DGW, EWI, IWI**
[50]	Italy	Heritage public building	Energy audit Thermographic survey Simulation	Bo, **DGW, GFI, IWI, RI**
[51]	UK	Traditional dwelling	Simulation	**TGW**
[52]	UK	Traditional dwellings	Simulation	Bo, **DGW, DP, GFI, IWI**, PV, **RI, SGW**
[53]	Denmark	Residential building	Experiment Simulation	Bo, **IWI, SGW**
[54]	Italy	Heritage buildings	Simulation	Bo
[55]	UK	Victorian house	Experiment	**DGWa, DP, GFI, IWI, RI, TGW**
[7]	UK	5 Traditional dwellings archetypal models	Simulation	Bo, **DGW, GFI, EWI, IWI**, L, **RI**, PV, **SGW, TGW**
[56]	Denmark	Listed residential building	Simulation	**DGW, IWI, RI**
[57]	UK	Victorian terrace house	Experiment Simulation	Bo, **DP, GFI, IWI, FR, SGW**
[58]	Turkey	Heritage public building	Simulation	Bo, **DGW, DP, GFI, IWI, RI**
[59]	UK and France	Social housing	Monitoring Simulation	**DGW, DP, EWI, GFI, RI**
[60]	Paraguay	Dwelling	Simulation Sensitivity analysis	**DGW, EWI, IWI, RI**
[61]	Denmark	Historic building	Simulation	**IWI**
[62]	Belgium	Traditional wall	Simulation Probabilistic analysis	**IWI**
[63]	UK	12 traditional heritage dwellings	Simulation	Be, Bo, **EWI, GFI, RI, SD, SGW**

Be: Behavioral measures; Bo: boiler upgrade; CWI: cavity wall insulation; DGW: double-glazed windows; DGWa: double-glazed windows—argon filled; DP: draught-proofing; EWI: external wall insulation; FR: fabric repair; GFI: ground floor insulation; IWI: internal wall insulation; L: energy-efficient lighting; PI: pipe insulation; PV: photovoltaic panels; RI: roof insulation; SD: shading devices; sDGW: slim double-glazed windows; SGW: secondary glazing windows; SP: solar panels; TGW: triple-glazed windows.

The studies conducted on heritage buildings constitute only a minority of the whole body of research on building energy efficiency. The literature review showed that the use of energy simulation was extensive in such studies; when simulation was the main method, the use of CSs allowed for a reliable validation tool because it allowed for a comparison of simulated and measured data, providing a triangulation of the findings and better reliability of the results generated. However, the studies on heritage buildings were often limited either to the investigation of a single retrofit intervention or a single CS of (often public) buildings, whereas little has been conducted on heritage buildings using multiple CS methods. The studies conducted by Ingram [64] and Moran [52] on traditional dwellings used a limited number of CSs and a limited validation strategy (energy consumption data for some of the cases). Organ et al. [7] and Wise et al. [63] attempted a more comprehensive approach to the problem of traditional heritage dwellings retrofit; however, their results had some limitations due to the use of steady-state models and archetypal dwellings, as well as the lack of a validation strategy based on measured data. Their methodology could be improved using multiple methods of data collection, a quality check and analysis and multiple CSs to enhance the validity and reliability of the results generated. Due to the wide range of uncertainties in the input data and in the assumptions necessary to model traditional dwellings, the potential of such models to accurately represent the thermal behavior of the real building depends on the quality of the data used for calibration. Only a few studies on traditional heritage buildings calibrated their results with real data, and they were frequently limited to a single CS [7,17,24,26,29,35–37,46,47,50,52,54,57,58,61,63].

2.2. Energy Retrofit for Buildings of Heritage Value: Guidance and Advice from Conservation Bodies

To make an informed decision concerning the retrofit measures available to improve the energy performance of TLDs, a wide range of studies were consulted, covering regulation, guidance and advice, as well as previous academic research on the energy retrofit of traditional buildings and buildings of heritage value.

When addressing energy improvement interventions for older buildings of heritage value, a special approach is advocated unanimously by conservation bodies; one that takes the fabric of such buildings in special account for their heritage value and thermo-hygrometric behavior, while aiming to improve their energy performance [3,6,65–84].

Traditional buildings are made of porous, breathable materials; their envelope can easily absorb moisture from the indoor and outdoor environment and release it when it dries out. Thick, solid masonry walls also provide them with high thermal inertia. These proprieties combined facilitate traditional buildings' ability to buffer humidity and heat fluctuations [6,83]. Sympathetic interventions for traditional dwellings must account for such a delicate balance in their thermo-hygrometric behavior [4,16,65,77,78]. Therefore, before planning any retrofit interventions on buildings of traditional construction, a clear understanding of their heat and moisture behavior is key to avoid unintended consequences, such as the problems of moisture condensation and timber decay [6,83].

A whole-building approach is often called for by conservation bodies, one that considers the building as a whole and in its context, to find "balanced solutions that save energy, sustain heritage significance, and maintain a comfortable and healthy indoor environment" [77] (p. 9). This can be achieved by ensuring that any adopted measure is responsible (respects the heritage features of the building), safe (accounts for the thermo-hygrometric equilibrium of its constructions, minimizing the risk of unintended consequences) and effective (improves energy performance).

The validity of a 'fabric-first' approach, widely recommended to date for energy improvement interventions, has recently been questioned due to the urgency of the net-zero target, which might be more quickly addressed at a large scale by first decarbonizing heating for existing dwellings [85]. Even more difficult is the application of the fabric-first approach for TLDs, whose envelopes are their most vulnerable and valuable features. Retrofit interventions in this case must aim to strike a balance between improving energy performance, respecting the special character of the dwelling and avoiding risks for the health of the building's fabric and occupants.

The least invasive measures are the most favorable to conservation bodies [75,77,78,80]; therefore, the systematic maintenance and repair of the building fabric and systems should be a priority before considering any more invasive interventions [6,77,83,84,86]. Thus, a whole-building retrofit approach tailored to TLDs should consider improving building services and controls and addressing the occupants' behavior; lastly, any alteration to their fabric should be addressed [57,66].

3. Methodology—Materials and Methods

3.1. The Research Framework

The body of literature recommends a sequential approach to address energy efficiency improvements and the management of change for heritage buildings [4,33,52,58,64,86]. The most common approach starts with establishing a baseline scenario for the evaluation of different retrofit solutions [25,87,88] and for heritage impact assessment [24,89,90]. It is then possible to identify potential changes to be introduced to the baseline scenario and measure their impact against the base case performance.

Our study adopted this framework, and hence a methodological sequential approach; this stems from identifying the heritage value recognized by the listing of selected 19th C TLD CSs and assessing the current energy performance and thermal behavior of the CSs investigated, then follows by devising appropriate retrofit solutions and assessing their effect on energy consumption and carbon emissions. The mixed methodology adopted in this study further develops this approach to include an analysis of the impact of such interventions on the thermo-hygrometric balance and heritage value of TLDs and comes up with a range of responsible, safe and effective energy retrofit scenarios.

Figure 2 describes the methodological approach of this study, the methods utilized (in the circles), the data collected or generated from them (bullet points) and the models created at each subsequent stage of research. After the critical review of the literature and the secondary data collection, primary data were gathered through visual and measured surveys, questionnaires, interviews and thermal imaging. The output of the first data collection stage was used to feed into the following data generation stages via dynamic energy simulation (DES) applications using the software IES-VE 2019 (for details, please see [91]). The first stage of simulation was run for the status-quo models; the energy consumption and indoor condition results from this stage were then compared with metered data (data collection 2 in Figure 2) to calibrate the models (For details, please see [92]. An example of the calibration process can be found in the Supplementary Materials S1). Next, the calibrated models were normalized to standardize any behavioral determinant of energy consumption. Finally, a baseline scenario of energy consumption was generated by upgrading all the normalized models by means of a high-energy-efficient boiler; this was aimed to facilitate a fair cross-case comparison where only heating energy consumption (HEC) and the envelope characteristics could be considered (for details, please see [93]). The baseline scenario was used as a benchmark in the following stages of research to assess the effectiveness of a range of passive retrofit solutions by weighing the HEC outcome of simulations pre- and post-interventions. Therefore, responsible passive retrofit measures were selected for the CSs (for details, please see [94]); being listed buildings, the responsible measures selected were those allowed by their listing, and hence were respectful of their heritage value. The measures were modelled and applied—individually and combined—to the baseline models, and further stages of simulations were run to investigate the potential risks to the thermo-hygrometric balance of their constructions (ensuring they are safe) and the changes in the energy efficiency of the selected CSs (aiding in the selection of the most effective measure). A parametric analysis of the proposed solutions permitted the development of responsible, safe and effective combinations of retrofit interventions; finally, a sensitivity analysis of the simulation results highlighted the areas of the envelope where appropriate combinations of interventions are more likely to generate the highest energy and carbon savings.

3.2. The Case Studies

The use of mixed methods, multiple units of analysis and multiple CSs enabled the triangulation of the findings, which enhanced the validity and reliability of the results generated. To maximize what could be learnt from the cases, the sample size had to cover the key variables in the population from which they were drawn. Such variables were identified as follows: (1) aspect (number of sides where windows are located; being flats in converted terraced houses, the CSs were either single-aspect-in the case of small properties opened only on one side-or dual-aspect-if opened with windows on two opposite sides; a few triple-aspect ones could be found at the end of a row of terraced houses, but they were not taken into account because of their rarity), (2) floor level and (3) orientation (for the double-aspect dwellings, the orientation of the main elevation was considered). Eight representative CSs, converted flats within grand Regency buildings, were chosen in the two earliest Regency developments in Brighton: Brunswick Town and Kemp Town (for a detailed account of the CS selection process and their existing constructions, please see [95]).

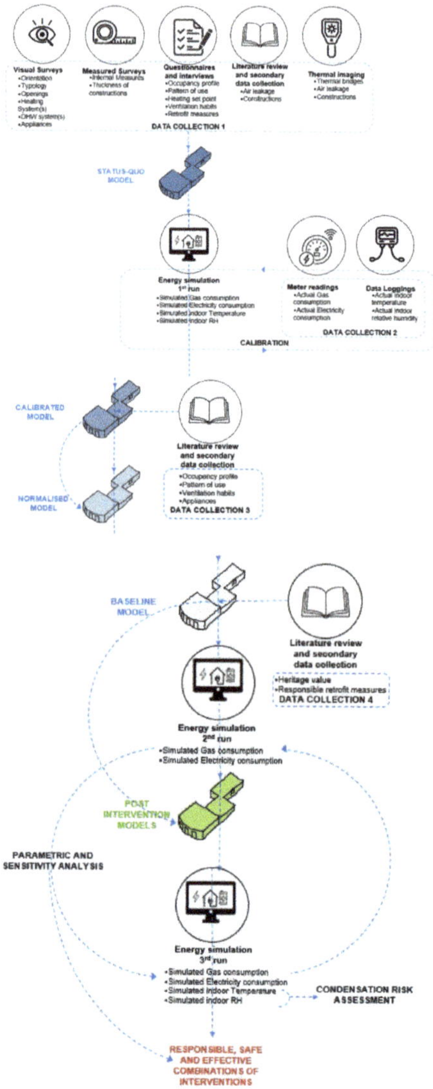

Figure 2. Research framework.

Tables 2 and 3 present a synthesis of the key variables for each CS, alongside the treated floor area (TFA-the heated floor area of the dwellings), form factor (calculated as the ratio of the thermal envelope to the treated floor area), windows-to-walls ratio (WWR), geometry and envelope constructions. The external walls in the dwellings investigated are made of uninsulated solid brickwork; most windows are (original or replacement) timber sashes, some are replacement casement windows and a few have secondary glazing or double glazing (in the original or replacement frame); the ground floors have solid or suspended timber constructions (uninsulated or insulated); and top-floor dwellings have pitched and/or flat (uninsulated or insulated) roofs.

Table 2. CSs 2–12 key variables, size, geometry and constructions.

CS Number		CS2	CS7	CS8	CS12
Conservation Area		Brunswick Town	Kemp Town	Montpellier and Clifton Hill	Kemp Town
Date		1st half 19th C	1st half 19th C	2nd half 19th C	1st half 19th C
Aspect		Dual aspect	Dual aspect	Single aspect	Dual aspect
Floor level		Middle floor	Middle and top floors	Middle floor	Ground floor
Orientation		West-facing	South-facing	West-facing	West-facing
TFAm^2		76.90	195.49	62.40	158.15
Form Factor		0.81	1.69	0.46	1.21
WWR		24	18	33	20
Geometry		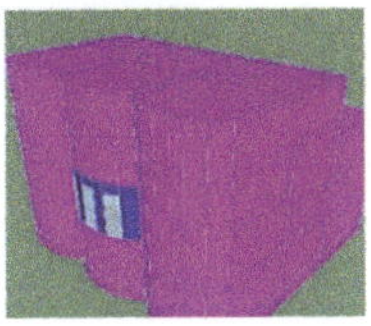			
Envelope Constructions	External wall—front	Lime render outside, brickwork (2 bricks), plaster-on-lath inside	3rd floor: Lime render outside, brickwork (1-and-half bricks), plaster-on-lath inside. Top floor: Cement render outside, Brickwork (1 brick), insulation and plasterboard inside	N.A.	Lime render outside, brickwork (2 bricks), plaster-on-lath inside
	External wall—back	Lime render outside, brickwork (1 brick), plaster on the hard inside	3rd floor: Lime render outside, brickwork (1, 1-and-half and 2 bricks), plasterboard inside Top floor: Cement render outside, Brickwork (1 brick), insulation and plasterboard inside	Lime render outside, brickwork (2 bricks), plaster inside	Lime render outside, brickwork (1 and 1-and-half bricks), plasterboard inside
	Windows—front	n.3 Timber sash—single glazing	3rd floor: n.3 Timber sash—double glazing n.3 Timber sash—double glazing + gas krypton Top floor: n.1 UPVC Sliding—double glazing—argon	N.A.	n.2 Timber sash—single glazing
	Windows—back	n.5 Timber sash—single glazing	3rd floor: n.3 Timber sash—double glazing n.1 Timber sash—secondary glazing n.1 Upvc Casement—double glazing—argon Top floor: n.1 Upvc Casement—double glazing—argon-filled	n.3 Timber sash—single glazing	Ground floor: n.1 Timber sash—single glazing n.2 Timber casement—double glazing n.1 Timber casement—single glazing 1st floor: n.2 Timber sash—secondary glazing n.1 Timber sash—single glazing n.1 Timber casement—single glazing
	Ceiling	Carpet floor, timber boarding, timber joists, plaster-on-lath ceiling	Carpet floor, timber boarding, timber joists, plaster-on-lath ceiling	Carpet floor, timber boarding, timber joists, plaster-on-lath ceiling	Carpet floor, timber boarding, timber joists, plaster-on-lath ceiling
	Ground floor	N.A.	N.A.	N.A.	Rear extension: Suspended timber floor—plastic tiles, timber boarding, timber joists, ventilation void
	Roof	N.A.	Timber pitched roof—insulation at loft level Timber flat roof—insulation between joists	N.A.	Rear extension: Timber flat roof—waterproof membrane, timber boarding, timber joists, ventilation void, insulation between joists

Table 3. CSs 13–17 key variables, size, geometry and constructions.

	CS Number	CS13	CS14	CS16	CS17
	Conservation Area	Kemp Town	Kemp Town	Brunswick Town	Brunswick Town
	Date	1st half 19th C	1st half 19th C	2nd half 19th C	1st half 19th C
	Aspect	Dual aspect	Dual aspect	Single aspect	Dual aspect
	Floor level	Middle floor	Lower ground floor	Lower ground floor	Ground floor
	Orientation	South-facing	East-facing	West-facing	East-facing
	TFAm^2	123.93	48.70	72.72	120.45
	Form Factor	1.25	1.94	1.63	0.89
	WWR	25	18	20	19
		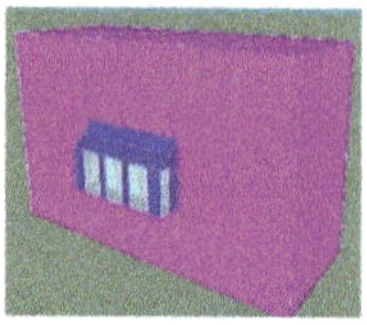	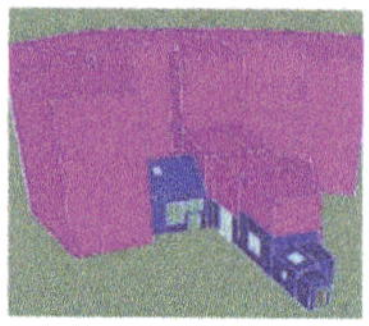	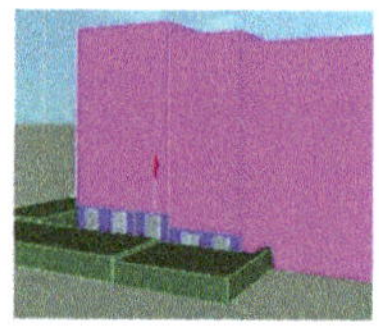	
Envelope Constructions	External wall—front	Lime render outside, brickwork (1-and-half bricks), plaster-on-lath inside	Lime render outside, brickwork (1 and 1-and-half brick), plasterboard inside	Cement render outside, brickwork (1-and-half bricks), plaster inside	Lime render outside, brickwork (1-and-half bricks), plaster-on-lath inside
Envelope Constructions	External wall—back	Lime render outside, brickwork (1 brick), plaster-on-lath inside	Painted brickwork (1 and 1-and-half bricks) outside, plasterboard and plaster	N.A.	Lime render outside, brickwork (1-and-half bricks), plaster on the hard and plaster-on-lath inside
Envelope Constructions	Windows—front	n.3 Timber casement—single glazing	n.2 Timber casement—single glazing n.1 Timber sash—single glazing n.2 Timber skylights—double glazing	n.5 Timber sash—single glazing	n.2 Timber sash—single glazing
Envelope Constructions	Windows—back	n.3 Timber casement—single glazing n.4 Timber sash—single glazing	n.4 Timber casement—single glazing n.1 Timber sash—single glazing n.1 Timber skylight—double glazing	N.A.	n.4 Timber sash—single glazing n.1 Timber casement—double glazing n.1 Timber casement—single glazing
Envelope Constructions	Ceiling/floor	Carpet floor, timber boarding, timber joists, plaster-on-lath ceiling	Carpet floor, timber boarding, timber joists, plaster-on-lath ceiling	Carpet floor, timber boarding, timber joists, plaster-on-lath ceiling	
Envelope Constructions	Ground floor	N.A.	Clay tiles and stone tiles, solid concrete floor	Suspended timber floor, timber boards	N.A.
Envelope Constructions	Roof	Above the kitchen and bedrooms: timber flat	Above the living room and bathroom: timber pitched—insulation between rafters	N.A.	Above the master bedroom: timber pitched

3.3. *The Selection and Simulation of Suitable Retrofit Measures*

Responsible retrofit measures for the CSs under investigation were devised following an in-depth review of the literature, regulation and guidance (for details, please see [94]). Further developing the list of interventions shaped by Historic England [77], the energy retrofit measures available were arranged into low- (green), medium- (amber) and high-risk (red) options (Table 4). Each of the measures available and permissible was then assessed against the level of impact it has on the heritage features needing protection (assessment of heritage significance [4]). Finally, a systematic approach was taken to shape the list of responsible measures to use in this study, subdividing the building envelope into six different areas of intervention (Table 4). The range of low-, medium- and high-risk interventions available for the CS models were then modeled in IES-VE to obtain a range of post-intervention scenarios.

Table 4. Responsible and safe retrofit options selected for each area of intervention.

Areas of Intervention		First Stage	Second Stage	
		Low-Risk Options	**Medium-Risk Options**	**High-Risk Options**
A1. Whole dwelling		L1. Draught-proofing	N.A.	N.A.
A2. Windows/ glazed doors	**A2a. Front**	L1. Heavy curtains and shutters	M1. Secondary glazing (single)	H1. Slim double glazing
			M2. Secondary glazing (double)	
	A2b. Back	L1. Heavy curtains and shutters	M1. Secondary glazing (single)	H1. Slim double glazing
			M2. Secondary glazing (double)	
A3. External doors	**A3a. Front**	N.A.	M1. Internal insulation board—wood fibreboard	N.A.
	A3b. Back	N.A.	M1. Internal insulation board—wood fibreboard	N.A.
A4. Ground floor	**A4a. Solid ground floor**	L1. Carpets—wool	M1. Thin insulation board—aerogel	H1. New limecrete floor
	A4b. Suspended timber ground floors	L1. Carpets—wool	M1. Thin insulation board—aerogel	H1. Insulation between joists—sheep wool
A5. Roof	**A5a. Pitched**	L1. Loft insulation—sheep wool	M1. Insulation at rafters' level between rafters—sheep wool	H1. Insulation at rafters' level above rafters—wood fibreboard
	A5b. Flat	N.A.	M1. Insulation between joists—sheep wool	H1. Insulation above flat roof—wood fibreboard
A6. Walls	**A6a. Internal face plaster-on-lath**	N.A.	M1. Internal blown insulation behind lath—cellulose	H1. Internal thin blanket insulation—aerogel
				H2. Internal board insulation—wood fibreboard
	A6b. Internal face solid or drylined		M1. New permeable plaster—cork	H1. Internal thin blanket insulation—aerogel
				H2. Internal board insulation—wood fibreboard

The colors assigned to the low-, medium- and high-risk measures is meant to highlight their respective level of risk, as it was also shown in the parametric tables of analysis (see Supplementary Materials S2).

The simulation of interventions was carried out in two stages. Firstly, it applied all the low-risk options available for each CS to the baseline model (B scenario) and generated the B_L scenario. The following stage involved the medium- and/or high-risk options applied to the B_L scenario. This strategy is unanimously recommended by conservation bodies [4,66,77,78,80,82,84] and previous research [6,83] which suggest the application of low-risk measures prior to medium- and/or high-risk ones. In fact, they are the most easily applicable and removable, and they imply lower costs, low levels of disruption, no need for planning application or listed building consent and the lowest risk of unintended consequences. To firstly apply low-risk interventions and in due course add other more expensive, delicate and disruptive ones is also a common practice, which was confirmed by interviews with the occupants.

To effectively manage the large number of simulations that resulted from the combination of the interventions selected and modeled in IES-VE, a coding system was devised. This way, each unique simulation's name indicated, for each CS, the variable (and its variation) or combination of variables (and their variations) associated with it. Table 4 summarizes the key areas of intervention (parameters in the sensitivity analysis), the measures (variations) for each area and their codes. The measures listed in this Table are allowable by regulation and are respectful of the heritage value of each specific CS (responsible).

The interventions proposed utilize breathable insulants and finishing materials, allowing for the movement of water vapor [4,78,83]. The hygroscopic properties of breathable insulation materials, together with the right amount of ventilation, are meant to store part of the vapor that can pass through the insulation from the inside, preventing it from reaching the cold masonry and condensing, while also permitting some moisture movement and evaporation both internally and externally [83]. All the modeled retrofitted constructions for the walls (A6) were also tested for condensation, utilizing the function available in IES-VE, to ensure that the newly created solutions were moisture safe. The IES condensation assessment is a steady state analysis, and is hence only capable of investigating vapor diffusion in each construction at a specific moment in time under steady boundary conditions. Like the Glaser method, it does not account for variations in the material properties due to changes in moisture content [21]. Therefore, the setting of boundary conditions that allow for the analysis of a worst-case scenario was considered extremely important to overcome the limitations that may be due to the use of a steady-state approach. The data input approach utilized in similar studies, together with what is suggested by standards (i.e., the Glaser method [96] and the dynamic hygrothermal simulation method [97], both described in BS 5250:2021 [98]) were considered to help make decisions regarding the boundary conditions to use for the analysis.

Because the IES analysis only assesses one specific moment in time, and because the highest difference between the indoor and outdoor temperature (leading to the highest risk of condensation) is recorded during the winter months for heated dwellings in this climatic area, the lowest monthly temperature value and the highest monthly relative humidity (RH) value were inputted in IES to account for the worst-case scenario in the external conditions. For the internal conditions, for each construction in this study, temperature set-points (21 °C in the living area and 18 °C in the rest of the dwelling) were used. For the indoor RH, the same value of 50% was used for all the constructions investigated. This reflects the average indoor RH based on the external daily mean temperature data, as recommended by BS EN 15026 [97], and was often used in precedent studies to assess the risk of condensation, even when a dynamic simulation method was adopted [37,44,61,99], to cite some). As a result of this analysis, wherever needed, the thickness of the insulation layer was decreased to decrease the temperature gradient between internal and external spaces and reduce the risk of condensation taking place within the envelope in order to generate a range of responsible and safe interventions. All the medium- and high-risk measures were finally applied to the B_L scenario, both individually and combined, aiming to achieve the target U-value given by the Building Regulations in place at the time of this study [11] for each thermal element. It was not always possible to achieve the target U-value due to heritage value considerations, condensation risk analysis, space constraints and the maximum thickness of insulation achievable with the material selected. Condensation risk was the main cause of the limits imposed on the thickness of the insulation layer added to wall constructions. This restricted the extent of U-value improvements. Table 5, which follows, shows, for each area of intervention, the area-weighted baseline U-value of the construction (in dark grey) and the U-values achieved as an outcome of the interventions (in light grey)-in bold those that met or exceeded the target U-value. The U-values post-retrofit ranged from 0.44W/m^{2}K (for solid walls-finished with plaster on the brickwork, as opposed to walls finished with plaster-on-lath internally-with aerogel insulation A6bH1-in CS7) to 0.92W/m^{2}K (for walls finished in plaster-on-lath and solid walls with wood fiberboard insulation—A6aH2 and A6bH2—in CS13).

Table 5. Envelope U-values pre- and post-interventions.

		Windows		Doors		Ground Floor		Roof		Walls	
	Target U-Value W/m^{2}K	**1.6** [1]		**1.8**		**0.25**		**0.18**		**0.30**	
		A2a Front	**A2b Back**	**A3a Front**	**A3b Back**	**A4a Solid**	**A4b Suspended**	**A5a Pitched**	**A5b Flat**	**A6a Plaster-on-Lath**	**A6b Solid/ Drylined**
CS2	U-value Baseline (B)	5.55	5.55							1.02	2.13
	U-value M1	2.35	2.35							0.55	0.84
	ΔU-value (B-M1)	3.20	3.20							0.47	1.29
	U-value M2	**1.44**	**1.44**							n.a.	n.a.
	Δ U-value (B-M2)	4.11	4.11							n.a.	n.a.
	U-value H1	n.a.	**1.58**							0.60	0.55
	Δ U-value (B-H1)	n.a.	3.97							0.42	1.58
	U-value H2	n.a.	n.a.							n.a.	0.68
	Δ U-value (B-H2)	n.a.	n.a.							n.a.	1.45
CS7	U-value Baseline (B)	1.63	2.22						0.27	1.18	0.95
	U-value M1	**1.48**	1.79						0.25	0.67	0.64
	Δ U-value (B-M1)	0.15	0.43						0.02	0.51	0.31
	U-value M2	**1.38**	**1.29**						n.a.	n.a.	n.a.
	Δ U-value (B-M2)	0.25	0.93						n.a.	n.a.	n.a.
	U-value H1	**1.48**	**1.44**						n.a.	0.66	0.44
	Δ U-value (B-H1)	0.15	0.78						n.a.	0.52	0.51
	U-value H2	n.a.	n.a.						n.a.	0.94	0.49
	Δ U-value (B-H2)	n.a.	n.a.						n.a.	0.24	0.46
CS8	U-value Baseline (B)		5.58								1.36
	U-value M1		2.25								0.69
	Δ U-value (B-M1)		3.33								0.67
	U-value M2		**1.31**								n.a.
	Δ U-value (B-M2)		4.27								n.a.
	U-value H1		**1.46**								0.71
	Δ U-value (B-H1)		4.13								0.65
	U-value H2		n.a.								0.57
	Δ U-value (B-H2)		n.a.								0.79
CS12	U-value Baseline (B)	5.40	3.68	2.20			0.91		0.21	1.03	1.40
	U-value M1	3.15	1.84	**1.73**			**0.24**		0.20	0.60	0.82
	Δ U-value (B-M1)	2.24	1.83	0.47			0.67		0.01	0.43	0.58
	U-value M2	2.52	**1.10**				n.a.		n.a.	n.a.	n.a.
	Δ U-value (B-M2)	2.87	2.58				n.a.		n.a.	n.a.	n.a.
	U-value H1	n.a.	**1.23**				0.27		**0.18**	0.61	0.75
	Δ U-value (B-H1)	n.a.	2.44				0.64		0.02	0.42	0.65
	U-value H2	n.a.	n.a.				n.a.		n.a.	n.a.	0.60
	Δ U-value (B-H2)	n.a.	n.a.				n.a.		n.a.	n.a.	0.80
CS13	U-value Baseline (B)	3.97	5.56						1.60	1.28	2.08
	U-value M1	2.30	2.31						0.33	0.95	0.95
	Δ U-value (B-M1)	1.49	3.26						1.27	0.33	1.13
	U-value M2	1.71	**1.38**						**0.18**	n.a.	n.a.
	Δ U-value (B-M2)	2.08	4.18						1.42	n.a.	n.a.
	U-value H1	n.a.	**1.53**						n.a.	0.78	0.78
	Δ U-value (B-H1)	n.a.	4.04						n.a.	0.49	1.30
	U-value H2	n.a.	n.a.						n.a.	0.92	0.92
	Δ U-value (B-H2)	n.a.	n.a.						n.a.	0.36	1.16

Table 5. *Cont.*

		Windows		Doors		Ground Floor		Roof		Walls	
	Target U-Value W/m^{2}K	1.6 [1]		1.8		0.25		0.18		0.30	
		A2a Front	**A2b Back**	**A3a Front**	**A3b Back**	**A4a Solid**	**A4b Suspended**	**A5a Pitched**	**A5b Flat**	**A6a Plaster-on-Lath**	**A6b Solid/ Drylined**
CS14	U-value Baseline (B)		5.09		2.20	0.68		0.28	0.28		1.39
	U-value M1		2.43		**1.73**	**0.24**		0.27	0.27		0.77
	Δ U-value (B-M1)		2.67		0.47	0.44		0.01	0.01		0.62
	U-value M2		1.66			n.a.		n.a.	n.a.		n.a.
	Δ U-value (B-M2)		3.43			n.a.		n.a.	n.a.		n.a.
	U-value H1		1.66			**0.25**		**0.18**	**0.18**		0.59
	Δ U-value (B-H1)		3.43			0.44		0.10	0.10		0.80
	U-value H2		n.a.			n.a.		n.a.	n.a.		0.65
	Δ U-value (B-H2)		n.a.			n.a.		n.a.	n.a.		0.74
CS16	U-value Baseline (B)	5.17					0.66				1.50
	U-value M1	2.16					**0.24**				0.72
	Δ U-value (B-M1)	3.01					0.42				0.77
	U-value M2	**1.31**					n.a.				n.a.
	Δ U-value (B-M2)	3.86					n.a.				n.a.
	U-value H1	**1.44**					**0.24**				0.70
	Δ U-value (B-H1)	3.73					0.41				0.80
	U-value H2	n.a.					n.a.				0.56
	Δ U-value (B-H2)	n.a.					n.a.				0.94
CS17	U-value Baseline (B)	5.60	5.20							1.39	1.56
	U-value M1	2.21	3.03							0.77	0.74
	Δ U-value (B-M1)	3.39	2.17							0.62	0.82
	U-value M2	1.25	2.41							n.a.	n.a.
	Δ U-value (B-M2)	4.35	2.79							n.a.	n.a.
	U-value H1	n.a.	1.97							0.71	0.50
	Δ U-value (B-H1)	n.a.	3.23							0.68	1.06
	U-value H2	n.a.	n.a.							0.67	0.53
	Δ U-value (B-H2)	n.a.	n.a.							0.72	1.03

[1] Whole window U-value.

The final stage of research assessed the effectiveness of the responsible and safe interventions applied by comparing the HEC output—and correlated CO_2 emissions—from the simulation to that of the B scenario to finally obtain combinations of responsible, safe and effective interventions. The results were assessed using parametric and sensitivity analyses where the areas of intervention constituted parameters, subject to variations as a result of potential retrofit interventions, that were reflected in the variance of their respective U-value. Section 4, which follows, reports and discusses the outcomes of the sensitivity analysis of the simulation results.

4. Results

The results of this study were produced in three stages as the outcomes of the following simulation scenarios:

1. Status-quo, normalized and baseline (B) scenario (for details, please see [93])
2. B_L scenario, obtained from the simulation of low-risk interventions applied to the B scenario (for details, please see [95])

3. Whole retrofit scenario, obtained from the simulation of medium- and/or high-risk interventions applied to the B_L scenario.

This paper presents stage 3 in the analysis to assess the impact of medium- and/or high-risk options—individually and combined—on the energy performance of the base-case models once all the low-risk interventions were implemented (B_L scenario [95]).

A total of 770 new models were generated for all combinations of medium- and/or high-risk options, which, added to the low-risk interventions simulations, totaled 800 simulations runs. The resulting annual gas and electricity consumption data were accessed in IES-VE utilizing Vista-pro and exported, firstly into Excel for parametric analysis, and then into SPSS 2020, for sensitivity analysis.

Figure 3 shows the HEC savings percentages after the application of medium- and high-risk interventions individually (OAAT) on the B_L scenario (B scenario upgraded with all the low-risk interventions applicable in each CS).

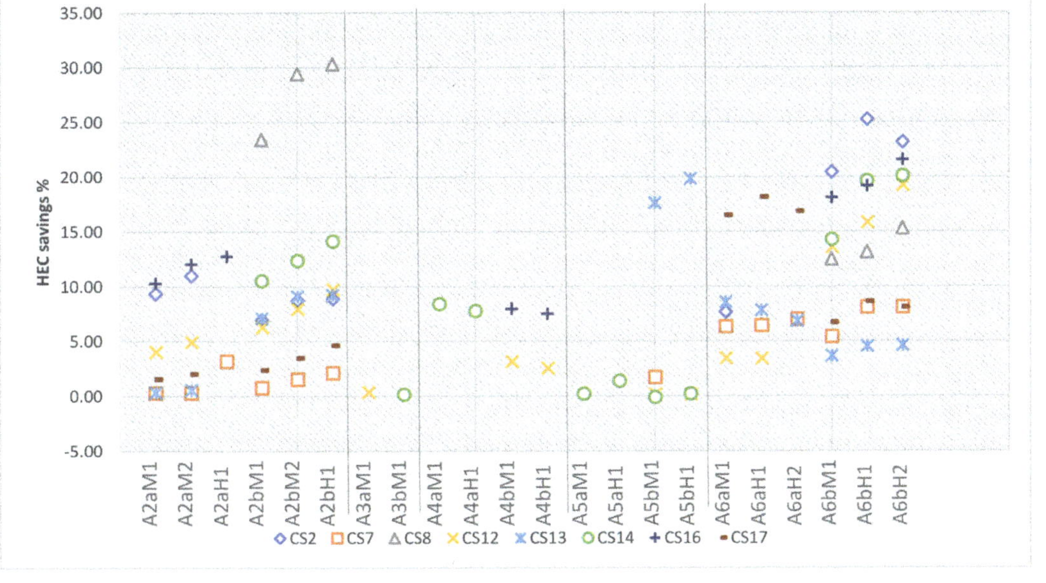

Figure 3. Annual HEC savings % [(kWh B_L-kWh post-intervention)/kWh B_L] achievable after the application of the medium- and high-risk interventions individually (OAAT) on the B_L scenario.

The savings percentages ranged from nearly 0 to more than 30% for different interventions. Interestingly, a wide range of results is also noticeable when looking at the same intervention across different CSs. This is likely due to the following:

- differences in the thermal envelope area addressed by the intervention
- differences in the baseline U-value of the same constructions across different CSs when they were subject to previous interventions aimed at improving their energy efficiency
- limitations imposed by heritage value, where the application of some interventions was not allowed or insulation was limited in thickness because heritage features could have been compromised
- limitations due to condensation risk, where the application of some interventions was not possible and might have required the use of a dehumidifier.

For these same reasons, there are intrinsic limitations in the comparison between the results from this study and from other studies, even when these were conducted in the same climatic conditions and on similar constructions.

The results of the simulations of all the combinations of measures were then used to produce a ranking matrix of combinations for each CS. To aid the assessment of the interventions selected, the tables of combinations highlight those producing the highest energy and carbon savings, alongside their level of risk, to come up with effective combinations that also imply the lowest risks to the thermo-hygrometric balance and heritage value of the envelope (Supplementary Materials S2).

Figure 4 shows a synthesis of the sensitivity analysis of the results from the simulation of the combinations of medium- and high-risk interventions, showing the parameters' % importance, and hence the impact that the retrofit measures—applied to each area of intervention—had on the reduction in HEC (and CO_2 emissions) in each CS.

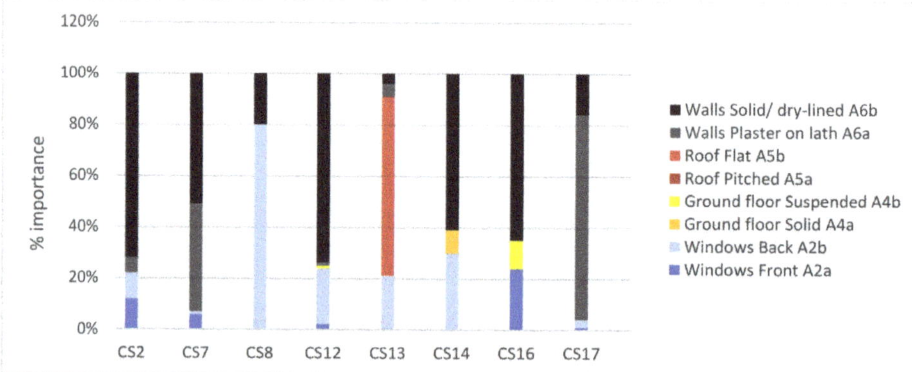

Figure 4. Parameter importance targeting HEC: % value of each area of intervention.

In six out of eight dwellings, the most important area of intervention was the external wall (in dark and light gray—A6), and more specifically solid walls (in dark gray—A6b). Solid (internally finished with plaster on the brickwork) walls also constituted the largest area of the thermal envelope in most CSs. Where walls finished in plaster-on-lath (in light gray in Figure 4—A6a) were found, they generally formed the front of the dwelling, often the most decorated part in Regency buildings. Therefore, this is where more limitations were likely imposed by the listing, which might hinder major alterations of the status-quo, limiting the fabric performance improvements achievable with the interventions. In all the dwellings except CS17, solid walls enclosed rooms to the back of the building, which were often the object of interventions over time. This might have led to the loss of the original plaster-on-lath, if there was any in place, hence allowing the application of a wider range of interventions.

Figure 5 represents a summary of the parameters' importance % of each area of intervention, obtained by summing up the % importance of the two sub-areas in which they are divided (e.g., A2 = A2a + A2b). It confirms that overall, walls (A6 = A6a + A6b, in gray) result as the most influential parameter targeting both HEC and CO_2 emissions, far outweighing windows (in blue), the ground floor (in yellow) and the roof (in red). Windows (in blue—A2) are generally poorly performing in the population investigated and offer a high potential for heat loss reduction. However, confirming the findings from the simulations of the medium- and high-risk interventions OAAT (Figure 3), windows only result as the most important parameter in CS8 (Figures 4 and 5). In this studio flat (62.4 m^2 of TFA), windows (A2b) take up 33% of the external envelope vs. 67% taken by walls (A6b), and the dwelling has the highest windows-to-walls ratio of all the CSs (33% vs. 21% average value of the other cases).

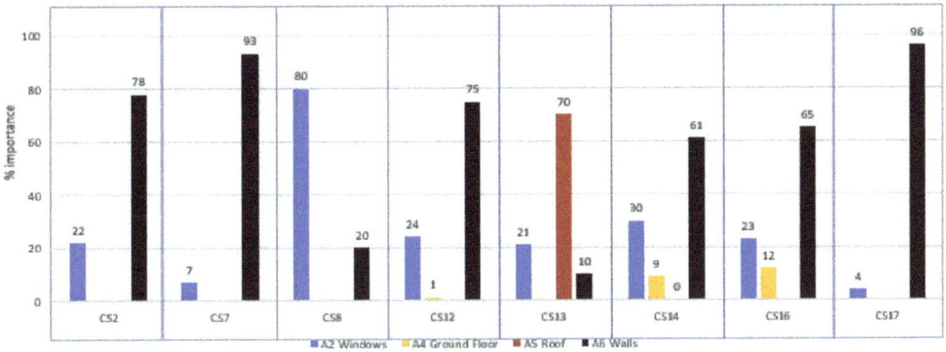

Figure 5. Parameter importance %: summing up all sub-areas of intervention.

The parameter importance results were then divided by their area in m_2 (Figure 6). Windows (in dark and light blue in Figure 6) resulted as the area of intervention having the highest importance/m_2 in five out of eight CSs, hence showing that retrofit interventions on them could be prioritized to maximize energy savings where a whole-building approach is not applicable.

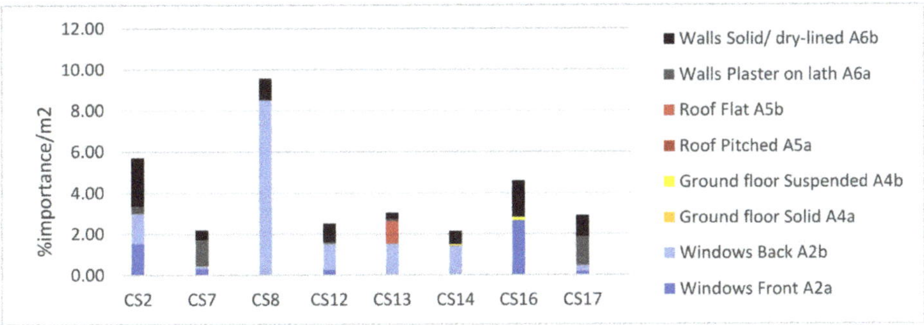

Figure 6. Parameter importance targeting HEC and CO_2 emissions per surface unit (m^2) of each area.

At this stage of our study, it was also considered necessary to assess how the importance of window interventions could change if including internal shading devices (this was considered a low-risk measure, and hence was part of the first stage of simulations of interventions, which was used to generate the BL scenario for the following stages of parametric and sensitivity analysis) as part of the window upgrades in the sensitivity analysis. A test was carried out on CS2. For the test, a new baseline scenario was created by excluding all internal shading devices from the B_L scenario. Then, the shading devices were added to the windows alongside each medium- and/or high-risk retrofit intervention. The new ranking obtained was the same as in the previous stage of analysis (where the impact of window interventions on the final HEC was estimated only based on secondary glazing and slim double-glazing interventions). Solid walls were confirmed to be the most important parameter; however, the percentage value of the parameter importance changed drastically, and windows resulted to be much more important when the interventions considered included the use of shading devices (from just above 10% to more than 30% for front windows and from nearly 10% to nearly 20% for back windows). The collective savings obtained from the front and back window (A2a + A2b) retrofit exceeded those obtained from the wall (A6a + A6b) retrofit.

The findings from this further stage of analysis confirm the need for a targeted study for each individual dwelling to devise the most responsible, safe and effective interventions. They contribute to the results of this study, shedding new light on the importance of a comprehensive approach to TLD retrofit, where the significance of retrofit interventions for windows should not be underestimated. This is especially true for the CSs where the area (m^2) of windows is higher than that of walls. Window interventions could then be prioritized in these cases where retrofitting internal wall insulation would noticeably reduce the usable internal space, result in higher condensation risks and/or impose higher risks to heritage value.

As previously noted by Historic England [7], original single-glazed window performance can sensibly be improved with the addition of double-glazed secondary glazing (A2aM2 and A2bM2 in Figure 3), resulting in windows becoming almost as energy efficient as if they were upgraded to slim double glazing (A2aH1 and A2bH1 in Figure 3). This can be particularly important for front windows, where the originals are often still in place, to ensure their conservation while improving performance. This is mostly true for the windows in CSs 2, 8, 12 and 16 (west-facing), as well as in CSs 7 and 13 (south-facing), which are exposed to the prevailing winds in this area (coming from the south-west). In fact, although double-glazed secondary glazing gives slightly lower energy savings compared to slim double-glazing in these CSs, their orientation is ideal to reduce the risk of condensation with this intervention [74,77,80]. Any intervention on windows in TLDs should, however, be negotiated with conservation officers due to the impact on their heritage value. In some dwellings, the risks imposed to the heritage value and the negative consequences for the usability of the window may outweigh the benefits of any medium- or high-risk interventions. In CS17, for instance, the savings obtainable by means of secondary glazing or slim-double glazing are minimal, ranging from 1.6% to 4.6% (Figure 3), and only the addition of internal shading devices permitted the achievement of similar savings (3.1%).

Further analyzing the parameter importance results presented in Figure 4, the flat roof (A5b) only results in the most important parameter for CS13, potentially producing up to 20% energy savings (Figure 3), far outweighing what can be achieved by means of wall insulation. This is certainly due, in this CS, to the large area ratio taken up by the flat roof construction (41%), uninsulated in the baseline scenario. In the other dwellings where roof constructions are found (CSs 7, 12, 14 and 17), these constitute a smaller area of the external envelope and are either suitable only for loft insulation (CSs 7 and 17) and/or have already been insulated in their baseline scenario (CSs 7, 12 and 14), which leads to a low potential of U-value improvement in the retrofitted scenarios.

In none of the ground floor-level dwellings investigated (CSs 12, 14 and 16) does the ground floor construction result in the most important parameter (Figures 4 and 5), and it is rather overshadowed by the importance of walls. In CS12, the energy savings achievable by means of retrofit options for the suspended timber ground floor are negligible (3.1% for aerogel insulation board above the ground floor construction—A4bM1—and 2.5% for sheep wool insulation between the joists—A4bH1—as shown in Figure 3) because only one room is in contact with the ground; hence, the area ratio of this construction is only 17% of the total thermal envelope. Anyway, in CSs 14 and 16, the results obtained for the solid ground floor (A4a, in CS14) and suspended ground floor (A4b, in CS16) show energy savings just below 10% and should not be disregarded (Figure 3). In these CSs, the whole dwelling area is at ground level, corresponding to 40% (in CS14) and 73% (in CS16) of the thermal envelope.

5. Discussion

5.1. Triangulation of Results Concerning Window and Wall Retrofit

The findings from this study confirm the results of most of the literature on building retrofit, both in the UK and elsewhere. Ben and Steemers [22] investigated non-traditional dwelling retrofit and found higher energy savings from cavity wall insulation (9%) than from secondary glazing (6%). The savings obtained from wall retrofit were smaller in this study than the ones reached in our study (approximately 5% to 25%—see Figure 3), which instead investigated traditional dwellings. However, the savings predicted from secondary glazing by Ben and Steemers were in the range of those calculated in our study. The study conducted by He et al. [38], simulating traditional and non-traditional dwellings in the north-east of England, obtained lower energy savings than our study but achieved a similar ranking of intervention effectiveness. Their study showed that the highest energy savings were achieved by means of solid wall insulation (heating demand reduction of 5.7%), which far outweighed those achievable by double glazing (heating demand reduction of 0.6%). A comprehensive review of the outcome of UK retrofit projects was conducted for the DEEP research paper [21] whose findings suggest that upgrading windows could reduce HEC by between 5% and 15% (slightly less than what was found in our study), while solid wall insulation could generate from 13% to 68% savings in HEC (within the same range as that obtained in our study), confirming the significant reductions in heat transfer achievable with solid wall insulation. The NEED analysis of domestic energy consumption [31] resulted in lower energy savings from solid wall insulation measures (median gas savings of 14%) compared to those achieved in our study. Nonetheless, solid wall interventions were potentially the most effective compared with the other measures evaluated in NEED (new condensing boiler, loft insulation, cavity wall insulation and PV panels). Other studies in cold climatic conditions came to different conclusions: from the analysis of an old Danish multi-family building retrofit, Morelli et al. [53] found the savings achievable with window interventions in the range of those achievable with interventions on walls—both about 20%. However, their best-performing intervention for windows used three panes of glass, which was applicable in our context due to the listing.

Most studies conducted in the Mediterranean area [24,32,35] came to similar conclusions to those achieved in most of the CSs in our research, showing the impact of energy-efficient windows to be less significant than the impact of thermal insulation on the HEC of the buildings investigated. The Mediterranean climate is the main cause for the low overall HEC savings post-intervention in the studies conducted in Turkey by Sahin [58] and in Portugal by Flores [33]. In both studies, the energy savings achievable with window retrofit interventions (for a public historic building and for historic dwellings, respectively) were estimated to be below those achievable through wall interventions. However, savings from each measure individually accounted for only below 5%. The study conducted on solid-walled residential building prototypes by Kolaitis et al. [44] showed a higher energy saving potential for internal wall insulation configurations when the modeled flat was simulated in the Mediterranean climate (65%) than when the same intervention was simulated in the colder Oceanic climate (47%) where the risks of condensation were higher.

There are also studies conducted in the Mediterranean area on historic public buildings, where window retrofit obtained the highest energy savings [17,50], showing opposite findings to those of our study. Ascione et al. [17] found that the heating energy demand was reduced by 37% with the application of new double-glazed windows (to achieve a U-value of 1.49 W/m^2K) vs. 9% with 5 cm of insulating plaster for most of the walls. Mancini et al. [50] showed potential HEC savings due to retrofit interventions for windows in the range of 28%, which far outweigh those achievable through other retrofit interventions, e.g., wall insulation (21%), loft insulation (9%) and floor insulation (7%).

Few studies refer to traditional dwellings in the UK, and even fewer address traditional dwellings of heritage value. Rhee Duverne and Baker [57], assessing the outcome of a range of retrofit interventions for a Victorian terraced house in the UK, conceded to the energy improvements achievable by reinstating old timber windows upgrading their glazing, which also aids in restoring the original character of the building. Their study showed that retrofit interventions for windows obtained a 43% heat loss reduction through the element; however, solid wall retrofit measures reduced heat loss through the wall by more than 80%. Wall insulation resulted in their study as the best-performing intervention (given the SAP rating). These data might not be directly comparable to the results of our study, but they confirm that interventions on walls can potentially generate the highest energy savings. The study conducted by Moran [52] in three 19th C terraced houses considered retrofit interventions applied progressively in packages. It included window retrofit in a first combination of measures involving draught-proofing, ground floor insulation and roof insulation, as well as boiler and appliance upgrades. The savings attributable to this package of interventions ranged from 8% to 51%, while those obtained when adding wall insulation to these measures ranged from 54% to 85%. Although in Moran's study it was not possible to directly compare the savings due to the walls' retrofit to those attributable to other parts of the building envelope, it is evident that external wall performance played an important role in determining the final HEC, and hence the CO_2 emissions, as shown for many of the CSs investigated here. A different conclusion was obtained in the study of a Victorian terrace house conducted by Neroutsou and Croxford [55]. Their study found a significant disproportion between the heating load reduction achievable with window retrofit interventions (30%) and that achievable with wall insulation (18%)—both aiming to meet the target U-value imposed by Building Regulations—despite the high percentage of the external envelope taken up by walls in their CS (based on an end of terrace house). When the building was simulated to meet EnerPHit standards (hence lower envelope U-values), the heating load reductions from retrofitting the walls (33%), roof (31%) and windows (32%) were similar.

5.2. Triangulation of Results Concerning Roof and Floor Retrofit

It is not possible to compare our results concerning roof retrofit with the majority of previous studies as most of them are either focused on retrofit interventions for walls exclusively [18,28,30,36,37,42,44,46,47,62], for windows only [40,41,45,51], or on combinations of multiple interventions, often including both active and passive measures [17,22,25,26,29,32–34,40,43,48–50,52,53,57,58] and frequently assessing them as a whole. Neroutsou and Croxford [55] found a 24% HEC saving potential from the pitched roof insulation between rafters in a traditional Victorian house when aiming to achieve a target U-value of $0.20 W/m^2K$, hence resulting in a similar scenario to the one found in our research. Similarly, the DEEP research paper [21] found that insulation at the rafters' level could reduce whole-house heat loss by 20%, as much as solid wall insulation. These data are not directly comparable with our results as the pitched roof constructions investigated were already insulated in the baseline scenario in our study.

Most of the previous studies did not include ground floor interventions or consider them together with a package of retrofit measures for the envelope. The DEEP literature review on domestic retrofit practices [21] found that suspended ground floor retrofits are rare in the UK despite there being around 10 million uninsulated floors. However, their study suggested that insulating these floors could reduce heat loss in homes by up to 20%, depending on how much infiltration is taking place through the ground floor. This value is much higher than that found in CS16 in our study, despite the large proportion of the thermal envelope taken up by the ground floor in that dwelling. This is possibly due to the combined effect of adding insulation and draught-proofing, which in our study are instead considered as two separate interventions (with draught-proofing as part of the first stage—low-risk interventions). The few studies that assessed the outcome of ground floor insulation for traditional dwellings generally showed lower savings than those found in our study. Wise et al. [63], investigating traditional heritage dwellings in the UK, obtained 1.5% and 1.7% HEC savings. This is likely because, in their study, the dwellings investigated presented a mixture of solid and suspended ground floor constructions, but only the suspended floors were insulated as part of the interventions proposed. Similarly, Neroutsou and Croxford [55] found a 3.5% HEC saving potential from the ground floor insulation in the Victorian two-story terraced house investigated. Rhee Duverne and Baker [57] found the energy savings achievable by floor insulation to be nine times smaller than those achievable by wall insulation (approximately 10 kWh/m^{2}yr vs. 90 kWh/m^{2}yr) for a Victorian terrace house and double those achievable by loft insulation. CSs 14 and 16 in our study showed similar energy savings obtained by ground floor insulation (from 7.7 kWh/m^{2}yr with limecrete floor—A4aH1—in CS14 to 8.9 kWh/m^{2}yr with thin aerogel board—A4bM1—for CS16). However, the maximum savings achievable by the walls in these CSs were below 20 kWh/m^{2}yr, hence just double the size of those achievable by ground floor insulation. These results are likely due to the thermal envelope area ratio taken by the ground floor constructions, which is certainly larger when assessing an individual dwelling (as in our study) than when assessing two-story terrace houses (like in Neroutsou and Croxford [55] and Rhee Duverne and Baker [57]).

5.3. Triangulation of Results Concerning the Whole Energy Retrofit Approach

A comparison of the results of our study with previous studies in the UK and elsewhere shows similarities but also many differences, even when considering the same population and similar climatic contexts. Such variances are a consequence of three main factors:

- differences in the specific baseline conditions and in the range of interventions, and hence in the achievable reduction in the U-values of the constructions
- differences in the surface areas of the envelope attributed to each intervention
- differences (if any) in the climatic conditions in which the studies were carried out.

This conforms to the suggestions made by heritage and conservation bodies (Historic England, 2012 and 2018a; Historic Scotland, 2013; STBA, 2012a) when it comes to traditional heritage building retrofit, articulating them further into the following conclusions:

- each building must be assessed individually
- each building is characterized by an individual baseline scenario
- it is essential to strike a balance between the following:
    - respect for heritage features
    - the thermo-hygrometric behavior of the construction; and
    - the target energy performance.
- These all play a special role in the selection of the most responsible, safe and effective retrofit measures and affect their outcome.

A wide range of energy and carbon savings were computed in this study based on different combinations of responsible and safe interventions. Figure 7 shows the HEC savings % achievable with the most effective combinations of the following:

- medium- and/or high-risk interventions against the B_L scenario (in blue)
- low-, medium- and/or high-risk interventions against the B scenario (in orange)
- boiler upgrade and low-, medium- and/or high-risk interventions against the normalized (N) scenario (in green).

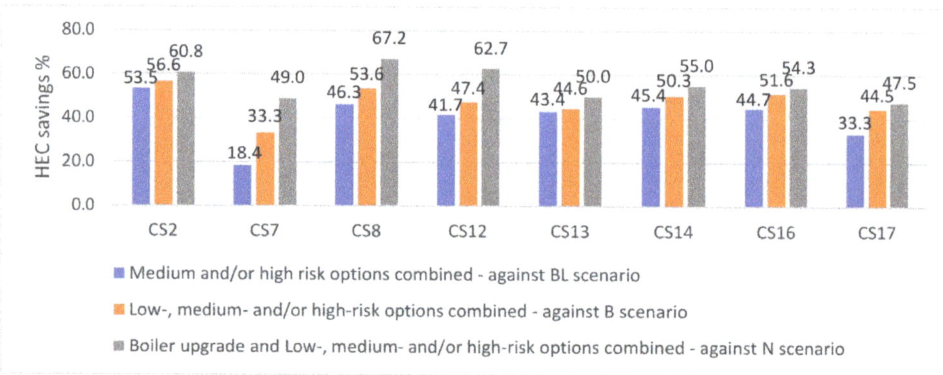

Figure 7. HEC savings % achievable with the most effective combinations of interventions.

The most effective combinations of medium- and/or high-risk interventions (in blue in Figure 7) for the selected CSs could provide energy savings ranging from 18% (CS7) to 54% (CS2) compared to the energy consumption in the B_L scenario. This is due to differences in baseline scenarios, where CS7 has already been partially retrofitted (added floor level, loft insulation and upgraded windows). In contrast, CS2, which has not been subject to any prior retrofit intervention, shows the highest energy savings potential despite its limitations due to heritage features and condensation risks. Furthermore, CS7 has a large form factor (1.69), which is double that of CS2 (0.81)—see Tables 2 and 3. Despite the windows-to-walls ratio (WWR) being higher in CS2—24% for CS2 vs. 18% for CS7—the windows can be retrofitted to higher levels of performance compared to the walls of CS2 (achieving a maximum ΔU-value of 4.11W/m^{2}K with secondary double glazing for both the front and back windows vs. a ΔU-value of 1.58W/m^{2}K with an aerogel blanket for solid walls). CSs 14 and 16 both show 45% energy savings with the best-performing combination of medium- and/or high-risk options, despite having large form factors (1.63 for CS16 and 1.94 for CS14—the largest of all CSs). This is likely because a large part of the thermal envelope of these dwellings is the ground floor construction (about 40% for the solid ground floor in CS14 and 61% for the suspended ground floor in CS16) and both dwellings are sheltered from the weather, being at a lower ground floor (with CS14 overlooking the internal courtyard of the building). Considering the impact of low-, medium- and/or high-risk interventions on the baseline scenario (orange bar in Figure 7), the savings range from a minimum of 33% for CS7 to a maximum of 57% for CS2. The savings due to low-risk options (the difference between the first and second bar for each CS in Figure 7) are highest for CS7 (where draught-proofing, internal shading devices and loft insulation can all be applied) despite the loft already being partially insulated. This is because the baseline air leakage rates are particularly high (ranging from 1.2 to 1.4ACH in most rooms vs. values for the other CSs ranging from 0.4ACH for CS14 to 1.1ACH for CS16); hence, CS7 has a larger potential for improvement by means of draught-proofing compared to the other dwellings (for details, please see Menconi et al. [95]).

When assessing these findings in the wider context of the literature, it is obvious that a unique pattern is very difficult to discern and probably less relevant because no two properties are the same. Nevertheless, this paper aims to present results that can be cross compared meaningfully not only between the CSs under investigation but also with those of other studies and can therefore be generalized with a satisfactory level of reliability both internally and externally. The energy and carbon savings potential of interventions are strictly related to the following:

- the geometry of each dwelling
- the thermo-physical characteristics of the envelope in its base case
- the area ratio of each part of the envelope to the whole thermal envelope
- the target performance to be achieved
- the limitations due to the thermo-hygrometric balance and heritage value of the constructions.

Bothwell et al. [25], simulating retrofit packages for the social housing stock in the UK, found a 25% HEC reduction when combining external wall, ground floor and loft insulation; triple-glazed windows; and draught-proofing. He et al. [38] reached an overall reduction of 13.9% in total heating demand by applying a range of passive retrofit options on a large sample of dwellings in the UK. Both UK studies resulted in energy savings predicted by the most effective combinations of interventions on dwellings (not of traditional construction), which were significantly lower than in our study.

Similar results to our study were obtained by Neroutsou and Croxford [55] and by Rhee Duverne and Baker [57] investigating passive retrofit measures for Victorian terrace houses (end-of-terrace and mid-terrace houses, respectively). They achieved approximately 50% and 43% energy savings, respectively, through the combination of all the measures selected.

When considering combinations of measures, a large part of the literature considers heating system upgrade alongside a range of passive measures applied to the envelope. To compare the findings of the current study with this part of the literature, the HEC savings achieved by the best-performing combinations of low-, medium- and/or high-risk passive measures together with boiler upgrade were considered (green bar in Figure 7); they ranged from 48% (CS17) to 67% (CS8). This resulted in a mean reduction in operational CO_2 emissions of 1800kg among the CSs selected. As expected, CS8 achieved the highest energy savings. This dwelling is heated by an old LPG heater and a small electric heater (in the bathroom) in its status-quo condition; hence, the change to a high-efficiency gas boiler certainly improved its energy performance. All the CSs showed considerable to significant improvements achievable by their heating system upgrade, resulting in an average 15% HEC savings potential across the CSs investigated. This finding further confirms the advice of conservation bodies concerning the importance of this intervention prior to any passive retrofit measure that can pose higher risks to the heritage value of the dwelling and to the thermo-hygrometric balance of its constructions [4,66,83,84], to cite but some.

The study conducted by Broström et al. [26] for dwellings of heritage value in Sweden resulted in annual energy savings ranging from just below 20% (for heating system upgrade, draught-proofing and loft insulation) to nearly 75% (when those measures were combined with external wall insulation and new triple-low-E-glazing windows). In the Mediterranean area, the study conducted by Dalla Mora et al. [29] demonstrated that the combination of all the retrofit interventions proposed for a traditional listed building (heating system upgrade, solar and photovoltaic panels, energy-efficient windows and internal wall insulation) achieved 88% HEC savings.

Amongst the UK studies, the NEED literature review [31] suggests that the HEC savings obtained by whole house retrofit studies in the UK range from 35 to 56%. Ben and Steemers's study [22] of a non-traditional heritage dwelling resulted in 30% energy savings achievable by means of the most effective combination of measures (including both passive measures and boiler upgrade). The study by Heat et al. [40] on a traditional heritage dwelling in Scotland utilized steady-state energy simulation software applications to simulate boiler upgrade together with a range of passive measures (double-glazed windows, ground floor insulation, loft insulation and internal wall insulation). They obtained a wide range of results, with average energy savings of approximately 65%. This high value can be attributed to the derelict conditions of the building in its baseline scenario, confirming the variability of the findings due to diversities in the status-quo conditions. Moran (2013) found savings ranging from 54% to 85%. This range of results is also a consequence of the range of baseline scenarios of the CSs (whose base-case energy use ranged from 499 kWh/m^2 to 197 kWh/m^2). The gap between these results and those of the current study is again due to different baseline scenarios (base-case energy use in the current study ranging from 231 kWh/m^2 to 40 kWh/m^2). Different U-value targets for the envelope—which were based on the passive house standards of Moran [52]—are another contributing factor to the differences between findings. Organ et al. [7] obtained HEC savings for the archetype models of a pre-1850 terrace house and two Victorian terrace houses ranging from 53% to 67% with a low-impact package of measures (including heating system upgrade alongside measures for the envelope). They also simulated the same models with a high-impact package of measures (including photovoltaic panels alongside all other measures, applied to a greater extent) and found HEC savings ranging from 68% to 75%. The Historic England report on carbon reduction scenarios in the built environment [7] concluded that out of the three sources of carbon savings evaluated (i. building fabric and air tightness improvements; ii. a shift away from fossil fuel-based heating; and iii. the decarbonization of the national electricity grid), building fabric improvements indicated the greatest potential for carbon reductions.

The results obtained in our study are in the range of those found in previous studies that investigated retrofitting traditional dwellings in the UK. These findings prove the significant energy and carbon saving potential for TLDs and in general for the pre-1919 dwelling stock. Our study showed that this such potential can be fulfilled by means of safe interventions that respect the heritage value of this part of the housing stock.

6. Conclusions

6.1. Summary of Results

This paper presented the results of our study aimed at devising a suitable whole house passive retrofit approach for TLDs. Each influential parameter analyzed in this study was considered and its contribution towards the change in HEC (hence, CO_2 emissions) assessed. Energy savings achievable through individual interventions and with the most effective combinations of interventions were discussed comparing the CSs transversally between each other and in the wider context of the literature. The analysis and discussion confirmed that each dwelling is unique, requiring individual assessment to devise the most effective, responsible and safe retrofit measures to improve energy performance and decrease carbon emissions while respecting heritage character and traditional construction. Nevertheless, the analysis highlighted some patterns in the results from this study that link the importance of each parameter to its corresponding area ratio of the envelope and to the actual change in its thermal performance as a result of interventions.

In this study, the most effective low-, medium- and high-risk interventions, combined with boiler upgrade, reduced operational CO_2 emissions by a mean of 1800kg among the CSs selected. As a rough estimate, if all the listed converted flats in Brighton (over 5000, as estimated from an analysis of the listed entries for Brighton and Hove—[15]) were retrofitted, approximately 9000 tCO_2 could be saved. If the whole number of traditional dwellings in Brighton (40,000 [13]) were retrofitted, the savings could increase to up to 72,000 tCO_2. The results of our study show that traditional heritage dwellings should not be overlooked in the effort to tackle the climate change crisis.

The following are wider implications that can be extrapolated from the findings of this study.

- Among the medium- and high-risk interventions, IWI generally outperforms all other measures. However, if combining window upgrade with draught-proofing and shading devices, interventions on windows can produce similar or higher HEC savings than those produced by IWI.
- IWI implies the risk of interstitial condensation. To avoid this, the insulation should be breathable, and the retrofitted construction should aim for a U-value ranging from 0.44 to 0.92 W/m^{2}K (for solid walls) and from 0.55 to 0.77 W/m^{2}K (for walls finished in plaster-on-lath). Hence, to be on the safe side, the target U-value suggested by AD Part L1B of the Building Regulations should not generally be aimed for dwellings of traditional construction.
- Among window measures, secondary slim double low-E glazing is the best performing and results in HEC savings higher than slim double glazing while implying less risk for the heritage value. To avoid risks of condensation, secondary double glazing should be prioritized, especially for windows facing south or south-west in Brighton, as these are exposed to its prevailing winds.
- Both secondary slim double low-E glazing and slim double low-E glazing allow the achievement of lower U-values than those asked for windows by AD Part L1B of the Building Regulations.
- External door retrofit produces only small HEC savings and is often discouraged in traditional dwellings for front doors due to heritage value considerations.
- Insulating ground floors can achieve nearly 10% HEC savings in both solid and suspended ground floors, especially when the ground floor construction takes up a large proportion of the external envelope.
- Both aerogel insulation blankets and sheep wool insulation (for suspended floors) or limecrete (for solid floors) help achieve similar energy savings for ground floors. The choice should be made by balancing the level of acceptable disruption. When there are no historic finishes to conserve, the option of aerogel blankets is preferable to limit the level of disruption; when historic finishes are in place, high-risk measures (i.e., sheep wool insulation or limecrete floor) are more suitable as they allow the conservation of the existing flooring material although causing more disruption.
- Roof insulation is an effective option when the roof is uninsulated and even more if it takes up a large proportion of the external envelope, e.g., in top-floor flats.
- For roofs, the choice between medium- or high-risk solutions mainly depends on the level of disruption acceptable and on the heritage value of the ceiling. When the ceiling is not decorated, sheep wool insulation is preferable if some disruption is acceptable. This may not be permitted if the occupants stay in the flat during the work. If there is a decorative ceiling, it is preferable to add insulation on top of rafters/joists when this does not lead to unacceptable changes in the uniformity of the roofs' height (i.e., in a row of listed terraces) and allowing for the higher costs associated with the removal and repositioning of the roof cladding.
- The highest HEC savings from the retrofit measures can be obtained, as expected, in the dwellings that have the highest thermal envelope-to-TFA ratio by combining low-, medium- and high-risk options with boiler upgrade.

6.2. Contributions of This Paper

The main novelty of this research lays in the rigorous, layered and systemic mixed-method approach taken to devise effective retrofit solutions for TLDs, applied and tested on multiple representative CSs, of which multiple units of analysis were accounted for. The critical review of the literature highlighted a major research gap indicating a lack of an all-inclusive methodology capable of ensuring that all participating factors (within the scope of the study) are considered, while the need for it has been repeatedly called for by academics. The methodological approach devised for this study builds upon that already taken by previous UK and international projects, e.g., CALEBRE, EFFESUS and RIBuild, as follows:

- it stems from a similar approach to retrofit to that of the CALEBRE project [100], aimed at improved air tightness and U-values of the external envelope
- it filters the range of measures selected through the identification of the specific heritage values to be protected in each CS
- it assesses the impact of each measure on such values, similarly to the EFFESUS [30] project, to come up with a list of responsible measures
- it applies a further filtering of the responsible retrofit measures selected, assessing the associated condensation potential, as in the CALEBRE and RIBuild [101] projects, to obtain a responsible and safe range of measures and determine in detail material build-ups for each of them
- it assesses the effectiveness of the interventions devised by measuring their impact on energy consumption and associated CO_2 emissions by means of DES, as in the CALEBRE project.

This strategy contributes to the novelty of this study, implementing what precedent studies had partially achieved, though they curtailed their ability to fully account for all the complex interrelated factors that characterize TLD retrofit. The strategy developed in this study, by contrast, aimed to take a holistic approach to address all these multiple aspects of the problem (HEC and carbon emission reduction, heritage value preservation and condensation risk). It does so by devising a trade-off between the need for individual solutions—accounting for the complexity of all factors involved in each dwelling—and the necessity of consistency in the rationale behind the choice of interventions and materials.

Due to its multifaceted, modular and adaptable approach, the methodology devised, used and tested for dwellings in South-East England is flexible and customizable; therefore, it could easily be applied to investigate the current behavior and formulate energy improvement solutions for similar or different buildings in similar or different contextual conditions.

The analysis and discussion of the results of this and similar studies showed that the solution to the problem of whole dwelling retrofit for TLDs cannot be easily generalized because each specific CS needs an individual assessment. Nevertheless, the results presented in this paper are a contribution to the body of research because a range of passive interventions were tested and studied in terms of their feasibility and effectiveness on selected TLD CSs to generate a framework of tested solutions. The need for decision making/assisting tools for retrofit has been highlighted in the literature to aid both users and designers in the choice of the most suitable and effective retrofit scenario. The envisaged results could, finally, provide a more reliable framework of tested solutions and could facilitate dialogue with conservation officers and public conservation bodies to extend the effective and healthy service life of heritage buildings, safely enhancing their energy performance while preserving their heritage values.

6.3. Research Limitations

6.3.1. U-Value Calculations

Invasive investigations (e.g., core sampling) were not possible in the CSs investigated due to the private ownership of the dwellings and their heritage value. Hence, the material build-ups for the whole envelope were based on the actual thickness of the construction, as measured in situ, the visual and tactile investigation of the internal and external surfaces, the thermal imaging survey and the literature review and conversations with local experts on the construction methods and materials of the time. The thermal performance of such envelopes, therefore, was subject to a certain degree of uncertainty. For this matter, it was also considered whether to collect primary data from the real cases to calculate the U-values of their envelopes using heat fluxmeters coupled with internal and external temperature sensors [102–104]. This could have been an option, adding to the credibility of the study, but did not prove to be the most practical one as it had cost implications and involved more time or repeated visits that might not have been welcomed by the participants. Furthermore, due to the range of materials used within the same envelope in traditional dwellings and to the varied thicknesses of one such envelope, the test should have been repeated in more than a few spots for each CS in order to reach credible fabric survey outcomes.

Therefore, the contingency plan was to use the calculated U-value for each building element according to the assumptions related to the material build-ups. A quality check of such values was achieved by comparing the U-values calculated by the software with those measured by similar research on traditional masonry buildings. The comparison showed U-values in the range of those measured [64,102–104] or calculated [105] by previous studies. A further confirmation of the assumptions made was given by the success of the calibration stage.

6.3.2. Condensation Risk Analysis

There are intrinsic limitations to the use of a steady-state method for condensation risk analysis. These were addressed in this study by the application of a worst-case scenario of boundary conditions. To assess how conservative the IES predictions were and whether this may have affected the results, a dynamic hygrothermal simulation test was carried out for a wall construction in its baseline scenario and post-interventions. The test confirmed the results given by the steady-state assessment carried out in IES.

6.4. Recommendations for Future Research

This study could be further developed to add other insights into this field of research, and the methodology applied here could also be further applied to other similar or different contexts to widen the impact of this research and allow for a wider breadth of results. Some areas where this research could be further developed, or the methodology could further be applied, are as follows.

6.4.1. Indoor Comfort

The retrofit measures selected and tested in this study were assessed against four main criteria, i.e., impact on heritage value, risk of condensation and potential reductions in both HEC and CO_2 emissions. The assessment of indoor comfort criteria pre- and post-intervention can add another useful output to this study and further contribute to its holistic approach. This could be important to address the risk of overheating for retrofitted dwellings and is particularly needed for dwellings retrofitted using internal wall insulation. A test run of thermal comfort and an overheating analysis using the TM59 methodology was carried out to ensure compliance with PAS 2035 [106] and can be used for future research. A further stage of simulation would then be needed to explore forthcoming climate scenarios using future weather projection averages.

6.4.2. Economic Implications of the Interventions

The cost of the interventions was excluded from this study. However, this could be a further decision factor to add to the investigation of suitable retrofit measures alongside other analysis criteria. The modularity and flexibility of the methodology devised in this study, however, prove effective and customizable to serve a wider analysis.

Supplementary Materials: The following supporting information can be downloaded at https://www.mdpi.com/article/10.3390/en18040850/s1: Table S1: CS2 Calibration stage 1-energy data; Figure S1: CS2 floor plan with annotated position of data loggers in the living area (L) and in the bedroom area (B1 and B2); Figure S2: CS2 Calibration stage 2a-graphic calibration-temperature data-Living room-5 May–1 July 2017; Table S2: CS2 Calibration stage 2b-NMBE and CV(RMSE) between simulated results and data logging-temperature data-Living room-5 May–1 July 2017; Figure S3: CS2 Calibration stage 2a-graphic calibration-relative humidity data-Living room-5 May–1 July 2017; Table S3: CS2 Calibration stage 2b-NMBE and CV(RMSE) between simulated results and data logging-relative humidity data-Living room-5 May–1 July 2017; Figure S4: CS2 Calibration stage 2a-graphic calibration-temperature data-Living room-14 December 2017–14 February 2018; Table S4: CS2 Calibration stage 2b-NMBE and CV(RMSE) between simulated results and data logging-temperature data-Living room-14 December 2017–14 February 2018; Figure S5: CS2 Calibration stage 2a-graphic-relative humidity data-Living room-14 December 2017–14 February 2018; Table S5: CS2 Calibration stage 2b-NMBE and CV(RMSE) between simulated results and data logging–relative humidity data-Living room-14 December 2017–14 February 2018; Figure S6: CS2 Calibration stage 2a-graphic -temperature data-Living room-30 May–1 August 2018; Table S6: CS2 Calibration stage 2b-NMBE and CV(RMSE) between simulated results and data logging-temperature data-Living room-30 May–1 August 2018; Figure S7: CS2 Calibration stage 2a-graphic calibration-relative humidity data-Living room-30 May–1 August 2018; Table S7: CS2 Calibration stage 2b-NMBE and CV(RMSE) between simulated results and data logging-relative humidity data-Living room-30 May–1 August 2018; Figure S8: CS2 Calibration stage 2a-graphic calibration-temperature data-Master bedroom-30 May–1 August 2018; Table S8: CS2 Calibration stage 2b-NMBE and CV(RMSE) between simulated results and data logging-temperature data-Master bedroom-30 May–1 August 2018; Figure S9: CS2 Calibration stage 2a-graphic calibration-relative humidity data-Master bedroom-30 May–1 August 2018; Table S9: CS2 Calibration stage 2b-NMBE and CV(RMSE) between simulated results and data logging-relative humidity data-Master bedroom-30 May–1 August 2018; Figure S10: CS2 Calibration stage 2a-graphic calibration-temperature data-Master bedroom-14 December 2017–14 February 2018; Table S10: CS2 Calibration stage 2b-NMBE and CV(RMSE) between simulated results and data logging temperature data-Master bedroom-14 December 2017–14 February 2018; Figure S11: CS2 Calibration stage 2a-graphic calibration-relative humidity data-Master bedroom-14 December 2017–14 February 2018; Table S11: CS2 Calibration stage 2b-NMBE and CV(RMSE) between simulated results and data logging-relative humidity data-Master bedroom-14 December 2017–14 February 2018; Table S12: Envelope U-values pre- and post-interventions TFA and thermal envelope area of the CSs; Figure S12: CS2 and areas of intervention A2 (windows) and A6 (wall); Figure S13: CS2 is a first-floor dwelling in the Brunswick Town Conservation Area; Table S13: Medium and high-risk interventions: individual ranking for CS2; Table S14: The combinations ranking matrix for CS2.

Author Contributions: Conceptualization, M.M., N.P. and P.P.; methodology, M.M., N.P. and P.P.; software, M.M.; validation, M.M.; formal analysis, M.M.; investigation, M.M.; resources, M.M.; data curation, M.M.; writing—original draft preparation, M.M.; writing—review and editing, M.M., N.P. and P.P.; visualization, M.M.; supervision, N.P. and P.P.; project administration, M.M. All authors have read and agreed to the published version of the manuscript.

Funding: This research received no external funding.

Data Availability Statement: The original data presented in the study are openly available at https://research.brighton.ac.uk/en/studentTheses/a-future-proof-cultural-heritage (accessed on 7 December 2024).

Conflicts of Interest: The authors declare no conflict of interest.

References

1. BEIS. 2020 UK Greenhouse Gas Emissions, Final Figures. 2024. Available online: https://assets.publishing.service.gov.uk/media/65c0d15863a23d0013c821e9/2022-final-greenhouse-gas-emissions-statistical-release.pdf (accessed on 10 June 2024).
2. CCC. *Progress in Reducing Emissions 2024 Report to Parliament*; Committee on Climate Change (CCC), 2024. Available online: https://gemserv.com/wp-content/uploads/2024/07/Gemserv-Summary-CCC-2024-Annual-Progress-Report-to-Parliament.pdf (accessed on 5 June 2024).
3. STBA. *Responsible Retrofit of Traditional Buildings*; Sustainable Traditional Buildings Alliance (STBA): London, UK, 2012. Available online: http://www.sdfoundation.org.uk/downloads/RESPONSIBLE-RETROFIT_FINAL_20_SEPT_2012.pdf (accessed on 25 October 2022).
4. Historic England. *Energy Efficiency and Historic Buildings—Application of Part L of the Building Regulations to Historic and Traditionally Constructed Buildings*; Historic Scotland: Edinburgh, UK, 2011.
5. STBA. *Performance and Energy Efficiency of Traditional Buildings: Gap Analysis Update 2020*; Historic England: London, UK, 2020.
6. May, N.; Griffith, N. *Planning Responsible Retrofit of Traditional Buildings*; STBA: London, UK, 2015.
7. Organ, S.; Wood, M.; Drewniak, D.; Lamond, J. *Carbon Reduction Scenarios in the Built Historic Environment: Final Report for Historic England*; University of the West of England: Bristol, UK, 2020.
8. Bottrill, C. *Homes in Historic Conservation Areas in Great Britain: Calculating the Proportion of Residential Dwellings in Conservation Areas*; University of Oxford: Oxford, UK, 2005.
9. Grosvenor. *Heritage & Carbon: How Historic Buildings Can Help Tackle the Climate Crisis*; Grosvenor: London, UK, 2021.
10. HM Government. *Adapting Historic Homes for Energy Efficiency: A Review of the Barriers*; HM Government, 2024. Available online: https://www.gov.uk/government/publications/adapting-historic-homes-for-energy-efficiency-a-review-of-the-barriers/adapting-historic-homes-for-energy-efficiency-a-review-of-the-barriers (accessed on 3 November 2024).
11. HM Government. *The Building Regulations 2010: Approved Document L1B—Conservation of Fuel and Power in Existing Dwellings*; HM Government: London, UK, 2010.
12. DCLG. *National Planning Policy Framework*; Department for Communities and Local Government (DCLG): London, UK, 2012. Available online: https://assets.publishing.service.gov.uk/government/uploads/system/uploads/attachment_data/file/6077/2116950.pdf (accessed on 1 April 2020).
13. BEST. *Brighton & Hove, Eastbourne, Hastings, Lewes, Rother, Wealden: Private Sector Housing Stock Condition Surveys-Sub Regional Overview Report*; Brighton and East Sussex Together, 2008. Available online: https://ww3.brighton-hove.gov.uk/sites/brighton-hove.gov.uk/files/Brighton%20&%20Hove%20Stock%20Condition%20Survey%202008.pdf (accessed on 21 January 2020).
14. Antram, N.; Morrice, R. *Brighton and Hove*; Yale University Press: New Haven, CT, USA, 2008.
15. BHCC. *Private Sector House Condition Survey 2008: Final Report*; Brighton and Hove City Council (BHCC), 2008. Available online: https://www.brighton-hove.gov.uk/sites/brighton-hove.gov.uk/files/Brighton%20&%20Hove%20Stock%20Condition%20Survey%202008.pdf (accessed on 15 September 2021).
16. Arregi, B.; Little, J. Hygrothermal Risk Evaluation for the Retrofit of a Typical Solid-walled Dwelling. *J. Sustain. Des. Appl. Res. Innov. Eng. Built Environ.* **2016**, *4*, 3.
17. Ascione, F.; de Rossi, F.; Vanoli, G.P. Energy retrofit of historical buildings: Theoretical and experimental investigations for the modelling of reliable performance scenarios. *Energy Build.* **2011**, *43*, 1925–1936. [CrossRef]
18. Atkinson, J.L. Evaluating Retrofitted External Wall Insulation. Ph.D. Thesis, Cardiff Metropolitan University, Cardiff, UK, 2015.
19. Avola, F.; Evola, G.; Marletta, L. Energy savings and summer thermal comfort for retrofitted buildings: A complex balance. In *Sustainability in Energy and Buildings: Proceedings of SEB 2019*; Springer: Singapore, 2019; pp. 281–293.
20. Banfill, P.; Simpson, S.; Haines, V.; Mallaband, B. Energy-led retrofitting of solid wall dwellings: Technical and user perspectives on airtightness. *Struct. Surv.* **2012**, *30*, 267–279. [CrossRef]
21. BEIS. *Demonstration of Energy Efficiency Potential (DEEP): Literature Review of Benefits and Risks in Domestic Retrofit Practice and Modelling*; Leeds Sustainability Institute, Leeds Beckett University: Great Britain, UK, 2021.
22. Ben, H.; Steemers, K. Energy retrofit and occupant behaviour in protected housing: A case study of the Brunswick Centre in London. *Energy Build.* **2014**, *80*, 120–130. [CrossRef]
23. Bianco, T.; Pollo, P.; Serra, V. Wood fibre vs synthetic thermal insulation for roofs energy retrofit: A case study in Turin, Italy. *Energy Procedia* **2017**, *111*, 347–356. [CrossRef]
24. Blecich, P.; Franković, M.; Kristl, Ž. Energy retrofit of the Krsan Castle: From sustainable to responsible design—A case study. *Energy Build.* **2016**, *122*, 23–33. [CrossRef]
25. Bothwell, K.; Saich, M.; Mallion, P. Retrofit of existing housing in the United Kingdom: The carbon reduction possibilities. In *PLEA 2011 Architecture & Sustainable Development*; Presses Universitaires de Louvain: Louvain-la-Neuve, Belgium, 2011.
26. Broström, T.; Eriksson, P.; Liu, L.; Rohdin, P.; Ståhl, F.; Moshfegh, B. A Method to Assess the Potential for and Consequences of Energy Retrofits in Swedish Historic Buildings. *Hist. Environ. Policy Pract.* **2014**, *5*, 150–166. [CrossRef]

27. Byrne, A.; Byrne, G.; Davies, A.; Robinson, A.J. Transient and quasi-steady thermal behaviour of a building envelope due to retrofitted cavity wall and ceiling insulation. *Energy Build.* **2013**, *61*, 356–365. [CrossRef]
28. Campbell, N.; Mcgrath, T.; Nanukuttan, S.; Brown, S. Monitoring the hygrothermal and ventilation performance of retrofitted clay brick solid wall houses with internal insulation: Two UK case studies. *Case Stud. Constr. Mater.* **2017**, *7*, 163–179. [CrossRef]
29. Dalla Mora, T.; Cappelletti, F.; Peron, F.; Romagnoni, P.; Bauman, F. Retrofit of an Historical Building toward NZEB. *Energy Procedia* **2015**, *78*, 1359–1364. [CrossRef]
30. Eriksson, P.; Hermann, C.; Hrabovszky-Horváth, S.; Rodwell, D. EFFESUS Methodology for Assessing the Impacts of Energy-Related Retrofit Measures on Heritage Significance. *Hist. Environ.* **2014**, *5*, 132–149. [CrossRef]
31. ESNZ; BEIS. *National Energy Efficiency Data-Framework (NEED): Summary of Analysis*; Department for Energy Security and Net Zero (ESNZ) and Department for Business, Energy and Industrial Strategy (BEIS): London, UK, 2023.
32. Fabbri, K.; Tronchin, L.; Tarabusi, V. Energy Retrofit and Economic Evaluation Priorities Applied at an Italian Case Study. *Energy Procedia* **2014**, *45*, 379–384. [CrossRef]
33. Flores, J.A.M. The Investigation of Energy Efficiency Measures in Traditional Buildings in the Oporto World Heritage Site. Ph.D. Thesis, Oxford Brookes University, Oxford, UK, 2013.
34. Fregonara, E.; Lo Versob, V.R.M.; Lisac, M.; Callegaria, G. Retrofit scenarios and economic sustainability. A case-study in the Italian context. *Energy Procedia* **2017**, *111*, 245–255. [CrossRef]
35. Galatioto, A.; Ricciu, R.; Salem, T.; Kinab, E. Energy and economic analysis on retrofit actions for Italian public historic buildings. *Energy* **2019**, *176*, 58–66. [CrossRef]
36. Galliano, R.; Wakili, K.G.; Binder, B.; Daniotti, B. Evaluation of Three Different Retrofit Solutions Applied to the Internal Surface of a Protected Cavity Wall. *Energy Procedia* **2015**, *78*, 848–853. [CrossRef]
37. Harrestrup, M.; Svendsen, S. Full-scale test of an old heritage multi-storey building undergoing energy retrofitting with focus on internal insulation and moisture. *Build. Environ.* **2015**, *85*, 123–133. [CrossRef]
38. He, M.; Lee, T.; Taylor, S.; Firth, S.; Lomas, K. Dynamic modelling of a large scale retrofit programme for the housing stock in the North East of England. In Proceedings of the 2nd International Conference in Urban Sustainability and Resilience, London, UK, 3–5 November 2014.
39. Heath, N.; Baker, P.; Menzies, G. *Historic Scotland Technical Paper 9: Slim-Profile Double Glazing Thermal Performance and Embodied Energy*; Historic Scotland: Edinburgh, UK, 2010.
40. Heath, N.; Pearson, G.; Barnham, B.; Atkins, R. *Historic Scotland Technical Paper 8: Energy Modelling of the Garden Bothy, Dumfries House*; Historic Scotland: Edinburgh, UK, 2010.
41. Heath, N.; Baker, P. *Historic Scotland Technical Paper 20: Slim-Profile Double-Glazing in Listed Buildings Re-Measuring the Thermal Performance*; Historic Scotland: Edinburgh, UK, 2013.
42. Hopper, J.; Littlewood, J.R.; Taylor, T.; Counsell JA, M.; Thomas, A.M.; Karani, G.; Geens, A.; Evans, N.I. Assessing retrofitted external wall insulation using infrared thermography. *Struct. Surv.* **2012**, *30*, 245–266. [CrossRef]
43. Hugoni, C.; Hugoni, M.E.; Causone, F.; Morello, E. A working methodology for deep energy retrofit of residential multi-property buildings. In *Sustainability in Energy and Buildings: Proceedings of SEB 2019*; Springer: Singapore, 2019; pp. 687–699.
44. Kolaitis, D.I.; Malliotakis, E.; Kontogeorgos, D.A.; Mandilaras, I.; Katsourinis, D.I.; Founti, M.A. Comparative assessment of internal and external thermal insulation systems for energy efficient retrofitting of residential buildings. *Energy Build.* **2013**, *64*, 123–131. [CrossRef]
45. Litti, G.; Audenaert, A.; Lavagna, M. Life cycle operating energy saving from windows retrofitting in heritage buildings accounting for technical performance decay. *J. Build. Eng.* **2018**, *17*, 135–153. [CrossRef]
46. Little, J.; Ferraro, C.; Arregi, B. *Historic Scotland Technical Paper 15: Assessing Risks in Insulation Retrofits Using Hygrothermal Software Tools*; Historic Scotland: Edinburgh, UK, 2015.
47. Little, J.; Arregi, B. Managing moisture—The key to healthy internal wall insulation retrofits of solid walls. In Proceedings of the 17th International Passive House Conference, Frankfurt, Germany, 19–20 April 2013.
48. Lo Cascio, E.; Ma, Z.; Borelli, D.; Schenone, C. Residential building retrofit through numerical simulation: A case study. *Energy Procedia* **2017**, *111*, 91–100. [CrossRef]
49. Lowe, R. Technical options and strategies for decarbonizing UK housing. *Build. Res. Inf. Int. J. Res. Dev. Demonstr.* **2007**, *35*, 412–425. [CrossRef]
50. Mancini, F.; Cecconi, M.; De Sanctis, F.; Beltotto, A. Energy Retrofit of a Historic Building Using Simplified Dynamic Energy Modeling. *Energy Procedia* **2016**, *101*, 1119–1126. [CrossRef]
51. Memon, S. Analysing the potential of retrofitting ultra-low heat loss triple vacuum glazed windows to an existing UK solid wall dwelling. *Int. J. Renew. Energy Dev.* **2014**, *3*, 161–174. [CrossRef]
52. Moran, F. Benchmarking the Energy Use of Historic Dwellings in Bath and the Role for Retrofit and LZC Technologies to Reduce CO_2 Emissions. Ph.D. Thesis, University of Bath, Bath, UK, 2013.

53. Morelli, M.; Ronby, L.; Mikkelsen, S.E.; Minzari, M.G.; Kildemoes, T.; Tommerup, H.M. Energy retrofitting of a typical old Danish multi-family building to a "nearly-zero" energy building based on experiences from a test apartment. *Energy Build.* **2012**, *54*, 395. [CrossRef]
54. Nastasi, B.; di Matteo, U. Innovative use of Hydrogen in energy retrofitting of listed buildings. *Energy Procedia* **2017**, *111*, 435–441. [CrossRef]
55. Neroutsou, T.I.; Croxford, B. Lifecycle costing of low energy housing refurbishment: A case study of a 7year retrofit in Chester Road, London. *Energy Build.* **2016**, *128*, 178–189. [CrossRef]
56. Rasmussen, T. Retrofitting listed buildings: Measures, savings and requirements. *Open Constr. Build. Technol. J.* **2011**, *5*, 174–181. [CrossRef]
57. Rhee-Duverne, S.; Baker, P. *A Retrofit of a Victorian Terrace House in New Bolsover: A Whole House Thermal Performance Assessment*; Historic Scotland: Edinburgh, UK, 2015.
58. Şahin, C.D.; Arsan, Z.D.; Tunçoku, S.S.; Broström, T.; Akkurt, G.G.; Humanistisk-Samhällsvetenskapliga, V.; Uppsala, U.; Historisk-Filosofiska, F.; Konstvetenskapliga, I. A transdisciplinary approach on the energy efficient retrofitting of a historic building in the Aegean Region of Turkey. *Energy Build.* **2015**, *96*, 128–139. [CrossRef]
59. Sdei, A.G.F.; Tittelein, P.; Lassue, S.; Hannad, P.; Beslay, C.; Gournet, R.; Mcevoy, M. Social housing retrofit strategies in England and France: A parametric and behavioural analysis. *Energy Res. Soc. Sci.* **2015**, *10*, 62–71. [CrossRef]
60. Silvero, F.; Rodrigues, F.; Montelpare, S. A Parametric Study and Performance Evaluation of Energy Retrofit Solutions for Buildings Located in the Hot-Humid Climate of Paraguay—Sensitivity Analysis. *Energies* **2019**, *12*, 427. [CrossRef]
61. Soulios, V.; De Place Hansen, E.J.; Peuhkuri, R. Hygrothermal simulation assessment of internal insulation systems for retrofitting a historic Danish building. *MATEC Web Conf.* **2019**, *282*, 2049. [CrossRef]
62. Vereecken, E.; van Gelder, L.; Janssen, H.; Roels, S. Interior insulation for wall retrofitting—A probabilistic analysis of energy savings and hygrothermal risks. *Energy Build.* **2015**, *89*, 231–244. [CrossRef]
63. Wise, F.; Moncaster, A.; Derek, J. Rethinking retrofit of residential heritage buildings. *Build. Cities* **2021**, *2*, 495–517. [CrossRef]
64. Ingram, V. Energy Performance of Traditionally Constructed Dwellings in Scotland. Ph.D. Thesis, Heriot-Watt University, Edinburgh, UK, 2013.
65. English Heritage. *Constructive Conservation in Practice*; English Heritage: London, UK, 2008.
66. English Heritage. *Energy Conservation in Traditional Buildings*; English Heritage: London, UK, 2008.
67. English Heritage. *Heritage Counts*; English Heritage: London, UK, 2008. Available online: https://historicengland.org.uk/content/heritage-counts/pub/previous/london/london_2008-pdf/ (accessed on 25 January 2018).
68. Historic England. *Energy Efficiency and Historic Buildings: Draught-Proofing Windows and Doors*; Historic Scotland: Edinburgh, UK, 2016.
69. Historic England. *Energy Efficiency and Historic Buildings: Insulating Flat Roofs*; Historic Scotland: Edinburgh, UK, 2016.
70. Historic England. *Energy Efficiency and Historic Buildings: Insulating Pitched Roofs at Ceiling Level*; Historic Scotland: Edinburgh, UK, 2016.
71. Historic England. *Energy Efficiency and Historic Buildings: Insulating Solid Ground Floors*; Historic Scotland: Edinburgh, UK, 2016.
72. Historic England. *Energy Efficiency and Historic Buildings: Insulating Solid Walls*; Historic Scotland: Edinburgh, UK, 2016.
73. Historic England. *Energy Efficiency and Historic Buildings: Insulating Suspended Timber Floors*; Historic Scotland: Edinburgh, UK, 2016.
74. Historic England. *Energy Efficiency and Historic Buildings: Secondary Glazing for Windows*; Historic Scotland: Edinburgh, UK, 2016.
75. Historic England. *Heritage Works: A Toolkit of Best Practice in Heritage Regeneration*; Historic Scotland: Edinburgh, UK, 2017.
76. Historic England. *Traditional Windows: Their Care, Repair and Upgrading*; Historic Scotland: Edinburgh, UK, 2017.
77. Historic England. *Energy Efficiency and Historic Buildings: How to Improve Energy Efficiency*; Historic Scotland: Edinburgh, UK, 2018.
78. Historic Scotland. *Short Guide—Fabric Improvements for Energy Efficiency in Traditional Buildings*; Historic Scotland: Edinburgh, UK, 2013.
79. SPAB. *Energy Efficiency in Old Buildings*; Society for the Protection of Ancient Buildings (SPAB): London, UK, 2014.
80. SPAB. *Briefing: Windows and Doors*; Society for the protection of ancient buildings (SPAB): London, UK, 2016.
81. STBA. *Performance and Energy Efficiency of Traditional Buildings: Gap Analysis Study*; STBA: London, UK, 2012.
82. STBA. *Planning Responsible Retrofit of Traditional Buildings*; Sustainable Traditional Buildings Alliance (STBA): London, UK, 2015.
83. Suhr, M.; Hunt, R. *The Old House Eco Handbook: A Practical Guide to Retrofitting for Energy-Efficiency & Sustainability*; Frances Lincoln Limited.: London, UK, 2013.
84. The Prince's Regeneration Trust. *The Green Guide for Historic Buildings: How to Improve the Environmental Performance of Listed and Historic Buildings*; Stationery Office: London, UK, 2010.
85. Eyre, N.; Fawcett, T.; Topouzi, M.; Killip, G.; Oreszczyn, T.; Enkinson, K.; Rosenow, J. Fabric first: Is it still the right approach? *Build. Cities* **2023**, *4*, 965–972. [CrossRef]

86. Changeworks. *Energy Heritage: A Guide to Improving Energy Efficiency in Traditional and Historic Homes;* Changeworks: Edinburgh, UK, 2008.
87. Charles, A. Factors Affecting the Transformation of Existing Dwellings to Low Carbon Homes. Ph.D. Thesis, Glasgow Caledonian University, Glasgow, UK, 2012.
88. Hong, S.H.; Oreszczyn, T.; Ridley, I. The impact of energy efficient refurbishment on the space heating fuel consumption in English dwellings. *Energy Build.* **2006**, *38*, 1171–1181. [CrossRef]
89. ICOMOS. *How to Improve Energy Performance in European and Mediterranean Architectural and Urban Heritage? Heritage: A Model for Sustainable Development 2010 Paris;* ICOMOS France, Euromed Heritage: Paris, France, 2010.
90. Morris, P.; Therivel, R. *Methods of Environmental Impact Assessment*, 3rd ed.; Routledge: Abingdon, UK, 2009.
91. Menconi, M.; Painting, N.; Piroozfar, P. Building Energy Simulation of Traditional Listed Dwellings in the UK: Data sourcing for a base-case model. In *Sustainability in Energy and Buildings: Proceedings of SEB 2019;* Springer: Singapore, 2020; pp. 295–307.
92. Menconi, M.; Painting, N.; Piroozfar, P. Responsible retrofit measures for traditional listed dwellings: An energy simulation validation strategy. In *Sustainable Ecological Engineering Design: Selected Proceedings from the International Conference of Sustainable Ecological Engineering Design for Society (SEEDS) 2019;* Springer International Publishing: Cham, Switzerland, 2020; pp. 379–391.
93. Menconi, M.; Painting, N.; Piroozfar, P. An energy performance baseline scenario for 19th C listed dwellings in the UK. *Int. J. Build. Pathol. Adapt.* **2024**, *42*, 133–152. [CrossRef]
94. Menconi, M.; Painting, N.; Piroozfar, P. Building Energy Simulation of 19th C Listed Dwellings in the UK: A strategy to propose and assess suitable retrofit interventions. In *Sustainability in Energy and Buildings 2020;* Springer: Singapore, 2020; pp. 139–150.
95. Menconi, M.; Painting, N.; Piroozfar, P. Modelling and simulation of low-risk energy retrofit measures for Traditional Listed Dwellings in the UK. *J. Build. Eng.* **2024**, *82*, 108346. [CrossRef]
96. *BS EN ISO 13788:2002;* Hygrothermal Performance of Building Components and Building Elements. Internal Surface Temperature to Avoid Critical Surface Humidity and Interstitial Condensation. Calculation Methods. British Standard Institute (BSI): London, UK, 2002.
97. *BS EN ISO 15026:2007;* Hygrothermal Performance of Building Components and Building Elements. Assessment of Moisture Transfer by Numerical Simulation. British Standard Institute (BSI): London, UK, 2007.
98. *BS 5250:2021;* Management of Moisture in Buildings. Code of Practice. British Standard Institute (BSI): London, UK, 2021.
99. Liu, F.; Jia, B.; Chen, B.; Geng, W. Moisture transfer in building envelope and influence on heat transfer. *Procedia Eng.* **2017**, *205*, 3654–3661. [CrossRef]
100. Loveday, D.; Vadodaria, K. *Project CALEBRE: Consumer Appealing Low Energy Technologies for Building REtrofitting—A Summary of the Project and Its Findings;* Loughborough University: Loughborough, UK, 2013.
101. Giorgi, M.; Favre, D.; Goulouti, K.; Lasvaux, S. Hygrothermal assessment of historic buildings' external walls: Preliminary findings from the RIBuild project for Switzerland. *J. Phys. Conf. Ser.* **2019**, *1343*, 012183. [CrossRef]
102. Baker, P. *Historic Scotland Technical Paper 2: In-Situ u Value Measurements in Traditional Buildings;* Historic Scotland: Edinburgh, UK, 2008.
103. Baker, P. *Historic Scotland Technical Paper 10: U-Values and Traditional Buildings: In-Situ Measurements and Their Comparisons to Calculated u-Values;* Historic Scotland: Edinburgh, UK, 2011.
104. Rye, C. *The SPAB Research Report 1. U-value Report;* SPAB: London, UK, 2010.
105. IES-VE. *Historic Scotland Technical Paper 5: Energy Modelling of a Mid-19th C Century Villa—Baseline Performance and Improvement Options;* Historic Scotland: Edinburgh, UK, 2009.
106. *PAS 2035/2030:2019+A1:2022;* Specification and Guidance for Retrofitting Dwellings for Improved Energy Efficiency. British Standards Institution (BSI): London, UK, 2019.

energies

MDPI

Article

Investigation of Energy-Efficient Solutions for a Single-Family House Based on the 4E Idea in Poland

Piotr Ciuman [1], Jan Kaczmarczyk [1,*] and Dorota Winnicka-Jasłowska [2]

1 Department of Heating, Ventilation and Dust Removal Technology, Faculty of Energy and Environmental Engineering, Silesian University of Technology, ul. Konarskiego 18, 44-100 Gliwice, Poland; piotr.ciuman@polsl.pl

2 Department of Design and Architectural Research, Faculty of Architecture, Silesian University of Technology, ul. Akademicka 7, 44-100 Gliwice, Poland; dorota.winnicka-jaslowska@polsl.pl

* Correspondence: jan.kaczmarczyk@polsl.pl; Tel.: +48-323272820

Abstract: The paper analyses multi-variant energy simulations carried out in IDA ICE 4.8 software for a newly designed single-family building within the framework of the 4E Idea. This idea assumes the use of energy-saving, ecological, ergonomic, and economic solutions in construction and building operation. Energy simulations were conducted to evaluate the annual energy-saving potential of the developed architectural house concept, which incorporates ergonomic analyses and cost-effective construction solutions. Analyses were conducted to optimise the non-renewable primary energy index by selecting mechanical ventilation system (CAV or VAV) with heat recovery; the configuration of photovoltaic module installation in terms of their location and orientation; the exposure and type of solar thermal collectors (flat and vacuum); and the use of two types of heat pumps (air- and ground-source). The most favourable energy performance of the building was achieved with an HVAC system equipped with a VAV mechanical ventilation system with heat recovery, an on-grid photovoltaic installation, vacuum solar thermal collectors, and a ground-source heat pump with a horizontal heat exchanger. This configuration resulted in a primary energy index value of 2 kWh/m^2/year. The results of the analyses carried out for the 4E building concept may serve as a reference point for future energy-efficient building designs aspiring to meet higher standards of sustainable development.

Keywords: renewable energy sources; energy performance; single-family building; heat pump; solar thermal collector; photovoltaics; energy analyses

1. Introduction

The global goal of reducing energy consumption in buildings and mitigating their environmental impact continues to challenge the architectural, engineering and construction sectors. Maintaining high-quality comfort in terms of building functionality and occupant satisfaction is an additional challenge and, therefore, requires the application of a person-centred design of physical and social environments [1]. These challenges apply to all types of buildings, but due to the prolonged time people spend indoors, residential buildings are particularly important.

In Poland, single-family buildings accounted for 97.5% of all new residential buildings constructed in the first quarter of 2024. Almost all of these buildings, i.e., 98.2%, were constructed using traditional, improved technologies [2].

Statistical data show that over many years the household sector in the European Union has consistently contributed significantly to the final energy consumption [3]. In 2021, households made up a large proportion of the national energy consumption in EU countries, ranging from 11.4% (Luxembourg) to 28.1% (Croatia). In Poland, this indicator was 20.2% and was higher than the EU average (18.4%) [4]. In recent years, residential buildings in Poland have consumed the most energy: approximately 66% for space heating, followed by domestic hot water preparation (17%), lighting and electrical applications (9%), and cooking (8%) [5].

The building sector in the EU is also a substantial contributor to greenhouse gas emissions in the EU. In 2022, it was responsible for 34% of energy-related emissions [6]. Analyses of the whole life cycle performed for a single-family house showed that the greatest contributor to the environmental effects was the energy consumed during building operation, which contributed to 58% to 90% of the CO_2 emissions [7]. Heating-related energy consumption is one of the main sources of greenhouse gas emissions and accounts for a large part of domestic energy use. Reducing heating-related energy consumption offers great potential to reduce Europe's greenhouse gas emissions. The analysis also showed that facilities using heat pumps are characterised by an environmental impact that is six times lower than that of facilities powered by coal combustion and electricity from the network. Similarly, the social costs associated with CO_2 emissions were significantly lower in the case of the use of renewable energy sources [7]. It should be noted that in Poland, despite the increased share of renewable sources in electricity production, hard coal still has a significant share. In 2023, hard coal accounted for just over 60% of total energy generation, while renewable energy sources contributed approximately 25% (wind 14%, photovoltaics (PV) solar 6.8%, and biofuels 4.7%) [8].

In the single-family home sector, a viable option is the production of electricity by photovoltaic modules and its efficient on-site use. Thus, the application of PV panels with heat pumps [9–11] and domestic hot water systems [12] is a promising solution. Several studies have proven that such a combination can increase monthly self-consumption from 7% to 18%, and annually up to 13% [13]. An advanced smart-grid-ready controlled PV-HP-battery system in a single-family household showed that, over one-year operation, self-consumption can be increased even further, up to 43% [14]. The performance evaluation of a combination of heat pump and PV system controlled by a novel algorithm and based on simulations showed that between 25.3% and 41.0% of the building's electricity consumption, including the heat pump, can be covered directly by the PV installation annually. It was noted that the characteristics of the heating system can significantly influence the results. New buildings with floor heating and low supply temperatures yield a higher PV self-consumption levels compared to buildings with radiator heating and higher supply temperatures. The addition of a battery to the system further increased the degree of PV self-consumption. It was also noted that due to the high investment costs of batteries, they do not pay off within a reasonable period [12]. Another study in a Danish single-family house with a heat pump and floor heating [15], experimentally demonstrated that load shifting may be feasible and cost-effective, even without energy storage, and that the current pricing scheme, which allows the grid operators to differentiate the end-user tariffs throughout the day, provides a stimulus for end-consumers to shift heating loads. In addition, it has been shown that space heating systems in Swedish houses equipped with heat pumps have the potential to increase the resilience of the power grid during major network disruptions by temporarily reducing the room temperature to an acceptable level for the users [16].

Energy used in a building, apart from heating and hot water production [17], is to a large extent intended for heating (or cooling) the ventilation air. In accordance with the requirements in Poland, the supply air volume flow rate in residential premises should be constant and should correspond to the exhaust air volume flow rate planned for the ventilation of kitchens and bathrooms. However, it should not be less than 20 m^3/h per person intended for permanent residence in the building design [18]. The installation of a Variable Air Volume (VAV) system working with the demand–control principle can noticeably reduce source energy and energy costs. Study measurements were carried out in a 140 m^2 single-family house occupied by two adults and two children, where the controls installed in the existing mechanical ventilation system showed that the ventilation operation can be reduced to a low rate of running 37% of the time without significant changes in the CO_2 concentration and moisture level in the house. Savings in electrical energy for running fans were estimated at 35% [19]. Another study suggested that the VAV system can be especially beneficial to houses in cooling-dominant climates [20]. In Poland, due to legal restrictions, the VAV solution is not used, but its effectiveness is worth investigating, especially if the air quality is not compromised.

Reduction in the amount of energy for ventilation air treatment can be achieved by using appropriate heat recovery solutions and their frost control [21] or application of a ground heat exchanger [22]. The operation of the ventilation system is essential for removing air pollutants generated in the room and to maintain appropriate indoor air quality. In residential buildings the primary pollutants are those related to occupants and their activity, mainly bioeffluents represented by CO_2 and humidity. Other indoor sources are building materials and equipment emitting VOCs [23], which is especially important in newly built houses. The pollution emission rate is strongly influenced by the occupancy patterns [24]. The emission from building materials is continuous, although it is strongest for new materials and decreases in time. One option to reduce emissions from the building is the use of green materials [25]. However, it is necessary to provide measures to mitigate risks and to remove the pollution. Studies performed in 25 energy-efficient residential buildings with mechanical ventilation showed lower concentrations of indoor PM10, PM2.5, CO_2, and VOCs compared to conventional apartments. Also, the prevalence of symptoms, particularly children's atopic dermatitis and allergic rhinitis, was significantly lower. Occupants showed a higher satisfaction level with their sleep quality, indoor air, and indoor humidity than in conventional apartments [26].

Addressing these challenges requires innovative design strategies and technologies that balance energy efficiency, environmental protection, economic viability, and user comfort—the cornerstones of the 4E Idea [27]. The main assumptions of the 4E Idea model single-family house project were:

- Energy saving;
- Ecology, including: management of rainwater and limiting its collection from the network, use of construction materials from recycling to the greatest extent possible and with the lowest possible carbon footprint;
- Economy, low investment and operational costs over the life cycle;
- Ergonomics, design solutions based on the analysis of the real needs of users in different age groups, a building designed based on the principles of ergonomics.

The second assumption of the project was IDEA 2+2+(1), which assumes surface and functional–spatial solutions for a family with a changing number of members at different stages of the building's life, i.e., two, four, or five family members. The project assumed space flexibility, which can easily be rearranged or expanded. For this purpose, pre-designed studies were carried out, which allowed for the definition of the guidelines for the project.

The topic of energy-efficient and environmentally friendly buildings, along with the search for appropriate technical solutions, has been the focus of numerous studies in recent years [28]. Most of these studies address individual aspects, such as the use of photovoltaic panels or the optimisation of heat pump performance. However, only a few studies consider multiple factors simultaneously. A novelty of the study presented in this manuscript lies in the comprehensive consideration of the building as a whole, including its construction and technical systems. The primary goal was to identify a combination of solutions that would enable a single-family house to achieve zero-energy status.

This paper focuses primarily on the problems of energy production and consumption and indicates solutions that include renewable energy sources and advanced HVAC systems, which will allow the goal of creating a zero-energy 4E Idea building to be achieved. In addition to demonstrating the impact of the type of PV installation, solar thermal collectors, and the type of heat pump, which was the subject of previous studies, the presented analyses also raised the issue of the justification for using a CO_2 concentration-controlled VAV mechanical ventilation system to reduce energy consumption and the non-renewable primary energy index. The analyses were conducted based on energy simulations of the building designed for the local climate of Poland. The calculations used the IDA ICE simulation tool, which has been previously used in many energy analyses of buildings, including analyses of the operation of renewable energy sources [12,29–31]. Although IDA ICE does not take into account the directionality effects of direct solar radiation [32], it has been proven to accurately predict building thermal behaviour [33,34]. As a result of the activities carried out, a concept of an energy-efficient building was developed, which in terms of architecture, construction, and the configuration of the HVAC system and renewable energy sources can be a starting point for future projects of energy-efficient single-family buildings. It can also contribute to the development of energy-efficient single-family homes that align with Poland's environmental goals and global sustainability standards.

2. Materials and Methods

2.1. Characteristics of the Designed Building and Assumptions for the Simulation

The analysed facility was a newly designed single-family building that is detached, single-storey, without a basement, and consisting of two cuboid-shaped parts covered with gable roofs intersecting in the central part of the building. The building was divided into four zones: the entrance zone—connecting the daytime and nighttime zones, as well as the garage and utility rooms; the daytime zone—an open space comprising the living room, kitchen, and access to the terrace; the nighttime zone—a private area containing bedrooms and a bathroom; and the utility zone. A non-habitable attic was located above the nighttime and utility zones. The usable floor area of the building was 147 m^2. Including the garage, the total area was 189.7 m^2, with a total volume of 583 m^3. Typically, the area of a house for a family of four in Poland is approximately 120–150 m^2, with an average of 132.6 m^2 [2]. This space typically includes two rooms for the children, a bedroom for the parents, a shared living room, a kitchen, a bathroom, a toilet, and an additional room.

The 1300 m^2 plot was initially divided into four zones: the northern zone, designated for entrance and utility functions; the eastern and south-eastern zone, allocated for the installation of a ground heat exchanger; and the southern, south-western, and western zones, intended for leisure and recreational purposes, which together constituted the largest part of the plot.

The daytime zone of the building, characterised by the largest glazed area on the external wall, was oriented southward to maximise solar heat gains and thereby reduce the building's energy consumption. Figure 1 presents the building's floor plan, showing the division into daytime and nighttime zones, along with the detailed layout and floor areas of the rooms. Figure 2 shows the visualisation of the building's southern façade and the interior of the living room with the kitchen.

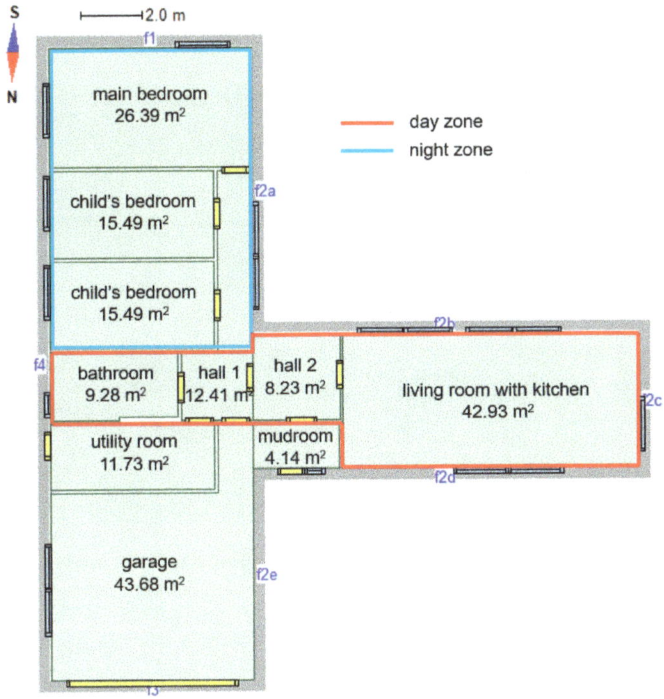

Figure 1. Floor plan of the designed building, created using IDA ICE 4.8 software.

(a) (b)

Figure 2. Visualisation of the building's southern façade (**a**) and interior of the living room with the kitchen (**b**).

The height of the rooms in the nighttime (private) zone was 2.95 m, while in the daytime (functional) zone, the height at the peak line of the roof was 4.90 m. The largest room in the daytime part of the building was the living room with the kitchen. This part of the building featured a gable roof, with exposed trusses inside, as well as large windows on the southern side. The glazing area accounted for 53% of the surface area of the southern external wall. The nighttime zone contained two children's bedrooms, the main (larger) master bedroom, and a bathroom. Additionally, the building included a garage with parking space for two cars and a utility room, which housed the domestic hot water storage tank, a heat recovery ventilation unit (HRV), and other HVAC and electrical system equipment. The connecting space between the two parts of the building was a corridor with large windows opening onto the terrace. In all rooms, except for the non-habitable attic, the air temperature was adjustable. Figures 3 and 4 depict the façades and a 3D view of the geometric model of the building, created using the IDA ICE 4.8 software.

(a) (b) (c)

Figure 3. Views of the designed building's façades—(**a**) northern façade, (**b**) eastern façade, (**c**) southern façade, created using IDA ICE 4.8 software.

Figure 4. Three-dimensional views of the designed building, created using IDA ICE 4.8 software.

The building structure consisted of an external and internal arrangement of load-bearing brick walls supported on a foundation slab, which incorporated a heating system installation and was covered by a trussed roof.

The heat transfer coefficient of individual building partitions was calculated based on the type and thickness of the construction materials comprising these partitions. The proper materials were preselected based on their thermal properties and costs. The partitions were designed to ensure that their heat transfer coefficient did not exceed the maximum value specified in the standard [34]. Table 1 presents their structure and heat transfer coefficients.

The building was designed with a heating, ventilation, and air conditioning (HVAC) system equipped with renewable energy sources. The heating system consisted of heating elements in the form of floor heating with operating parameters of 35 °C/30 °C. These parameters are widely used due to their compatibility with low-temperature heating sources, such as heat pumps. They are commonly recommended in industry guidelines and standards for residential heating systems of well-insulated buildings. In the garage, a wall-mounted panel radiator was installed.

The heat source for the heating and domestic hot water systems was an air-source heat pump with a nominal heating capacity of 6.1 kW and a coefficient of performance (COP) of 5.2 (for an air inlet temperature of 7 °C and a hot water temperature at the heat pump outlet of 35 °C). The system was supplemented with a buffer tank. The heating power of the radiators and the heat pump was selected based on the calculated heat demand of the building and individual rooms. The heating system was equipped with an automatic air temperature regulation system for rooms, based on the set minimum air temperature values: 24 °C for bathrooms, 20 °C for rooms designated for regular occupancy, 16 °C for the utility room, and 8 °C for the garage [35]. The domestic hot water system was designed with a 300-litre storage tank. In heating mode, the heat pump was activated when indoor air temperature fell below the heating setpoint. Heat was transferred to the buffer tank, which supplied energy to the space heating system. The heat pump prioritised domestic hot water production when the hot water tank temperature dropped below a specific setpoint (55 °C). The system switched to domestic hot water mode when hot water demand was detected, with priority over space heating to ensure uninterrupted hot water supply. The heat pump operated at optimal efficiency by modulating its output based on outdoor air temperature and load requirements. During mild weather, the heat pump ran at lower capacity, while at peak demand, it operated at full capacity.

Table 1. Construction of external building partitions and their heat transfer coefficients. The layers are presented in order from the inner side to the outer side of the building.

Building Partition	Construction	Heat Transfer Coefficient U, W/m²K
External wall	- cement–lime plaster - brick wall made of perlite blocks in SYSTEM 3E, 35.2 cm thick - mineral wool (insulation), 8 cm thick - cement–lime plaster	0.14
Roof	- plasterboard - vapour barrier foil - polyurethane foam PUR/PIR - mineral wool, 36 cm thick - pine wood class C27 - roofing membrane - flat roof tile Actua 10	0.08
Ground floor slab	- glass tiles - primer, 6 cm thick - damp proofing - reinforced concrete slabs, 50 cm thick - XPS Styrofoam, 20 cm thick - substrate made of gravel, gravel and coarse sand, 50 cm thick	0.14
Ceiling under the unheated attic	- cement–lime plaster - SMART type compressed plates, 15 cm thick - mineral wool, 30 cm thick	0.11
Windows		0.9
External door		1.3

The mechanical supply and exhaust ventilation system was equipped with a cross-flow heat recovery unit (HRV) with an efficiency of 80%, along with an air heater with a glycol heat exchanger and a ground heat exchanger (GHE) that helps save electricity by cooling the supply air in the summer while pre-heating it in the winter. The type of mechanical ventilation system, with variable air volume (VAV) or constant air volume (CAV) flow rates, was selected based on the analysis results presented in Section 3. The cross-flow heat exchanger in the HRV recovered heat from exhaust air to preheat incoming fresh air during cold seasons. In the summer, the heat recovery function operated in reverse to avoid overheating the incoming air. The air heater (connected to the horizontal GHE) was used to condition the supply air further when the preheated air did not meet the desired temperature. The heater was activated based on real-time indoor air temperature and demand. The GHE preconditioned incoming fresh air by exchanging heat with the ground. In the winter, it preheated cold outdoor air to reduce the load on the ventilation system and the air heater. In the summer, it cooled the warm outdoor air, providing passive cooling and reducing the need for active cooling systems.

The building was also planned to be equipped with an on-grid photovoltaic system aimed at covering the building's electricity demand, including the operation of HVAC components, such as the heat pump. A self-consumption rate of 40% of the produced electricity was assumed [36,37]. The location of the photovoltaic modules was also determined based on the results of the analysis presented in Section 3. In IDA ICE, the energy generated by the PV system depends on solar radiation data from a weather file. This file contains hourly values for global and diffuse solar radiation, which is essential for simulating localised energy output. Users can select or define a specific weather file, enabling real-time calculations based on solar radiation, panel efficiency, environmental conditions, and system losses.

Additionally, two rainwater collection tanks were planned on the plot around the building to support the toilet flushing system and the garden irrigation system.

Figure 5 shows the schematic diagram of the designed HVAC and photovoltaic systems within the building.

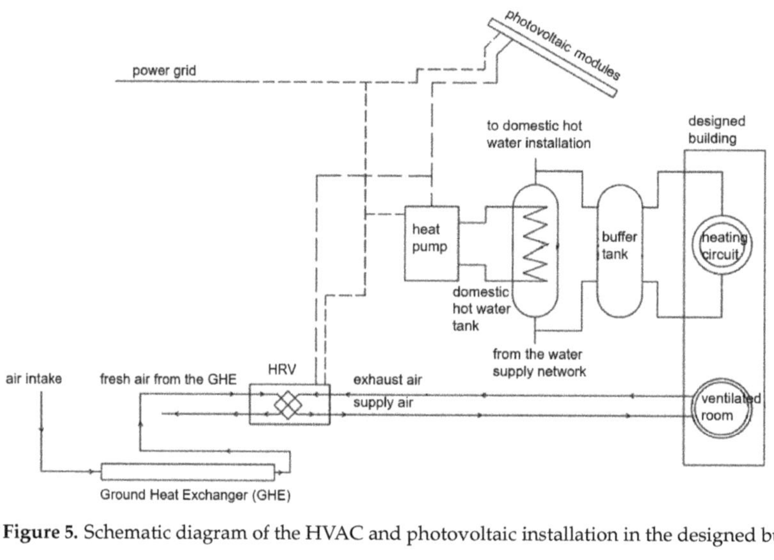

Figure 5. Schematic diagram of the HVAC and photovoltaic installation in the designed building.

2.2. Research Methodology

Year-round energy demand analyses for various computational variants were carried out using the IDA ICE 4.8 software [38]. This is an advanced software tool for conducting energy simulations and thermal comfort analysis in buildings. It is used by designers, engineers, and architects to assess the energy efficiency of buildings and ensure optimal thermal comfort for occupants. The software allows for detailed building modelling, considering the geometric and material properties of the building as well as the parameters of the HVAC system. With the ability to simulate various climatic conditions, IDA ICE enables the evaluation of the impact of year-round climatic conditions on energy consumption and occupants' comfort in buildings.

A geometric model of the designed building defining the construction of the building's envelope and elements, and orienting it according to the cardinal directions, was created in the IDA ICE 4.8 software in line with the design assumptions. The building was located in Katowice, Poland, for which climatic data from the ASHRAE database [39] were available. Default values for thermal bridges were assumed. The minimum outdoor air exchange rate was set to 0.5 L/h [35]. The numerical model included heat gains from occupants, electronic equipment, and lighting. It was assumed that the building would house four people: two adults and two children. For each zone (room), an occupancy schedule was set depending on the room's intended use. In bedrooms, occupants were assumed to be present during nighttime hours (22:00–06:00), while in the daytime zones, higher occupancy was assumed during the afternoon and evening hours (15:00–20:00), reflecting typical household routines after returning home from work or school. A clothing insulation value of CLO = 0.5 and an activity level of MET = 1.0 were assumed, in accordance with [40]. The software automatically adjusts the clothing of occupants within a range of ±0.25 CLO, depending on internal environmental conditions. Lighting was assumed to consist of LED bulbs with a power consumption of 3 W and luminous efficacy of 90 lm/W each (total of 81 W). The number of bulbs in each zone was determined based on the size and intended use of the room. Lighting schedules were aligned with individual room occupancy patterns. Electrical appliances, such as laptops, televisions, and household electronics, were included, and their power consumption was set for each zone based on the intended use and manufacturer data (total of 5.3 kW). For each room, an equipment usage schedule was set according to the time occupants spent in the building and used each device: in bedrooms, laptops (3 × 40 W) were assumed to be in use from 16:00 to 23:00 daily; in a bathroom, a washing machine (1450 W) operated three hours a week; in a living room and kitchen, a television (175 W) operated from 18:00 to 21:00 daily, a coffee machine (200 W)—one hour per day, a microwave (500 W)—two hours per week, a dishwasher (1450 W)—three hours per week, an oven (1450 W)—three hours per week.

Domestic hot water consumption was set at 80 litres per person per day [41], and its usage was scheduled. The total losses in the HVAC system were assumed to be 0.2 W/m^2 of floor area with 50% of the lost heat penetrating into the rooms.

Year-round energy analyses were conducted to select the most energy-efficient ventilation system and thermal and electrical energy sources for the newly designed building, through the optimisation of the non-renewable primary energy index. The analyses were carried out in the following stages:

1. Analysis and selection of the mechanical ventilation system in the building—Constant Air Volume (CAV) or Variable Air Volume (VAV).
2. Analysis and selection of the photovoltaic modules' configuration depending on location and exposure (southern and eastern roof surface and on the ground in the southern direction), which determined the tilt angle and possible maximum area and power of photovoltaic modules.
3. Analysis and selection of solar thermal collectors, considering two types (flat and vacuum) and two orientations (south and east) of collectors.
4. Replacement of the air-source heat pump with a ground-source heat pump.

The first two stages of the research were carried out simultaneously, while the subsequent stages were based on the most energy-efficient solutions from earlier stages.

In the first two stages, six computational variants were defined, which are presented in Table 2. These variants differed in the type of mechanical ventilation system (CAV or VAV) as well as the exposure, location, and power of the photovoltaic installation. Variants 1 and 4 differed in the exposure of the photovoltaic installation, which was oriented to the east, unlike the other variants with southern exposure. Variants 2 and 3, as well as 5 and 6, differed in the size of the photovoltaic installation. In the variants with a freestanding ground-mounted structure, the surface area was larger compared to the rooftop-mounted installation, which was limited by the roof's surface area.

For variants 4–5, with the CAV ventilation system, the supply air volume flow rate was set according to the intended use of each room, as specified in [42]. The total ventilation air volume flow rate for the CAV system was 450 m^3/h.

In variants 1–3, the mechanical ventilation system VAV was equipped with an automatic ventilation flow rate control system using CO_2 sensors to adjust the supply air volume flow rate according to the number of occupants in each room. The diffuser efficiency was set to 50% of the required hygienic airflow for a given room when occupants were absent, and 100% of the ventilation airflow when their presence was detected, based on the increased CO_2 levels in the room. A minimum CO_2 concentration of 400 ppm was assumed in the absence of occupants, with a maximum CO_2 concentration of 1000 ppm. The maximum calculated ventilation air volume flow rate for the VAV system was 318 m^3/h.

Table 2. Computational variants.

Variant	Ventilation System	PV Modules Surface Area, m^2	Tilt Angle of PV Modules, °	PV Modules Exposure and Location	PV Installation Power, kW_p
1	VAV	70	44	east, roof	6.6
2	VAV	40	55	south, roof	4.3
3	VAV	100	45	south, ground	10.7
4	CAV	70	44	east, roof	6.6
5	CAV	40	55	south, roof	4.3
6	CAV	100	45	south, ground	10.7

The three PV system variants analysed were selected to evaluate the impact of surface area and tilt angle to reflect diverse architectural and spatial conditions, as well as orientation and placement to address practical constraints and optimise energy yield based on location-specific solar irradiance conditions. This approach was deliberately focused on assessing how these parameters affect the annual energy-saving potential and primary energy index of the building. The analysis aimed to establish a foundational understanding of optimal PV module placement within the context of a holistic building energy performance evaluation.

2.3. Methodology for Calculating the Non-Renewable Primary Energy Index Value

For all variants, the calculations of the non-renewable primary energy index value were carried out in accordance with Equations (1)–(6), presented in [43].

$$EP = Q_P/A_f, \quad kWh/m^2/year \tag{1}$$

where

Q_P—year-round demand for non-renewable primary energy for technical systems, kWh/year

A_f—an area of rooms with regulated air temperature, m^2

$$Q_P = Q_{P,H} + Q_{P,W} + Q_{P,C} + Q_{P,L}, \quad kWh/year \tag{2}$$

where

$Q_{p,H}$—year-round demand for non-renewable primary z energy for the heating system, kWh/year
$Q_{p,W}$—year-round demand for non-renewable primary energy for the preparation of domestic hot water, kWh/year
$Q_{p,C}$—year-round demand for non-renewable primary energy for the cooling system, kWh/year
$Q_{p,L}$—year-round demand for non-renewable primary energy for the lighting installation, kWh/year

$$Q_{p,H} = Q_{f,H} \cdot w_H + E_{el,aux,H} \cdot w_{el}, \quad \text{kWh/year} \tag{3}$$

$$Q_{p,W} = Q_{f,W} \cdot w_W + E_{el,aux,W} \cdot w_{el}, \quad \text{kWh/year} \tag{4}$$

$$Q_{p,C} = Q_{f,C} \cdot w_C + E_{el,aux,C} \cdot w_{el}, \quad \text{kWh/year} \tag{5}$$

$$Q_{p,L} = Q_{f,L} \cdot w_{el}, \quad \text{kWh/year} \tag{6}$$

where

$Q_{f,H}$—year-round demand for final energy supplied to the building for the heating system, kWh/year
$Q_{f,W}$—year-round demand for final energy supplied to the building for preparation of domestic hot water, kWh/year
$Q_{f,C}$—year-round demand for final energy supplied to the building for the cooling system, kWh/year
$Q_{f,L}$—year-round demand for final energy supplied to the building for the lighting installation, kWh/year
w_i—non-renewable primary energy input factor for production and delivery of [43]:

energy for the heating system (w_H factor, for solar energy (energy from PV installation) equal 0), -
energy for the preparation of domestic hot water (w_W factor, for solar energy (energy from PV installation) equal 0), -
energy for the cooling system (w_C factor), -
electrical energy (w_{el} factor, for system power grid equal 2.5), -

$E_{el,aux,H}$—year-round demand for auxiliary final energy supplied to the building for the heating system, kWh/year
$E_{el,aux,W}$—year-round demand for auxiliary final energy supplied to the building for preparation of domestic hot water, kWh/year
$E_{el,aux,C}$—year-round demand for auxiliary final energy supplied to the building for the cooling system, kWh/year

The values of year-round demand for final energy and auxiliary energy supplied to the building for the heating system and preparation of domestic hot water were obtained through simulations conducted using the IDA ICE 4.8 software. Based on these results, the value of the non-renewable primary energy index was calculated.

Lighting installation calculations were excluded, as lighting requirements for single-family houses are not addressed in the energy analysis for the EP index, in accordance with [43].

3. Results and Discussion

To select the heating energy source for the designed building, heat gains and losses, as well as the use of thermal energy in the building, were calculated using the IDA ICE

4.8 software. Based on the results of the electrical energy usage calculations, a photovoltaic installation was chosen. Additionally, an analysis of the free energy gain from the photovoltaic system and the air heat recovery unit was conducted for each variant. For each calculation variant, the year-round non-renewable primary energy index values were calculated. Based on these, the most energy-efficient variant was selected. Subsequently, further possibilities for optimising the building's energy performance were analysed.

3.1. Heat Balance of the Designed Building

To better understand the building's energy performance, the analysis of the building's heat balance is essential. This approach helps assess how various factors influence the overall heating and cooling requirements. Figure 6 presents the year-round distribution of heat gains and losses for the analysed building for Variants 1–3 with the VAV ventilation system. The total year-round heat losses in the building amounted to 18,504 kWh, while the heat gains were 15,335 kWh. The heat balance includes heat losses through the external building envelope and for heating the ventilation supply air, as well as heat gains from electronic devices, lighting, and occupants. The year-round heat losses through the external walls and thermal bridges were 13,139 kWh. The heat losses for heating the supply air were 5352 kWh. Additionally, the balance accounted for heat losses in the HVAC system, which amounted to 14 kWh. Heat gains from electronic devices were 2919 kWh, from lighting 139 kWh, and from occupants 1954 kWh. Heat gains due to air infiltration were also considered, amounting to 12 kWh. The heat balance also includes heat losses and gains through windows due to radiation and penetration through the window frame. During the winter period (November–January), heat losses through the windows were 400 kWh, while during the summer and transitional periods (February–October), heat gains through the windows were 10,688 kWh.

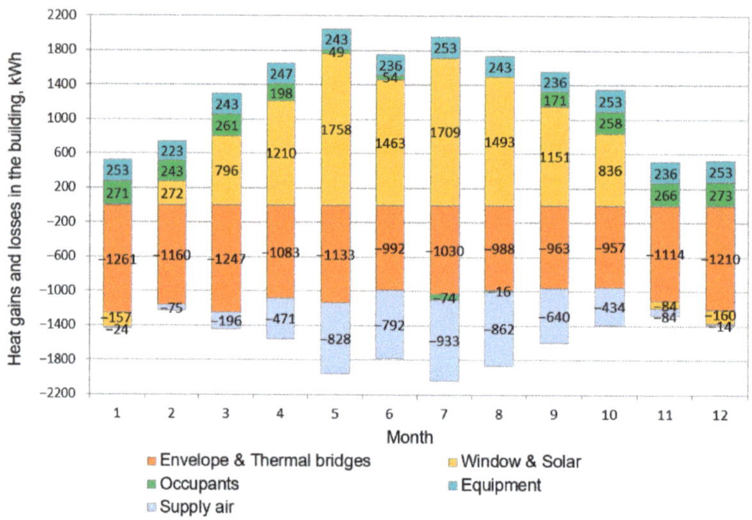

Figure 6. Year-round heat gains and losses in the building for Variants 1–3.

For Variants 4–6 with the CAV ventilation system, the heat losses for heating the supply air were 33% higher compared to Variants 1–3, amounting to 7130 kWh.

The obtained results highlight the advantages of the energy-efficient design of the building. Due to the significant glazing of the building's envelope, particularly the southern external wall, especially in the living room and kitchen area, as well as the external walls in the nighttime part of the building, where floor-to-ceiling windows were installed, the heat gains from solar radiation were maximised, accounting for 70% of the total heat gains. As a result, total heat gains balanced 83% of the building's year-round heat losses. Furthermore, the well-insulated building envelope contributed to stabilising the monthly heat losses throughout the year, preventing an increase in heat losses during the winter period. At the same time, the increase in the building's thermal insulation increases the risk of overheating during the summer period, which is mitigated by higher energy consumption for mechanical ventilation, aimed at maintaining the set indoor air temperature.

IDA ICE heat losses, presented in Figure 6, are calculated dynamically based on detailed models of building components and systems. Heat loss through building envelopes depends on the U-values of materials, surface areas, and the temperature difference between indoor and outdoor environments. Thermal bridges are accounted for using linear and point thermal transmittance coefficients. For windows, both glazing and frame properties, as well as edge effects, are considered. Ventilation heat losses are determined by airflow rates, temperature differences, and air properties, with adjustments for heat recovery if a system is installed. These calculations adapt to hourly variations in weather and system conditions, providing a comprehensive analysis of heat loss.

3.2. Thermal Energy Consumption in the Building

To further evaluate the energy performance of the building, it is essential to examine how thermal energy is distributed across different operational needs throughout the year. This analysis helps understand the contributions of various energy-consuming systems to the building's overall performance. Figure 7 shows the year-round distribution of thermal energy consumption in Variants 1–3 for space heating, domestic hot water preparation, and ventilation supply air heating. The total amount of heat delivered to the floor heating system (considering losses associated with heat generation and distribution) amounted to 3274 kWh. The total heat delivered to the heating coil in the HRV unit was 1022 kWh. From May to October, no energy was required for heating. The total heat delivered to the domestic hot water system was 6827 kWh per year, with an average of 569 kWh per month. The distribution of heat used for domestic hot water heating remained constant throughout the year due to the assumptions made regarding hot water usage.

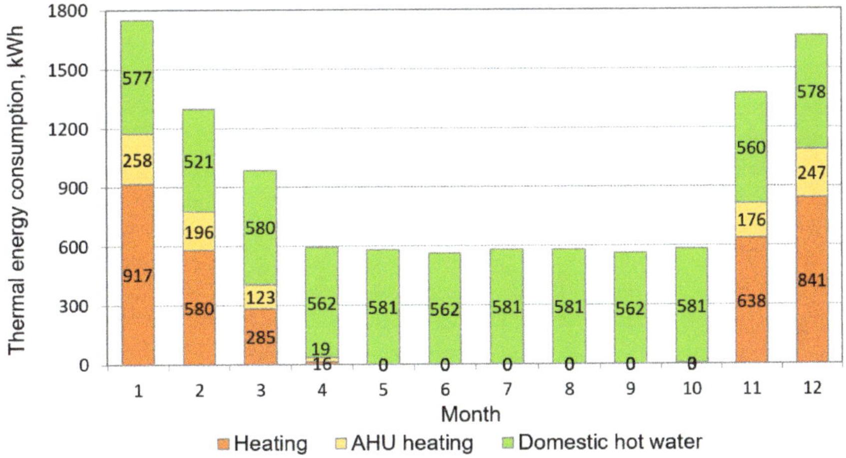

Figure 7. Year-round thermal energy consumption in the analysed building for Variants 1–3.

In Variants 4–6, the heat consumption for heating the supply air was 75% higher than in Variants 1–3, amounting to 1784 kWh.

3.3. Electrical Energy Consumption in the Building

To fully evaluate the energy demand of the building, it is essential to consider the electricity consumption associated with various systems and equipment. This includes energy use for both essential services and auxiliary components. Figure 8 shows the year-round electrical energy consumption for electronic equipment, lighting, heat pump compressor, and auxiliary HVAC system energy (i.e., circulation pumps, fans in the HRV unit) in Variants 1–3. The total energy consumption for electronic equipment was 1954 kWh, for lighting 114 kWh, for the heat pump 2815 kWh, and for auxiliary energy 657 kWh. The total year-round electricity demand for the building was 5540 kWh.

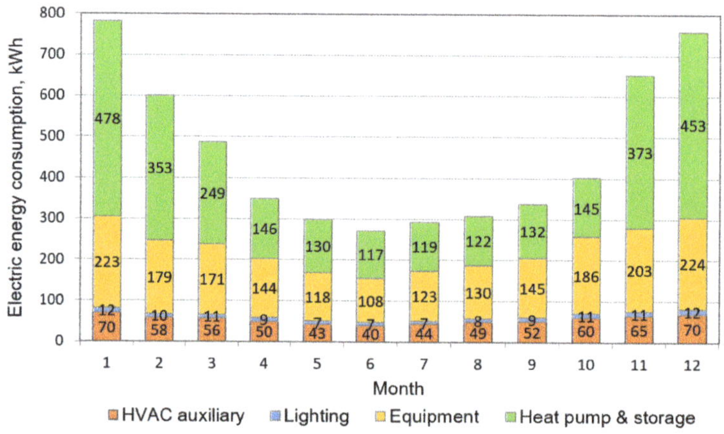

Figure 8. Year-round electrical energy consumption in the analysed building for Variants 1–3.

In Variants 4–6, the auxiliary electrical energy use in the HVAC system was 99% higher, and the energy demand for the heat pump compressor was 9% higher than in Variants 1–3, amounting to 1321 kWh and 3068 kWh, respectively. This was due to the higher ventilation air volume flow rate, which the fan had to handle in the CAV system.

3.4. Generated Free Energy

The performance of the PV system is influenced by factors such as the orientation and location of the PV modules. These factors are integral to determining the overall electrical energy generation efficiency. Figure 9 presents the year-round distribution of the generated electrical energy in the PV system, depending on the orientation and location of the PV modules in the different computational variants. The largest amount of energy, amounting to 10,797 kWh/year, was generated in Variants 3 and 6, in which the PV modules were located on a free-standing structure on the ground and oriented towards the south. The location of the PV modules on the ground, without the limitation of the roof area, allowed for the largest surface area of the PV modules, which was 100 m^2. In Variants 1 and 4, the PV modules, with a surface area of 70 m^2, were located on the eastern roof surface, generating 5881 kWh/year, which was 83% lower than in Variants 3 and 6. In Variants 2 and 5, the PV modules, with a surface area of 40 m^2, were located on the southern roof surface, generating 4247 kWh/year, which was 154% lower than in Variants 3 and 6. Analysing the monthly distribution, it can be observed that there is a sharp decrease in the amount of energy generated during the winter months. For example, in Variant 1, in December, the energy generated was more than seven times lower than in May. This is primarily due to the significant shortening of the day, meaning less time for the PV modules to be exposed to sunlight, and the lowest position of the sun above the horizon during this period.

The efficiency of heat recovery in ventilation systems significantly contributes to reducing the overall energy consumption in the building. The process of heat recovery from ventilated air is essential for optimising energy use, particularly in buildings with mechanical ventilation systems. Table 3 presents the year-round distribution of recovered energy in the building through the heat recovery process of ventilated air. This value depended on the ventilation air volume flow rates in the VAV and CAV systems in the respective variants. In Variants 4–6, equipped with the CAV system, 63% more energy was recovered compared to Variants 1–3 with the VAV system. This difference resulted from a 42% higher ventilation air volume flow rate in the CAV system variants compared to the maximum value in the VAV system variants. In the summer, the ground heat exchanger was utilised to cool the ventilation supply air, ensuring thermal comfort in the building without the use of air conditioning.

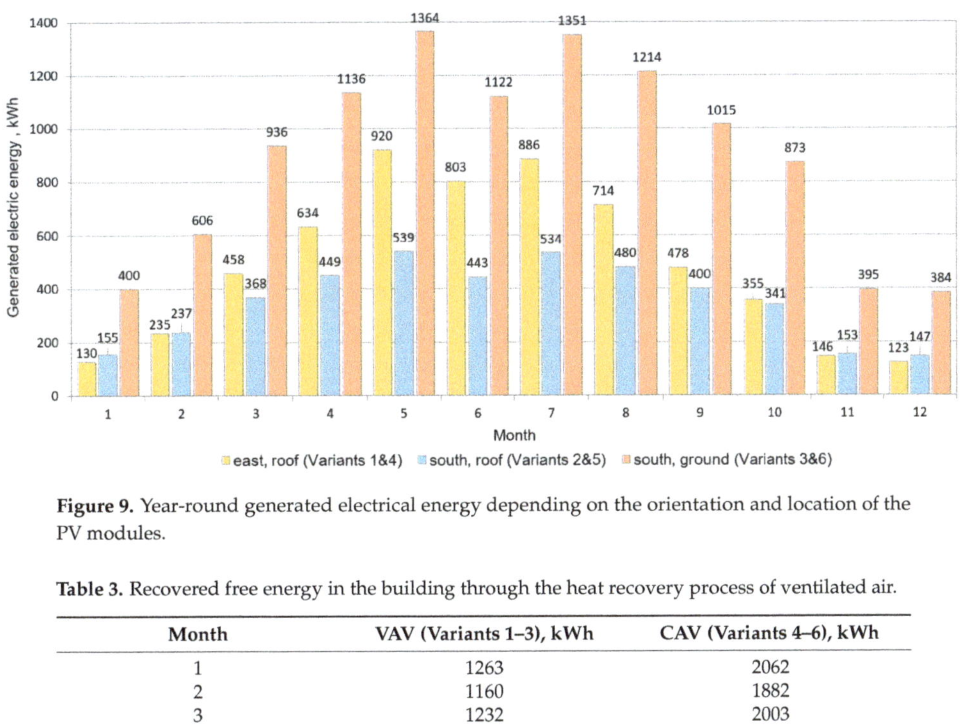

Figure 9. Year-round generated electrical energy depending on the orientation and location of the PV modules.

Table 3. Recovered free energy in the building through the heat recovery process of ventilated air.

Month	VAV (Variants 1–3), kWh	CAV (Variants 4–6), kWh
1	1263	2062
2	1160	1882
3	1232	2003
4	817	1323
5	496	809
6	350	576
7	259	425
8	276	453
9	550	893
10	764	1215
11	1154	1873
12	1248	2043
Total	9568	15,558

3.5. Analysis of the Non-Renawable Primary Energy Index

The analysis of the non-renewable primary energy (EP) index value is essential for evaluating the environmental impact of a building's energy consumption. This index reflects the amount of non-renewable energy needed to operate the building's systems and provides a benchmark for comparing different energy strategies.

Table 4 presents the following for each computational variant:

- The values of electrical energy demand for powering the building's technical systems, i.e., lighting, heating, and domestic hot water system (operation of the heat pump compressor) as well as auxiliary energy for the HVAC system (operation of HRV fans and circulation pumps).
- The amount of electrical energy supplied to the technical systems by the photovoltaic installation (assuming a self-consumption rate of 40%) and the amount of electrical energy supplied by the power grid.
- The values of the year-round non-renewable primary energy index value.

Table 4. Primary energy index values for Variants 1–6.

Variant	Lighting, kWh/Year	Technical System HVAC Aux, kWh/Year	Heat Pump, kWh/Year	Energy from the PV Installation, kWh/Year	Energy from the System Power Grid, kWh/Year	Primary Energy Index, kWh/m²/Year
1	114	657	2815	2352	1234	16
2	114	657	2815	1699	1887	25
3	114	657	2815	4319	0	0
4	114	1321	3068	2352	2151	28
5	114	1321	3068	1699	2804	37
6	114	1321	3068	4319	0	0

Variants 1 and 2, equipped with the VAV ventilation system, exhibited lower primary energy index values (16 kWh/m^2/year and 25 kWh/m^2/year, respectively) compared to the corresponding Variants 4 and 5 with the CAV system (28 kWh/m^2/year and 37 kWh/m^2/year, respectively). This difference was attributed to higher electricity consumption for the powering and auxiliary energy of the HVAC system in the CAV variants, caused by the larger volume flow rate of the ventilation air. Variants 3 and 6 demonstrated the lowest primary energy index value, equal to 0 kWh/m^2/year. This was due to the largest photovoltaic module surface area installed on a ground-mounted structure. The absence of space constraints, as opposed to roof-mounted systems in other variants, allowed for the highest electricity production in these cases. All computational variants met the criterion of the maximum value of the non-renewable primary energy index for single-family buildings, which must not exceed 70 kWh/m^2/year [18].

Based on the results of the analyses conducted, it was determined that the most advantageous solution in terms of the EP index value for the designed building is the application of the VAV ventilation system. Regarding the photovoltaic installation on the roof, a south-facing orientation is the most favourable. In the analysed building, the unit value of electricity generated by the PV installation mounted on the south-facing roof surface was 42 kWh/m^2/year, compared to 34 kWh/m^2/year for the east-facing roof surface. However, due to the east-facing roof surface being 30 m^2 larger, allowing for more PV modules to be installed, mounting the modules on the east-facing surface is the more advantageous solution for this building. The most favourable option for PV installation among all variants is its placement on the ground. With a PV module area of 100 m^2, the building required no electricity from the grid, thus achieving a zero-energy building status. However, this option demands a significant plot area, potentially limiting its feasibility for many single-family homes, especially considering the space required for a horizontal ground heat exchanger installation. Additionally, such a solution may increase investment costs due to factors such as the need for foundation construction, while the modules themselves may be more susceptible to damage. For these reasons, it was decided to examine how the EP index value could be reduced in Variants 1 and 2 with the VAV ventilation system, in which PV modules were mounted on the roof, to achieve a building performance as close as possible to that of a zero-energy building—one of the criteria of the 4E Idea.

3.6. Optimalisation of the Non-Renewable Primary Energy Index

To further enhance the building's energy performance, two variants of solar thermal collector installations—flat-plate and vacuum—were analysed. They were mounted on the southern and eastern roof surfaces, depending on the location of the PV modules in Variants 1 and 2. Subsequently, for the selected solar thermal installation variant characterised by the highest energy efficiency, an analysis of the use of a ground-source heat pump instead of an air-source heat pump was conducted.

The selection of solar thermal collectors was carried out using the T*SOL 2023 software [44] based on the year-round demand for domestic hot water calculated in the IDA ICE 4.8 software. The following calculation variants were established as a result of this selection:

- Variant 1.1: HVAC and PV installation from Variant 1 + two flat-plate solar thermal collectors mounted on the southern roof surface. Parameters of a single thermal collector: aperture area of 2.33 m^2, efficiency of 78%, linear heat loss coefficient of 4.14 $W/m^2{\cdot}K$, and square heat loss coefficient of 0.0145 $W/m^2{\cdot}K$.
- Variant 1.2: HVAC and PV installation from Variant 1 + two vacuum solar thermal collectors mounted on the southern roof surface. Parameters of a single thermal collector: aperture area of 1.6 m^2, efficiency of 77%, linear heat loss coefficient of 1.256 $W/m^2{\cdot}K$, and square heat loss coefficient of 0.005 $W/m^2{\cdot}K$.
- Variant 2.1: HVAC and PV installation from Variant 2 + two flat-plate solar thermal collectors mounted on the eastern roof surface, with parameters as in Variant 1.1.
- Variant 2.2: HVAC and PV installation from Variant 2 + two vacuum solar thermal collectors mounted on the eastern roof surface, with parameters as in Variant 1.2.

Flat-plate and vacuum solar thermal collectors were selected to reflect the two widely used collector technologies, which differ in efficiency and suitability under varying climatic conditions. The location of PV modules and solar thermal collectors was varied to assess the influence of orientation and roof geometry on energy yield. This accounts for real-world scenarios where building constraints or aesthetic considerations might dictate the placement of these systems.

Figure 10 shows the arrangement of solar thermal collectors and PV modules in Variants 1.1, 1.2, 2.1, and 2.2.

To assess the performance of the solar thermal collectors, the amount of solar radiation energy absorbed is an important factor. These data allow for a comparison of the energy contributions from solar thermal systems across different design variants. Figure 11 presents the amount of solar radiation energy absorbed by the solar thermal collectors in Variants 1.1–2.2. Over a year, the highest amount of solar energy was absorbed by vacuum tube solar thermal collectors oriented southwards (Variant 1.2), amounting to 2908 kWh/year. The second-highest value was absorbed by vacuum tube solar thermal collectors oriented eastwards (Variant 2.2) at 2090 kWh/year. Flat-plate solar thermal collectors exhibited lower levels of absorbed solar energy. For flat-plate collectors oriented southwards (Variant 1.1), the absorbed energy was 1668 kWh/year, while those oriented eastwards (Variant 2.1) absorbed 1077 kWh/year. Thus, it can be concluded that vacuum tube solar thermal collectors with southern exposure were more efficient in terms of absorbed solar energy.

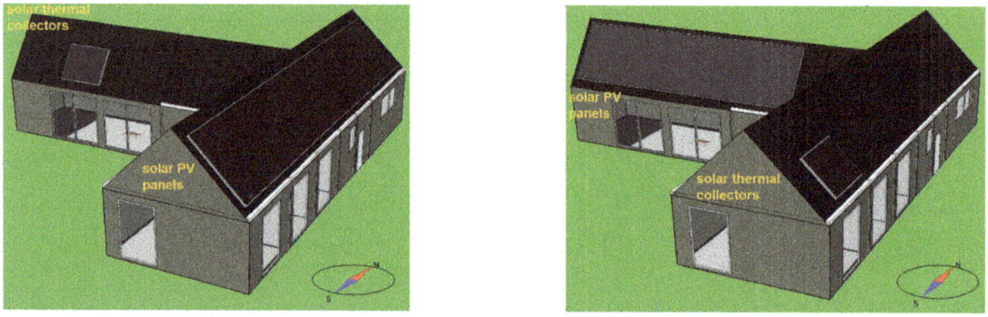

(a) (b)

Figure 10. The arrangement of solar thermal collectors and photovoltaic panels in Variants: (a) 1.1 and 1.2, (b) 2.1 and 2.2.

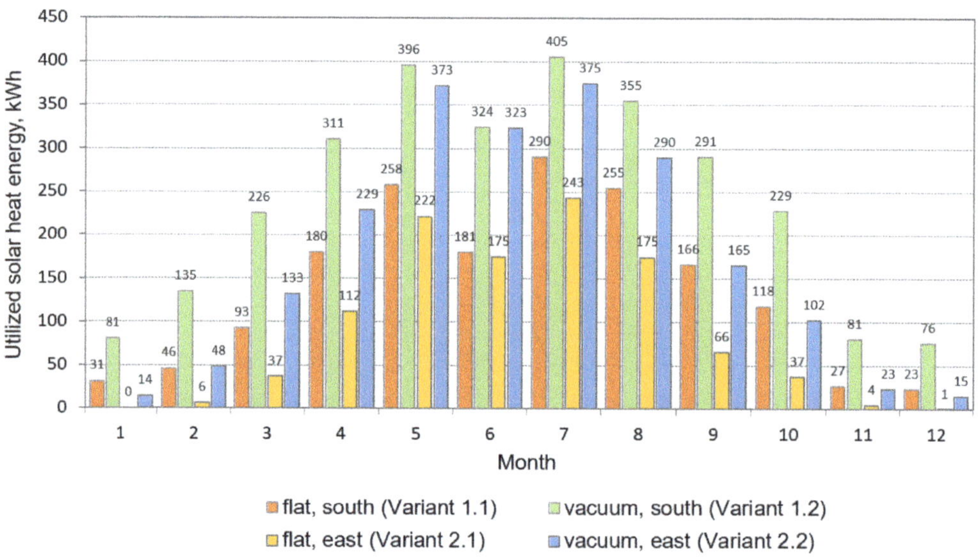

Figure 11. Solar radiation energy absorbed by solar thermal collectors in Variants 1.1–2.2.

Analysing the year-round distribution of solar energy gained by the solar thermal collectors reveals that vacuum tube solar collectors were particularly more effective than flat-plate solar collectors during winter months when the angle of solar radiation is the lowest. For instance, in January, the solar energy yield of the vacuum tube solar system in Variant 1.2 was 161% higher than that of the flat-plate solar system in Variant 1.1. During the summer, this difference decreased but remained significant; for example, in July, the yield of vacuum tube solar collectors (Variant 1.2) was 40% higher compared to flat-plate solar collectors (Variant 1.1).

The distribution of power consumption by the heat pump compressor is a crucial aspect in evaluating the overall energy efficiency of the building. The comparison between variants with and without solar thermal collectors highlights the impact of renewable energy integration on the system's performance. Figure 12 shows the heat pump compressor power distribution in Variant 1, without solar thermal collectors, and Variants 1.1 and 1.2, equipped with flat-plate and vacuum-tube solar thermal collectors, respectively. It can be observed that the greatest reduction in the operational load of the heat pump compressor due to the solar thermal installation occurred between early May and late September. This resulted from the fact that during this period the system worked mainly for domestic hot water preparation. For part of this period, on days with high solar energy gains, the heat pump did not need to operate at all. Vacuum-tube solar collectors contributed to a greater reduction in heat pump operation compared to flat-plate solar collectors. During the remaining part of the year, when the heat pump also operated for heating purposes, the reduction in compressor load due to the solar thermal installation was minimal. However, it is notable that even during the winter months, solar collectors occasionally reduced the workload of the heat pump compressor, depending on the prevailing weather conditions. The figure also highlights that the selected heat pump was equipped with an inverter compressor. This type of compressor operates by adapting its power and speed to varying conditions and individual user needs.

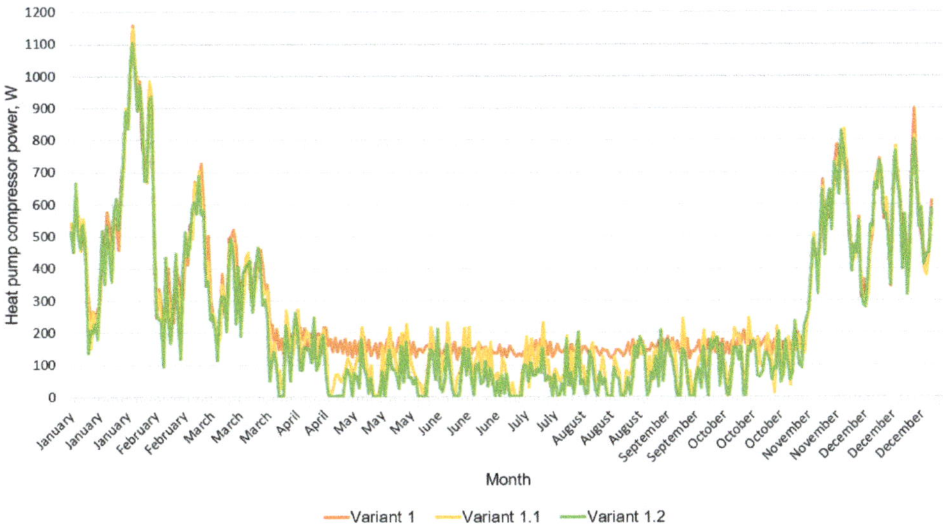

Figure 12. Year-round distribution of heat pump compressor power in Variants 1.1–1.2.

The integration of renewable energy systems, such as solar thermal collectors, can significantly affect the performance and efficiency of heating installations. Combining various technologies, such as heat pumps and solar systems, can considerably reduce overall energy consumption. Table 5 presents the impact of retrofitting the heat pump installation with a solar thermal system. The installation of solar collectors on the southern roof surface resulted in a reduction in energy consumption by the heat pump by 497 kWh/year for flat-plate solar collectors (Variant 1.1) and 680 kWh/year for vacuum solar collectors (Variant 1.2). In the case of solar collectors installed on the eastern roof surface, the energy consumption by the heat pump was reduced by 233 kWh/year for flat-plate solar collectors (Variant 2.1) and by 399 kWh/year for vacuum solar collectors (Variant 2.2). Thus, the installation of south-facing solar collectors resulted in a greater reduction in the heat pump compressor power, by 113% for flat-plate solar collectors and 70% for vacuum solar collectors, compared to the east-facing variants. At the same time, energy consumption by auxiliary devices increased due to the addition of a circulation pump. The most advantageous variants in terms of optimising the primary energy index were the installation of south-facing vacuum and flat-plate solar collectors, with EP index values of 8 kWh/m^2/year (Variant 1.2) and 10 kWh/m^2/year (Variant 1.1), respectively. Retrofitting Variant 2 with solar collectors resulted in a smaller reduction in the EP value compared to the variant without solar collectors than in the case of retrofitting Variant 1. Therefore, although Variant 2 exhibited a higher EP index value, the most favourable variant from this group, Variant 2.2, was characterised by an EP index value that was higher by 12 kWh/m^2/year than the corresponding Variant 1.2.

In the final stage of the analysis of the possibility of optimising the EP index value for the analysed building, an analysis was carried out regarding the use of a ground-source heat pump instead of an air-source heat pump. A ground-source heat pump with a heating capacity of 6 kW and a COP of 5.5 was selected. The thermal efficiency of the ground was assumed to be 20 W/m^2, in accordance with the guidelines [45], which recommend adopting this value when the parameters of the ground are unknown. It was assumed that the horizontal ground heat exchanger would be placed at a depth of 1.5 m. At this depth, the ground temperature is at least 6 °C [45] and, according to it the COP value for the ground-source heat pump, this was determined based on the manufacturer's data. The surface area of the ground heat exchanger was calculated according to the guidelines [45] and was 241 m^2. The ground temperature was assumed to be constant throughout the year, set at 8 °C. Propylene glycol was selected as the heat carrier fluid in the ground heat exchanger installation. The energy analysis was carried out based on Variant 1.2, which had the lowest EP index value.

Table 5. The primary energy index values for Variants 1-2.2.

Variant	Lighting, kWh/Year	Technical System HVAC Aux, kWh/Year	Heat Pump, kWh/Year	Energy from the PV Installation, kWh/Year	Energy from the System Power Grid, kWh/Year	Primary Energy Index, kWh/m^2/Year
1	114	657	2815	2352	1234	16
1.1	114	708	2318	2352	788	10
1.2	114	708	2135	2352	605	8
2	114	657	2815	1699	1887	25
2.1	114	708	2582	1699	1705	22
2.2	114	708	2416	1699	1539	20

Table 6 presents the impact of replacing the air-source heat pump (Variant 1.2) with the ground-source heat pump (Variant 7). The energy consumption by the compressor decreased by 458 kWh/year as a result of using a device with a higher COP value and adopting a constant ground temperature, as opposed to a variable outdoor air temperature throughout the year. As a result, the use of the ground-source heat pump led to a reduction in the EP index value by 6 kWh/m^2/year compared to the variant with the air-source heat pump. Ultimately, the designed building, with a VAV ventilation system with heat recovery, an on-grid photovoltaic installation, vacuum solar thermal collectors, and a ground-source heat pump, was characterised by an EP index value of 2 kWh/m^2/year. If an off-grid photovoltaic system with an energy storage battery had been used in the building, it could be assumed that the designed building would have achieved a zero-energy building characteristic, as the self-consumption of electricity would have been nearly 100%, as opposed to the 40% in the analysed variant.

Table 6. The primary energy index values for Variants 1.2 and 7.

Variant	Lighting, kWh/Year	Technical System HVAC Aux, kWh/Year	Heat Pump, kWh/Year	Energy from the PV Installation, kWh/Year	Energy from the System Power Grid, kWh/Year	Primary Energy Index, kWh/m²/Year
1.2	114	708	2135	2352	605	8
7	114	740	1645	2352	147	2

4. Conclusions

The paper presents the concept of a newly designed single-family building based on the 4E Idea, which assumes that the building should be energy-efficient, ecological, economical, and incorporate ergonomic design solutions. The energy efficiency criterion stipulates that the building should generate zero energy consumption or a positive year-round energy balance. The aim of the paper was to conduct multi-variant energy simulations of the proposed building, the results of which were intended to enable the selection of an HVAC system and a renewable energy sources concept that would ensure that the building's energy performance is as close as possible to a zero-energy building. It was analysed how the following factors influenced the reduction in the non-renewable primary energy index value: the type of mechanical ventilation system (CAV and VAV); the orientation and power of the photovoltaic installation; the exposure and type of solar thermal collectors (flat-plate and vacuum); and the type of heat pump (air and ground source). Based on the obtained results, the following conclusions were drawn:

- The architectural and structural concept of the building enabled the maximisation of heat gains and the reduction in heat losses through the external building envelope. Due to the significant glazing of the building envelope, the solar heat gains accounted for 70% of the total heat gains. As a result, the total heat gains covered 83% of the building's year-round heat losses. Furthermore, the well-insulated building envelope contributed to stabilising the monthly heat losses throughout the entire year.
- The analysis of the type of supply-and-exhaust mechanical ventilation with heat recovery showed that the VAV system is more energy-efficient than the CAV system. The variant in which an air-source heat pump and a VAV ventilation system were applied exhibited half the energy consumption compared to the corresponding variant with a CAV system (657 kWh/year for the VAV system and 1321 kWh/year for the CAV system). This difference resulted from the reduced ventilation air volume flow rate, as the VAV system adjusted the air supply to the current CO_2 concentration levels in the rooms. Consequently, the heat pump compressor consumed 253 kWh/year less energy, as it was required to supply less heat to warm the ventilation air.
- The analysis of the photovoltaic modules' location showed that a southern exposure is generally the most favourable for solar installations. The specific energy yield of the PV modules installed on the southern roof surface was 42 kWh/m^2/year, compared to 34 kWh/m^2/year for the eastern roof surface. However, due to the eastern roof's 30 m^2 larger surface area, which allowed for the installation of more PV modules, installing the modules on this surface proved to be the more advantageous solution for the analysed building. The variant featuring a VAV ventilation system, an air-source heat pump, and PV modules installed on the eastern roof surface (with a total module area of 70 m^2 and a capacity of 6.6 kWp) was characterised by an EP index value of 16 kWh/m^2/year. In contrast, the southern roof surface installation (with a total module area of 40 m^2 and a capacity of 4.3 kWp) had an EP index value of 25 kWh/m^2/year. In the case of the variant in which the PV modules were installed on a ground-mounted structure with a southern orientation and a surface area of 100 m^2,

the building achieved zero energy characteristics. However, this solution requires a significant plot area, involves higher investment costs, and presents a greater risk of damage, making it less suitable for widespread implementation.

- The analysis of the location and type of solar thermal collectors showed that the most favourable variant for optimising the PE index value was the installation of vacuum tube solar collectors on the southern roof surface. This configuration resulted in an EP index value of 8 kWh/m^2/year, and thus a two-fold reduction in the EP index value compared to the variant without solar thermal collectors. The installation of solar thermal collectors alleviated the workload of the heat pump compressor for domestic hot water preparation and contributed to a reduction in the heat pump compressor's electricity consumption by 629 kWh/year. This shows that under variable climatic conditions, vacuum tube solar thermal collectors exhibit higher efficiency compared to flat-plate solar thermal collectors.
- The analysis of the type of heat pump used indicated that the variant incorporating a ground-source heat pump with a horizontal ground heat exchanger demonstrated superior efficiency. This variant was characterised by a four-fold reduction in the EP index value compared to the variant with an air-source heat pump. This resulted from the higher COP value of the ground-source heat pump, which was constant throughout the year due to the stable ground temperature, in contrast to the fluctuating temperature of atmospheric air. Additionally, the installation of solar thermal collectors contributed to reducing the heat pump's workload and improving ground regeneration, particularly during the summer months.
- For the analysed building equipped with a VAV mechanical ventilation system with heat recovery, an on-grid photovoltaic installation, vacuum solar thermal collectors, and a ground-source heat pump with a horizontal heat exchanger, a primary energy index value of 2 kWh/m^2/year was achieved. The self-consumption of the generated electrical energy in the building was assumed to be 40%. It can be expected that implementing an off-grid photovoltaic system could result in achieving a zero-energy building characteristic.

The study has several limitations that should be noted. The simulations were conducted for a specific single-story building with particular construction characteristics, which may limit the generalizability of the results. Fixed parameters for occupant behaviour, such as device usage schedules and comfort temperatures, were assumed, which do not fully reflect variability in real-world scenarios. Additionally, local climatic conditions were used, which may not account for future climate changes, such as shorter periods of low temperatures. The analysis also excluded economic factors due to the dynamic nature of energy prices and equipment costs, which are influenced by social and political factors. Despite these limitations, the findings provide a valuable starting point for further research and analysis. As a result of the conducted analysis, a concept for an energy-efficient building was developed, which, in terms of architectural and construction design, as well as the configuration of the HVAC system and renewable energy sources installation, can serve as a starting point for future energy-efficient single-family building designs. Further research should be carried out for the variant incorporating electric energy storage to achieve further optimisation of the PE index value.

Author Contributions: Conceptualization, P.C., J.K. and D.W.-J.; methodology, P.C. and J.K.; software, P.C.; formal analysis, P.C. and J.K.; writing—original draft preparation, P.C.; writing—review and editing, J.K. and P.C.; visualisation, D.W.-J. All authors have read and agreed to the published version of the manuscript.

Funding: This research received no external funding.

Data Availability Statement: The original contributions presented in the study are included in the article, further inquiries can be directed to the corresponding author.

Acknowledgments: The work is supported by the Polish Ministry of Education and Science within the research subsidy.

Conflicts of Interest: The authors declare no conflicts of interest.

References

1. van Staalduinen, W.; Dantas, C.; Ferenczi, A.; Klimczuk, A.; Freitas, A.; Abreu Cordeiro, B.; Guley Goren Soares, B.; Pineda Revilla, B.; Hilario, C.; Vassiliou, C.; et al. White Paper: Designing the Perfect New European Bauhaus Neighbourhood. 2024. Available online: https://open.icm.edu.pl/items/0501e238-886e-40b0-81eb-c259afd03fc1 (accessed on 1 December 2024).
2. GUS Budownictwo w 1 Kwartale 2024 Roku. Available online: https://stat.gov.pl/obszary-tematyczne/przemysl-budownictwo-srodki-trwale/budownictwo/budownictwo-w-1-kwartale-2024-roku,13,22.html (accessed on 13 November 2024).
3. EU Energy Statistical Pocketbook and Country Datasheets. Available online: https://energy.ec.europa.eu/data-and-analysis/eu-energy-statistical-pocketbook-and-country-datasheets_en (accessed on 18 November 2024).
4. GUS Energy Consumption in Households in 2021. Available online: https://stat.gov.pl/en/topics/environment-energy/energy/energy-consumption-in-households-in-2021,2,6.html (accessed on 1 December 2024).
5. GUS Energy Efficiency in Years 2012–2022. Available online: https://stat.gov.pl/en/topics/environment-energy/energy/energy-efficiency-in-years-20122022,5,21.html (accessed on 1 December 2024).
6. Greenhouse Gas Emissions from Energy Use in Buildings in Europe. Available online: https://www.eea.europa.eu/en/analysis/indicators/greenhouse-gas-emissions-from-energy (accessed on 1 December 2024).
7. Kania, G.; Kwiecień, K.; Malinowski, M.; Gliniak, M. Analyses of the Life Cycles and Social Costs of CO_2 Emissions of Single-Family Residential Buildings: A Case Study in Poland. *Sustainability* **2021**, *13*, 6164. [CrossRef]
8. Poland—Countries & Regions. Available online: https://www.iea.org/countries/poland/energy-mix (accessed on 18 November 2024).
9. Zator, S.; Skomudek, W. Impact of DSM on Energy Management in a Single-Family House with a Heat Pump and Photovoltaic Installation. *Energies* **2020**, *13*, 5476. [CrossRef]
10. Niekurzak, M.; Lewicki, W.; Drożdż, W.; Miązek, P. Measures for Assessing the Effectiveness of Investments for Electricity and Heat Generation from the Hybrid Cooperation of a Photovoltaic Installation with a Heat Pump on the Example of a Household. *Energies* **2022**, *15*, 6089. [CrossRef]
11. Hesaraki, A.; Madani, H. *Energy Performance of Ground-Source Heat Pump and Photovoltaic/Thermal (PV/T) in Retrofitted and New Buildings: Two Case Studies Using Simulation and On-Site Measurements*; SINTEF Academic Press: Cambridge, MA, USA, 2020; ISBN 978-82-536-1679-7.
12. Kemmler, T.; Thomas, B. Design of Heat-Pump Systems for Single- and Multi-Family Houses Using a Heuristic Scheduling for the Optimization of PV Self-Consumption. *Energies* **2020**, *13*, 1118. [CrossRef]
13. Pater, S. Increasing Energy Self-Consumption in Residential Photovoltaic Systems with Heat Pumps in Poland. *Energies* **2023**, *16*, 4003. [CrossRef]
14. Baraskar, S.; Günther, D.; Wapler, J.; Lämmle, M. Analysis of the Performance and Operation of a Photovoltaic-Battery Heat Pump System Based on Field Measurement Data. *Sol. Energy Adv.* **2024**, *4*, 100047. [CrossRef]
15. Thorsteinsson, S.; Kalaee, A.A.S.; Vogler-Finck, P.; Stærmose, H.L.; Katic, I.; Bendtsen, J.D. Long-Term Experimental Study of Price Responsive Predictive Control in a Real Occupied Single-Family House with Heat Pump. *Appl. Energy* **2023**, *347*, 121398. [CrossRef]
16. Nalini Ramakrishna, S.K.; Björner Brauer, H.; Thiringer, T.; Håkansson, M. Social and Technical Potential of Single Family Houses in Increasing the Resilience of the Power Grid during Severe Disturbances. *Energy Convers. Manag.* **2024**, *321*, 119077. [CrossRef]
17. Drozd, W.; Kowalik, M. Analysis of Renewable Energy Use in Single-Family Housing. *Open Eng.* **2019**, *9*, 269–281. [CrossRef]
18. Obwieszczenie Ministra Rozwoju i Technologii z Dnia 15 Kwietnia 2022 r. w Sprawie Ogłoszenia Jednolitego Tekstu Rozporządzenia Ministra Infrastruktury w Sprawie Warunków Technicznych, Jakim Powinny Odpowiadać Budynki i Ich Usytuowanie. Available online: https://isap.sejm.gov.pl/isap.nsf/DocDetails.xsp?id=WDU20220001225 (accessed on 10 December 2024).
19. Nielsen, T.R.; Drivsholm, C. Energy Efficient Demand Controlled Ventilation in Single Family Houses. *Energy Build.* **2010**, *42*, 1995–1998. [CrossRef]

20. Lu, D.B.; Warsinger, D.M. Energy Savings of Retrofitting Residential Buildings with Variable Air Volume Systems across Different Climates. *J. Build. Eng.* **2020**, *30*, 101223. [CrossRef]
21. Bai, H.Y.; Liu, P.; Justo Alonso, M.; Mathisen, H.M. A Review of Heat Recovery Technologies and Their Frost Control for Residential Building Ventilation in Cold Climate Regions. *Renew. Sustain. Energy Rev.* **2022**, *162*, 112417. [CrossRef]
22. Cavazzini, G.; Zanetti, G.; Benato, A. Analysis of a Domestic Air Heat Pump Integrated with an Air-Geothermal Heat Exchanger in Real Operating Conditions: The Case Study of a Single-Family Building. *Energy Build.* **2024**, *315*, 114302. [CrossRef]
23. Halios, C.H.; Landeg-Cox, C.; Lowther, S.D.; Middleton, A.; Marczylo, T.; Dimitroulopoulou, S. Chemicals in European Residences—Part I: A Review of Emissions, Concentrations and Health Effects of Volatile Organic Compounds (VOCs). *Sci. Total Environ.* **2022**, *839*, 156201. [CrossRef]
24. Poirier, B.; Guyot, G.; Geoffroy, H.; Woloszyn, M.; Ondarts, M.; Gonze, E. Pollutants Emission Scenarios for Residential Ventilation Performance Assessment. A Review. *J. Build. Eng.* **2021**, *42*, 102488. [CrossRef]
25. Khoshnava, S.M.; Rostami, R.; Mohamad Zin, R.; Štreimikienė, D.; Mardani, A.; Ismail, M. The Role of Green Building Materials in Reducing Environmental and Human Health Impacts. *Int. J. Environ. Res. Public Health* **2020**, *17*, 2589. [CrossRef] [PubMed]
26. Lim, A.-Y.; Yoon, M.; Kim, E.-H.; Kim, H.-A.; Lee, M.J.; Cheong, H.-K. Effects of Mechanical Ventilation on Indoor Air Quality and Occupant Health Status in Energy-Efficient Homes: A Longitudinal Field Study. *Sci. Total Environ.* **2021**, *785*, 147324. [CrossRef]
27. Winnicka-Jasłowska, D.; Jastrzębska, M.; Kaczmarczyk, J.; Łaźniewska-Piekarczyk, B.; Skóra, P.; Kobiałko, B.; Kołodziej, A.; Mól, B.; Lasyk, E.; Brzęczek, K.; et al. Projekt koncepcyjny domu mieszkalnego opartego na Idei 4E. Project-based learning realizowany w Politechnice Śląskiej. *Builder* **2023**, *307*, 12–19. [CrossRef]
28. Pan, Y.; Zhu, M.; Lv, Y.; Yang, Y.; Liang, Y.; Yin, R.; Yang, Y.; Jia, X.; Wang, X.; Zeng, F.; et al. Building Energy Simulation and Its Application for Building Performance Optimization: A Review of Methods, Tools, and Case Studies. *Adv. Appl. Energy* **2023**, *10*, 100135. [CrossRef]
29. Tihana, J.; Ali, H.; Apse, J.; Jekabsons, J.; Ivancovs, D.; Gaujena, B.; Dedov, A. Hybrid Heat Pump Performance Evaluation in Different Operation Modes for Single-Family House. *Energies* **2023**, *16*, 7018. [CrossRef]
30. Xue, T.; Jokisalo, J.; Kosonen, R. Cost-Effective Control of Hybrid Ground Source Heat Pump (GSHP) System Coupled with District Heating. *Buildings* **2024**, *14*, 1724. [CrossRef]
31. Sankelo, P.; Ahmed, K.; Mikola, A.; Kurnitski, J. Renovation Results of Finnish Single-Family Renovation Subsidies: Oil Boiler Replacement with Heat Pumps. *Energies* **2022**, *15*, 7620. [CrossRef]
32. Mazzeo, D.; Romagnoni, P.; Matera, N.; Oliveti, G.; Cornaro, C.; De Santoli, L. Accuracy of the Most Popular Building Performance Simulation Tools: Experimental Comparison for a Conventional and a PCM-Based Test Box. In Proceedings of the Building Simulation 2019—16th IBPSA International Conference and Exhibition, Rome, Italy, 2–4 September 2019; pp. 4530–4537.
33. Mazzeo, D.; Matera, N.; Cornaro, C.; Oliveti, G.; Romagnoni, P.; De Santoli, L. EnergyPlus, IDA ICE and TRNSYS Predictive Simulation Accuracy for Building Thermal Behaviour Evaluation by Using an Experimental Campaign in Solar Test Boxes with and without a PCM Module. *Energy Build.* **2020**, *212*, 109812. [CrossRef]
34. Schaffer, M.; Bugenings, L.A.; Larsen, O.K. Validating a Building Performance Simulation Model of a Naturally Ventilated Double Skin Facade. *J. Phys. Conf. Ser.* **2023**, *2654*, 012092. [CrossRef]
35. *EN 12831-1:2017*; Energy Performance of Buildings. Method for Calculation of the Design Heat Load Space Heating Load, Module M3-3. European Committee for Standardization: Brussels, Belgium, 2017.
36. Allouhi, A. Solar PV Integration in Commercial Buildings for Self-Consumption Based on Life-Cycle Economic/Environmental Multi-Objective Optimization. *J. Clean. Prod.* **2020**, *270*, 122375. [CrossRef]
37. Analysis of Self-Consumption of Energy from Grid-Connected Photovoltaic System for Various Load Scenarios with Short-Term Buffering | Discover Applied Sciences. Available online: https://link.springer.com/article/10.1007/s42452-019-0432-5 (accessed on 15 November 2024).
38. IDA ICE—Simulation Software | EQUA. Available online: https://www.equa.se/en/ida-ice (accessed on 9 December 2024).
39. American Society of Heating, Refrigerating and Air Conditioning Engineers. *ASHRAE Handbook*; Fundamentals (SI Edition); American Society of Heating, Refrigerating and Air Conditioning Engineers: Atlanta, GA, USA, 2013.
40. *ASHRAE 55:2010*; Thermal Environmental Conditions for Human Occupancy. ASHRAE: Atlanta, GA, USA, 2010.
41. Rozporządzenie Ministra Infrastruktury z Dnia 14 Stycznia 2002 r. w Sprawie Określenia Przeciętnych Norm Zużycia Wody (Dz.U. 2002 Nr 8 Poz. 70). Available online: https://isap.sejm.gov.pl/isap.nsf/DocDetails.xsp?id=WDU20020080070 (accessed on 10 December 2024).
42. *PN-83/B-03430/Az3:2000*; Wentylacja w Budynkach Mieszkalnych Zamieszkania Zbiorowego i Użyteczności Publicznej—Wymagania. Wydawnictwa Normalizacyjne PKNiM: Warszawa, Poland, 1983.
43. Rozporządzenie Ministra Rozwoju i Technologii z Dnia 28 Marca 2023 r. Zmieniające Rozporządzenie w Sprawie Metodologii Wyznaczania Charakterystyki Energetycznej Budynku Lub Części Budynku Oraz Świadectw Charakterystyki Energetycznej.

(Dz.U. 2023 Poz. 697). Available online: https://isap.sejm.gov.pl/isap.nsf/DocDetails.xsp?id=WDU20230000697 (accessed on 10 December 2024).
44. T*SOL Online | Free Solar Thermal Calculator. Available online: https://online.tsol.de/en (accessed on 9 December 2024).
45. Polska Organizacja Rozwoju Technologii Pomp Ciepła. *Wytyczne Projektowania, Wykonania i Odbioru Instalacji z Pompami Ciepła;* Część 1. Dolne Źródła Do Pomp Ciepła; PEMP, Wydanie Drugie: Kraków, Poland, 2021.

energies

MDPI

Article

Application of Machine Learning Techniques for Predicting Heating Coil Performance in Building Heating Ventilation and Air Conditioning Systems

Adam Nassif, Pasidu Dharmasena and Nabil Nassif *

Department of Civil and Architectural Engineering and Construction Management, University of Cincinnati, Cincinnati, OH 45220, USA; nassifam@mail.uc.edu (A.N.)
* Correspondence: nassifnl@ucmail.uc.edu

Abstract: Heating systems in a building's mechanical infrastructure account for a significant share of global building energy consumption, underscoring the need for improved efficiency. This study evaluates 31 predictive models—including neural networks, gradient boosting (XGBoost), bagging, and multiple linear regression (MLR) as a baseline—to estimate heating-coil performance. Experiments were conducted on a water-based air-handling unit (AHU), and the dataset was cleaned to eliminate illogical and missing values before training and validation. Among the evaluated models, neural networks, gradient boosting, and bagging demonstrated superior accuracy across various error metrics. Bagging offered the best balance between outlier robustness and pattern recognition, while neural networks showed strong capability in capturing complex relationships. An input-importance analysis further identified key variables influencing model predictions. Future work should focus on refining these modeling techniques and expanding their application to other HVAC components to improve adaptability and efficiency.

Keywords: HVAC systems; machine learning; neural network; gradient boosting; bagging; XGBoost

1. Introduction

1.1. Issue at Large

Building occupants require a thermally comfortable environment to perform daily tasks efficiently and enhance overall productivity [1–3]. To achieve this, various HVAC systems are designed to regulate indoor temperatures based on occupant needs and external conditions. However, the energy consumption of these systems has significantly increased operational costs, making efficiency a critical concern [4].

Improving overall HVAC system efficiency requires a thorough evaluation of actual equipment performance under real-world operating conditions. A critical component of HVAC systems is the air handling unit (AHU), which conditions and distributes supply air through heating and cooling processes. The AHU consists of several key components, including heating and cooling coils, fans, air filters, and dampers, all of which contribute to indoor air quality and thermal comfort. The air filter removes particulates from the incoming air before conditioning, while the fan ensures proper air distribution through the ductwork. Figure 1 illustrates a typical AHU configuration with its major components.

The heating coil is supplied with hot water from a gas-fired or electric boiler, whereas the cooling coil is connected to a chiller, cooling tower, or an air-source system, depending on the building's design and climate conditions. These components play a crucial role in regulating indoor temperature and occupant comfort [5,6].

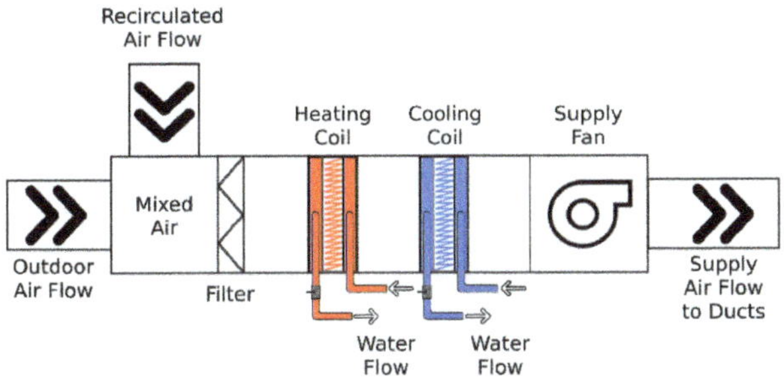

Figure 1. Typical AHU and major components.

According to the U.S. Energy Information Administration (EIA), space heating accounts for the largest single energy end-use in commercial buildings, representing approximately 32% of total building energy consumption in the United States [7]. This high energy demand translates to substantial operational costs, particularly in colder climates, where heating loads are more significant. Additionally, the majority of heating energy is supplied by natural gas, a primary contributor to global CO_2 emissions, highlighting the urgent need for energy-efficient heating solutions [8].

Heating coil plays a vital role in satisfying building's heating load by transferring thermal energy from the hot water loop to the supply air. This process occurs as air is forced over the coil and distributed into occupied spaces, ensuring thermal comfort. The coil's heat transfer characteristics significantly impact on the overall performance and efficiency of the HVAC system. Therefore, developing reliable and accurate energy models to predict heating coil performance is essential. Such models not only enhance system performance evaluation but also serve as a critical tool for optimization applications, improving energy efficiency and reducing operational costs [9].

1.2. Industry Trends

Modern HVAC research and industry trends emphasize decarbonization, electrification, and AI-driven optimization, increasing the demand for accurate heating coil models. The transition from fossil fuel-based heating systems to electric heat pumps and district heating networks necessitates precise performance modeling of heating components to maximize energy efficiency and minimize operational costs [10–13]. Furthermore, AI-driven optimization and digital twin technologies are being increasingly utilized to predict HVAC performance in real time, allowing for adaptive control strategies and improved fault detection [13–15]. Additionally, stringent energy codes such as ASHRAE 90.1 (American Society of Heating Refrigeration and Air-Conditioning Engineers) [16], international energy conversation codes [17], and global carbon reduction policies [18], are driving the need for more energy-efficient heating solutions, reinforcing the importance of accurate heating coil models. Current research has yet to establish a consistently reliable method for predicting HVAC system performance functions, in a way that directly enhances overall system efficiency. Developing new models will increase the likelihood of finding an optimal model to improve overall system efficiency.

2. Adoption of Data-Based Models to Predict Heating Coil Performance

2.1. Performance Prediction of the Heating Coil

Despite the growing emphasis on HVAC efficiency, a review of past literature reveals that limited attention has been given to accurately predict heating coil performance using laboratory tested or real-world data. Early research by Barney and Florez focused on temperature prediction models for heating system control [19]. Asamoah and Shittu explored methods to predict heating and cooling loads in residential buildings using building energy models [20]. More recently, researchers have explored AI-based optimization for HVAC system operations [21] and machine learning techniques for air conditioning load prediction [22]. Although advanced data analysis methods such as deep learning for indoor temperature prediction [23], AI-assisted HVAC controls [24], thermal comfort, and energy efficiency improvements [25] are gaining traction, much of this work relies on simulation-based data rather than actual equipment-tested data [26].

One of the key challenges in accurately modeling heating coil performance is the lack of experimental datasets. Most research has focused on cooling coil loads [27–29], damper controls [30], thermal comfort improvements [31], fault detection [15,32], HVAC control optimization [33], and broader energy efficiency measures [34], etc. However, heating loads are just as critical, yet research on heating coil performance remains limited due to data scarcity. The absence of comprehensive experimental datasets restricts the generalizability of data-driven models, making it difficult to develop reliable performance predictions [35]. Additionally, the thermal response of heating coils is highly dependent on airflow rates, water temperature fluctuations, and environmental conditions, which further complicates the development of static models capable of capturing actual performance [36].

Given the significant role of heating coils in building energy consumption, developing accurate and adaptable models is crucial for optimizing HVAC system efficiency. The key motivation for developing accurate heating coil performance models lies in their practical applications. Such models can enhance performance prediction under varying operating conditions, optimize energy efficiency to reduce operational costs, detect HVAC system faults to prevent failures, and provide real-time feedback when integrated with building automation systems (BAS). This, in turn, enables adaptive control strategies that improve overall system reliability and efficiency [16,37].

In the context of developing advanced data analysis models, it is essential to examine the data analysis methods commonly used in the industry. Understanding these established techniques provides a foundation for selecting and refining models that best suit the specific application.

2.2. Multiple Linear Regression Models (MLR)

Multiple linear regression (MLR) is the most straightforward and widely used method for predictive modelling, particularly suited for datasets with simple relationships [38]. MLR operates by formulating a simple equation that predicts a single output based on multiple input variables. By adjusting one input while holding the rest constant, the model estimates the impact of each variable on the output. The coefficients in the equation quantify these effects, with larger coefficients indicating a stronger influence on the predicted outcome. MLR remains interpretable and computationally efficient, making it a practical choice to prepare straightforward models for various applications. The following equation shows the general format of MLR.

$$y = x_1 b_1 + x_2 b_2 + \ldots + x_n b_n \tag{1}$$

Here, y is the dependent variable; a is the output intercept; $b_1, b_2, \ldots b_n$ are the coefficients; and $x_1, x_2, \ldots x_n$ are independent variables also known as the known parameters.

The primary limitation of MLR arises from the assumptions it imposes on data. It assumes a linear relationship between variables, independence among inputs and normally distributed errors with constant variances [39,40]. If these conditions are met, MLR offers a simple yet effective model for accurate predictions. However, if these assumptions are violated, model predictions become unreliable, potentially leading to significant errors in analysis [41].

2.3. Neural Networks

Neural networks are a powerful method for simulating decision-making processes based on data inputs. They work particularly well, compared to other models, at recognizing complex patterns with multiple inputs and outputs. Neural networks often outperform simpler models that struggle to capture such relationships [42].

At its core, neural networks consist of multiple layers of interconnected nodes, that process and transmit information. These layers contain a predefined number of nodes, where each node represents a specific aspect of the data. These nodes perform computations based on an associated equation that applies weights to the inputs received from the previous layer. Structurally, this computation may resemble an MLR equation. However, in this case, the intercept is replaced by a bias term, which adjusts the activation threshold and coefficients, function as weights that regulate the influence of preceding node outputs, and the independent variables correspond to the values received from the previous layer.

Each node processes its inputs using an activation function, such as the hyperbolic tangent or sigmoid function, which determines how the output passes through to the subsequent nodes. These functions generate a value that controls the significance of a node's contribution to the next layer. Figure 2 visualizes a selection of these activation functions, although there are many more variations exist, each with distinct properties suited for different modeling applications. A specific one to look at is the rectified linear unit or ReLU function, max(0, x), which gives a value of 0 to negative numbers and returns the exact same input for positive numbers.

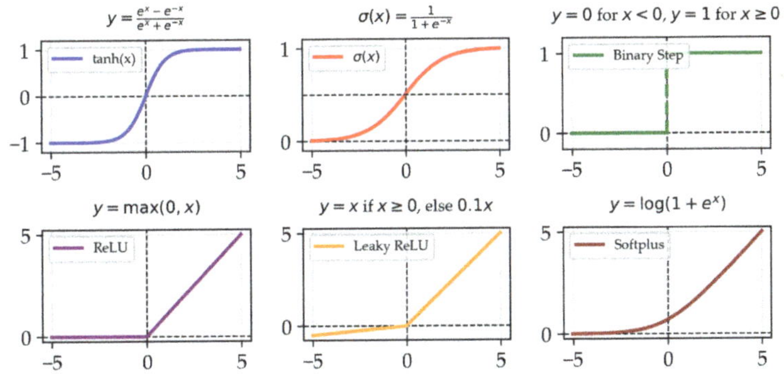

Figure 2. Common activation functions in neural networks.

Nodes in the subsequent layer interpret the activation function's output value to determine the appropriate weight to assign on each node, or whether to consider it at all. The activation threshold is inversely related to the node's importance; a lower activation threshold increases the node's influence [43].

Neural networks refine their predictive capabilities by iteratively processing data and adjusting the weight of connections between nodes. This improvement occurs through a training process known as backpropagation, which propagates errors backwards from the output layer to the input layer. By analyzing the difference between predicted and actual values, backpropagation adjusts the connection weights to minimize the loss function, which quantifies prediction errors. A high loss value indicates significant deviation from the true value. The optimization process known as gradient descent iteratively adjusts the network parameters to move toward the closest minimum point of the loss function. Figure 3 shows how the calculated loss decreases over successive iterations, demonstrating the network's improving accuracy [44].

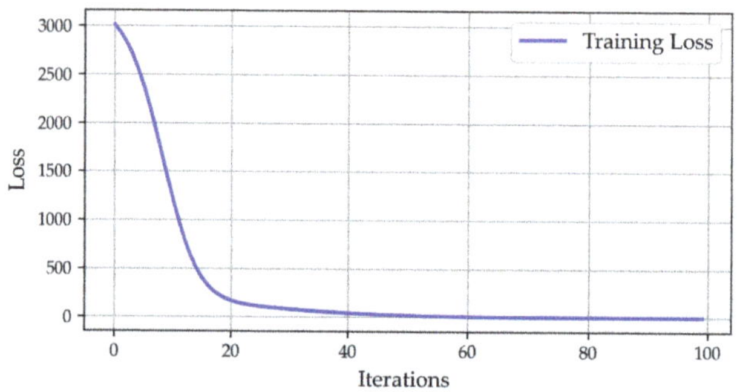

Figure 3. Neural network loss function over iterations.

During backpropagation, the network adjusts the weights of individual nodes based on predefined parameters. These parameters include bias calculations, learning rate, network architecture, iteration count, etc., contributing to the adaptability of neural networks [45]. Different network architectures such as right triangle (Figure 4a), left triangle (Figure 4b), diamond (Figure 4c), and block structures (Figure 4d) can significantly impact model performance. Architecture also relates to the number of nodes and layers, which influence the model's capacity to learn complex patterns in data.

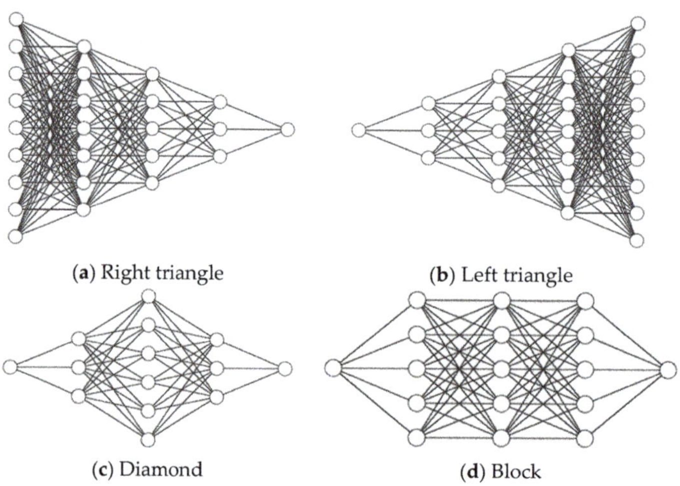

Figure 4. Common neural network architectural shapes.

Increasing the number of nodes and layers enhances the network's ability to capture intricate relationships but also raises the risk of overfitting, where the model memorizes training data instead of learning generalizable patterns. Overfit models perform exceptionally well on training data but fail to generalize unseen data. On the other hand, an insufficient number of nodes or layers may lead to underfitting, where the model oversimplifies patterns, reducing accuracy. Striking a balance between model complexity and generalization is crucial in neural network optimization, as it is across all machine learning paradigms. Given the vast number of possible configurations, optimizing a neural network often requires a combination of trial-and-error testing and advanced mathematical techniques [46]. By leveraging dynamic parameters and optimization techniques, neural networks provide a flexible and effective approach to pattern recognition, making them an essential tool in modern machine learning applications.

2.4. Bootstrap Aggregation (Bagging)

Bootstrap aggregation, also known as bagging, is a data modeling method that enhances predictive accuracy by training multiple models on randomly selected subsets of the data [47]. Bagging usually employs one of the following model types: decision trees, regressors, or classifiers. Individual instances of the selected models are trained on distinct subsets of the data. The aggregation of these models uses what is known as an ensemble technique. The ensemble ensures the final aggregated model is a proper accumulation of the individual models developed. Choice of the ensemble method must align with the base models used to ensure effective aggregation. Figure 5 highlights the process.

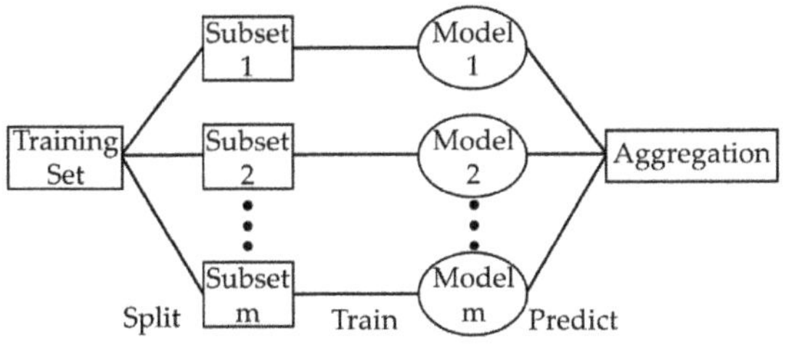

Figure 5. Schematic of the bagging process.

Generally, all models in the ensemble are typically assigned equal weight regardless of their individual performance. Consequently, bagging relies on the "wisdom of the crowd," a belief that while any single model may not perfectly represent the entire dataset, combining multiple models leads to a more accurate and reliable prediction.

The performance of the model depends significantly on the number of models analyzed, which is directly correlated to the number of subsets. A larger number of subsets helps mitigate the influence of outliers but may narrow the data analysis of the models, increasing the risk of overfitting. A relatively lower number of subsets will generalize the data more but will be influenced by outliers. Consequently, the tuning of the sample size is essential to finding the optimal number of models [48].

2.5. Gradient Boosting

Gradient boosting is another ensemble technique that improves predictive models through an iterative process whereby each model is trained independently to correct the errors of its predecessors [49]. Typically, decision trees are used as the base models, and each new model is added to the ensemble to enhance overall performance. The process is shown in Figure 6 below.

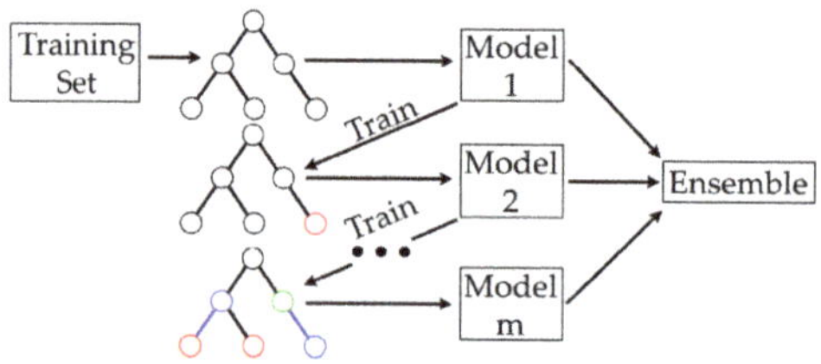

Figure 6. Schematic of the gradient boosting process (color changes indicates updates at each stage).

The first model developed is relatively simple, and each subsequent iteration refines it by focusing on residual errors to identify unrecognized patterns that the previous model failed to capture. Each addition to the ensemble is analyzed using a loss function, which determines the best adjustments needed to minimize prediction errors. This optimization process, known as gradient descent, aims to reduce the loss function. A loss function analyzes residual errors by comparing predicted and actual values, using techniques such as mean squared error to quantify discrepancies. Figure 7 shows a gradient descent path with two possible adjustment directions, labeled as x and y. While real-world models involve many possible adjustments, this figure conceptually represents how the algorithm identifies optimal modifications to reach a local minimum [50].

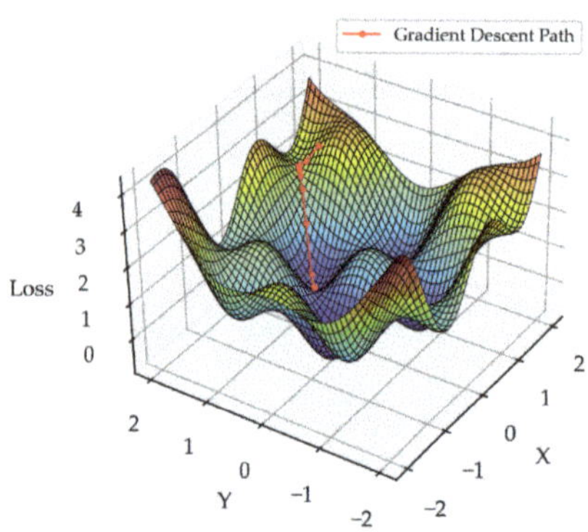

Figure 7. Gradient descent on a 3D loss surface.

Each model in the ensemble finds its local minimum, and models with lower loss values are weighted more heavily, as they contribute more reliable predictions. This weighting is necessary because early models in the process tend to overgeneralize the data. Moreover, as the ensemble evolves, attempts to correct errors may lead to overfitting. By prioritizing models that minimize loss effectively, gradient boosting ensures that only the most accurate models contribute significantly to the final prediction [51]. While the fundamental principles of gradient boosting remain consistent, different adaptations prioritize various aspects, such as handling missing data or improving computational efficiency. Therefore, testing and fine-tuning different gradient-boosting variants are essential to achieve optimal performance.

2.6. Application of Advanced Data Analysis on Heating Coil Performance Prediction

This research aims to apply machine learning models, such as bagging, neural networks, and gradient boosting, to predict heating coil performance in building mechanical systems—specifically, to develop models that estimate supply air temperature based on the heating coil's operation. Model selection is based on two key criteria: error analysis and model functionality. A chosen model must demonstrate acceptable predictive accuracy during both training and testing phases, with error thresholds determined by comparing the performance of all tested models. Since multiple models may meet these thresholds, it is also crucial to assess their underlying mechanisms and strengths.

In practical applications, dynamic models that effectively handle errors are essential. Given that HVAC systems operate with continuously changing data and require real-time decision-making, models must be capable of making logical adjustments based on all available information.

3. Methodology

3.1. Overview of the Methodology

Figure 8 illustrates the overall workflow followed in this research. The process began with equipment testing to ensure compliance with basic standards and data accuracy. Once verified, data were collected in a large spreadsheet and cleaned to remove inconsistencies. The cleaned dataset was then formatted for implementation in the developed code. Next, various libraries, including matplotlib to create the figures, were utilized to train multiple machine learning models, which were evaluated based on their error metrics and predictive performance. Finally, the analysis focused on how these models could be applied in real-world scenarios. The figure also highlights key libraries and tools used at each stage of the process.

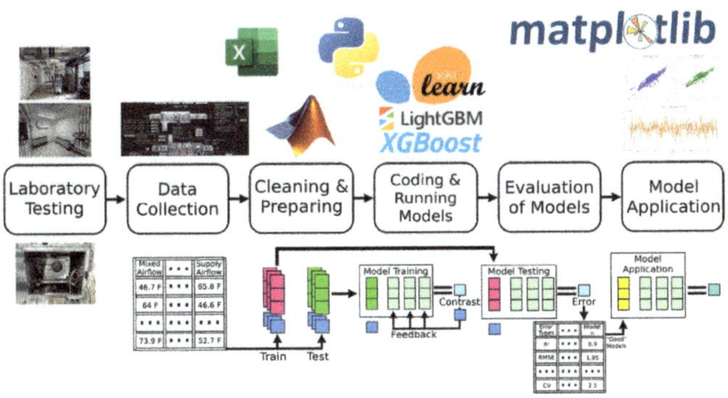

Figure 8. Overall workflow of the research.

3.2. Laboratory Testing and Data Collection

3.2.1. Overview of the Testing Facility

The experimental testing was conducted at the Building Energy Assessments, Solutions, and Technologies (BEAST) Laboratory (Figure 9a), located at the University of Cincinnati's Victory Parkway campus. This facility is equipped with various HVAC systems, including variable refrigerant flow systems, water-based air handling units, and direct expansion units connected to an electric heater and an air-cooled chiller. Additionally, the lab features three well-insulated and controlled spaces (Figure 9b) designed to simulate actual building zones, each equipped with a variable air volume (VAV) box. For this study, testing was performed on a water-based air-handling unit (Figure 9c).

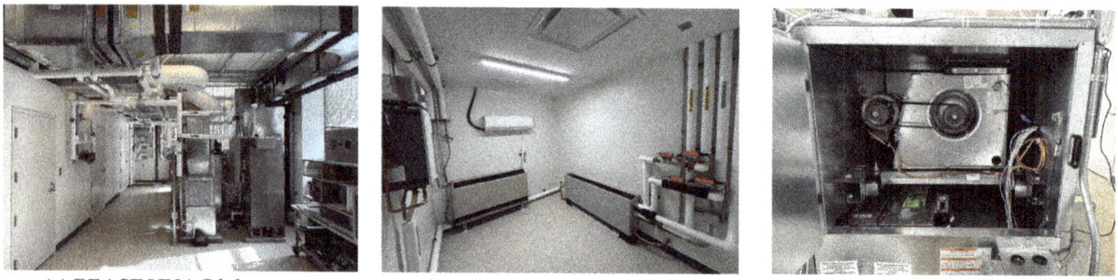

(a) BEAST HVAC laboratory (b) Example controlled space (c) Tested air handling unit

Figure 9. BEAST laboratory facility and tested equipment.

The laboratory's centralized building automation system (BAS) was utilized to monitor and control equipment parameters, minimizing errors associated with manual adjustments.

3.2.2. Laboratory Testing Procedure

Laboratory testing of the heating coil was conducted during the winter season. Throughout the testing period, the cooling coil and chiller remained inactive, with all associated chilled water valves being fully closed. Testing spanned for one week, during which the laboratory's BAS recorded system performance data at one-minute intervals, resulting in the collection of over 10,000 data points in imperial units.

To capture a broad range of operating conditions, the heating coil valve position and supply fan speeds were systematically adjusted every two hours. The heating coil valve position was incrementally modified from 20% (minimum valve position allowing at least 0.1 gallons per minute (GPM)) to 100% (maximum water flow) in 10% increments. A similar stepwise approach was applied to the supply fan speed, varying from 10% to 100% in 10% increments.

Precautions were taken to maintain the air handling unit's internal temperature between 20 °F (−6.67 °C) and 95 °F (37.78 °C) to prevent freezing, which could affect the heating and cooling coils, and to avoid excessive heat buildup that could damage electrical switches and sensors. Additionally, all other HVAC components, including the cooling coil, VAV box heating coil, and humidifier, were deactivated to ensure they did not interfere with the air conditioning system. The three lab zones remained unoccupied and were not subjected to external influences throughout the testing period.

By adjusting the coil valve position incrementally, a comprehensive dataset was obtained covering a wide range of hot water flow conditions. The control signal modifications resulted in hot water flow variations ranging from 0.1 GPM (0.000006 m^3/s) to a maximum of 1.6 GPM (0.0001 m^3/s). While the valve position influenced the hot water flow rate, it did not alter the hot water supply temperature, which remained constant at approximately 120 °F (48.89 °C). The impact of supply fan speed variations on heating coil performance was also examined by assessing changes in supply air temperature under different airflow conditions. At 100% fan speed, the system achieved a maximum airflow of approximately 1300 cubic feet per minute (CFM) (2208.7 m^3/h), representing the upper limit of the system's capacity.

The testing covered a wide range of operating scenarios, from worst-case heating conditions to optimal performance cases. This approach ensured that extreme conditions were accounted for in the dataset, improving the robustness of the model predictions.

3.2.3. Hot Water Loop and Airflow Control Conditions

The air handling unit's heating coil was supplied with hot water from an electric boiler connected through a closed-loop hot water system. The return water temperature varied based on system conditions. The supply air temperature was further influenced by outdoor and return air damper modulation, with the mixing air temperature maintained between 20 °F (−6.67 °C) and 95 °F (35 °C) to prevent freezing or overheating inside the unit.

During the testing period, outdoor temperatures dropped as low as −10 °F (−23.33 °C). It was ensured that no other conditioning sources influenced the supply air temperature besides the heating coil. The VAV damper positions remained fully open, and total supply airflow modulation was controlled solely by the supply fan.

This testing methodology enabled the collection of a diverse dataset that accurately represents real-world heating coil performance under various operational conditions, providing a strong foundation for predictive model development.

3.3. *Basis of Developing Predictive Models and Data Cleaning*

3.3.1. Data Cleaning

Following the completion of laboratory tests, key parameters influencing heating coil performance were identified. Initial results indicated that the supply air temperature conditioned through the heating coil was primarily influenced by the following variables:

1. Mixed air temperature (°F);
2. Supply water temperature of the heating coil (°F);
3. Hot water flow rate through the heating coil (GPM);
4. Total supply airflow (CFM).

Variations in these parameters directly impacted the supply air temperature. Based on this observation, the most influential factors determining heating coil performance were mixed air temperature, supply water temperature, hot water flow rate, and total supply airflow. Given this relationship, the supply air temperature was selected as the dependent variable, while the other identified parameters served as independent variables for model development.

Once the relevant parameters were finalized, data collected from the BAS underwent a structured cleaning process to ensure accuracy and reliability:

5. Handling missing data: Any data entries with missing values, typically marked as "---", were removed to prevent inconsistencies in the dataset.
6. Eliminating transition period data: Since data collection was continuous, readings captured during input transition periods were excluded to ensure only stabilized values influenced the analysis. To do this, any fluctuations of 0.5 GPM (0.00003 m^3/s) or greater in water flow and any fluctuations of 250 cfm (424.8 m^3/h) or greater in airflow were removed.

In total, 260 data points were dropped. The cleaned dataset was then used to develop predictive models, with these models aiming to accurately represent the heating coil's actual performance under varying operating conditions.

3.3.2. Development of Predictive Models

Predictive model development was conducted through Python 3.12.8 with the support of various libraries. Initially, MATLAB was considered due to its widespread use in engineering applications and its ability to produce visually appealing graphs. However, Python was ultimately chosen for its greater prevalence in machine learning applications, ease of use, and efficient computational capabilities. Python's performance advantage lies in its ability to handle computationally intensive data analysis tasks by leveraging libraries that execute complex mathematical operations in C or C++ while maintaining user-friendly, readable code. With the help of selected libraries, which simplify intricate operations, the predictive models were successfully developed.

3.3.3. Splitting Data

To effectively train and evaluate the models, the dataset was divided into training and testing subsets. Typically, models perform well on training data but may struggle with unseen data during testing. The choice of data split ratio depends on the dataset size, with 80/20 and 70/30 being common practices. In this research, both split ratios yielded nearly identical results, indicating that the dataset was sufficient for model training. Consequently, a 70/30 split was used for figures and tables in this study. Additionally, a random state was set to ensure reproducibility in data partitioning. Since computers generate pseudo-random splits, the seed value determines the starting point for randomness. Various random states were tested, all producing consistent results, as expected. For this research, a random state of 0 was used in all figures and tables.

3.3.4. Library Selection and Model Evaluation

A total of 31 machine learning models were developed and tested in this study, utilizing key libraries such as scikit-learn, XGBoost, LightGBM, Pandas, and NumPy [52]. Pandas and NumPy serve as foundational libraries that support data processing and mathematical operations. Pandas facilitates data manipulation, handling structured data such as tables and time series [53], while NumPy provides Python with capabilities for vectorized computations, indexing, and managing large, multidimensional arrays and matrices [54]. These libraries, in combination with various machine learning frameworks, allowed for the implementation and fine-tuning of predictive models.

The primary objective of this study was to identify the most effective approach for predicting supply air temperature in a HVAC system. Given the variability of environmental conditions such as climate, humidity, and seasonal fluctuations, a single predictive formula would be insufficient. Therefore, models were evaluated based on performance metrics, adaptability, and suitability for real-world applications.

When multiple models demonstrated comparable predictive accuracy, a more detailed analysis was conducted to assess their individual strengths and limitations. The goal was to develop a robust and adaptive set of models capable of recognizing localized patterns and learning system-specific behaviors.

3.4. Error Metrics Calculation

Equation (2) through (6) define the error metrics used to assess model performance. While multiple error methods were analyzed, R^2 (coefficient of determination) is the primary metric presented in most figures, as it measures how well the model's predictions align with actual values. An R^2 value of 1 indicates a perfect fit, while values below 0 suggest that the model performs worse than the mean value.

Mean absolute error (MAE) calculates the average absolute difference between predicted and actual values, treating all errors equally [55]. This method is effective when all data points hold equal importance. Mean squared error (MSE) determines the squared differences between predictions and actual values, penalizing larger errors more heavily [56]. This makes it useful when outliers significantly impact accuracy. Root mean squared error (RMSE) is similar to MSE but applies a square root to return the error to the original unit, making it easier to interpret. However, because RMSE applies a square root to the error, differences between models may appear less pronounced. For example, if one model has an MSE of 100 and another has an MSE of 81, the difference seems significant. But after applying the square root, their RMSE values become 10 and 9, making the difference appear smaller. Coefficient of variation (CV) evaluates whether a model is overfitting or excessively generalizing by analyzing its consistency across the dataset [2,57].

$$R^2 = 1 - \frac{\text{Sum of Squares}_{\text{residual}}}{\text{Sum of Squares}_{\text{Total}}} \tag{2}$$

$$MAE = \frac{\sum_{i=1}^{Number\ of\ Observations} \|(Observed\ Values) - (Predicted\ Values)\|}{Number\ of\ Observations} \tag{3}$$

$$MSE = \frac{\sum_{i=1}^{Number\ of\ Observations} \|(Observed\ Values) - (Predicted\ Values)\|^2}{Number\ of\ Observations} \tag{4}$$

$$RMSE = \sqrt{\frac{\sum_{i=1}^{Number\ of\ Observations} \|(Observed\ Values) - (Predicted\ Values)\|^2}{Number\ of\ Observations}} \tag{5}$$

$$\text{CV} = 100 \times \frac{\text{RMSE}}{\text{OutputMean}} \tag{6}$$

4. Results

4.1. Analysis of Models

Figure 10 compares the error values of all 31 tested models, providing an overview of their general performance. Refer Appendix A.1 for a better visualization of the R^2 values. The results indicate that most approaches achieved similar accuracy in predicting supply air temperature. While a few underperformed, the majority produced strong predictions, suggesting that the dataset was not overly complex for selected algorithms.

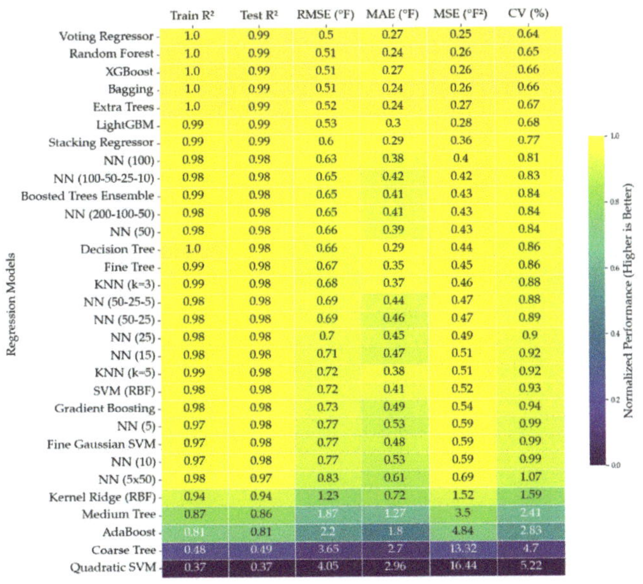

Figure 10. Training and testing R^2 for evaluated models.

Since different predictive techniques yielded comparable accuracy, selecting the best-performing ones requires more than just minimizing error values. Instead, additional practical considerations relevant to the field of study must be considered. For instance, when implemented in an actual system, predictive tools will inevitably encounter outliers, just as seen in the original dataset. Those capable of handling such anomalies effectively are preferred as they maintain performance stability during transitional periods when input conditions change. Another crucial aspect is adaptability. Given that not all HVAC systems are identical, models that can be adapted to different environments are more valuable. Based on these criteria, three models were selected: neural network, XGBoost, and bagging. Additionally, MLR was included as a baseline for comparison.

4.2. Model Selection

Four selected models offer strong potential for further development in machine learning applications. Neural networks, XGBoost, and Bagging demonstrated robust performance making them suitable for this research. However, it is reasonable to question why the voting regressor and random forest models, which achieved the lowest error values, were not selected. Several factors contributed to this decision.

While these models performed marginally better in terms of R^2, their performance across other error metrics was either equivalent to or worse than the selected models. Additionally, when tested on uncleaned data, the bagging model outperformed both the voting regressor and random forest, as is shown in Figure 10, across the analyzed error metrics [58,59]. Table 1 represents the error values obtained without data cleaning.

Table 1. Error values from uncleaned dataset.

	Error Values					
Models	**R^2 Train**	**R^2 Test**	**RMSE**	**MAE**	**MSE**	**CV**
Bagging	0.997854	0.977027	0.761906	0.291603	0.580500	0.980804
Random Forest	0.997867	0.976929	0.763534	0.291076	0.582985	0.982900
Voting Regressor	0.994969	0.976603	0.768907	0.321640	0.591218	0.989816
XGBoost	0.997564	0.974697	0.799610	0.327223	0.639376	1.029341

Bagging and XGBoost, similar to neural networks, offer greater potential due to their ability to handle errors effectively and the availability of extensive tuning and adjustment options. Since all models performed well based on evaluated error indices, the selection prioritized models known for their reliability and adaptability in real-world applications.

4.3. Analysis and Discussion of the Selected Models

4.3.1. MLR

MLR was chosen as a baseline due to its simplicity and transparency. As a straightforward method, it provides a clear mathematical equation, making it useful for establishing benchmark error values and identifying key input variables that influence predictions. Comparing MLR's insights with those from advanced models helps assess whether complex approaches capture the same underlying relationships or offer meaningful improvements beyond a linear framework. Given that many advanced techniques function as black boxes, MLR serves as a valuable reference for evaluating their effectiveness [60,61].

Equation (7) presents the developed MLR equation, illustrating how each input variable contributes to the predicted output. Among the independent variables, the mixed air temperature (x_3) exerts the greatest influence, as indicated by its larger coefficient.

$$Y = 77.77 + 0.54 \times x_1 + 5.42 \times x_2 + 7.30 \times x_3 - 1.92 \times x_4 \quad (7)$$

where y = supply air temperature (°F); x_1 = hot water supply temperature (°F); x_2 = water flow (gpm); x_3 = mixed air temperature (°F); and x_4 = supply fan air flow (cfm).

Figure 11 displays the testing and training scatter plots, along with residual plots, to analyze the accuracy of the model's predictions. Figure 11a,b reveal that many data points deviate significantly from the ideal predicted values, forming noticeable clusters away from the regression line. Similarly, Appendix A.2 visualizes these discrepancies through a line plot. These observations suggest that while MLR effectively captures general trends, it struggles with accuracy and reliability, particularly when modeling intricate or nonlinear patterns in the data. The residual plots in Figure 11c,d further highlight this limitation, as a significant number of data points are far from the central regression line, demonstrating that MLR frequently fails to provide precise predictions.

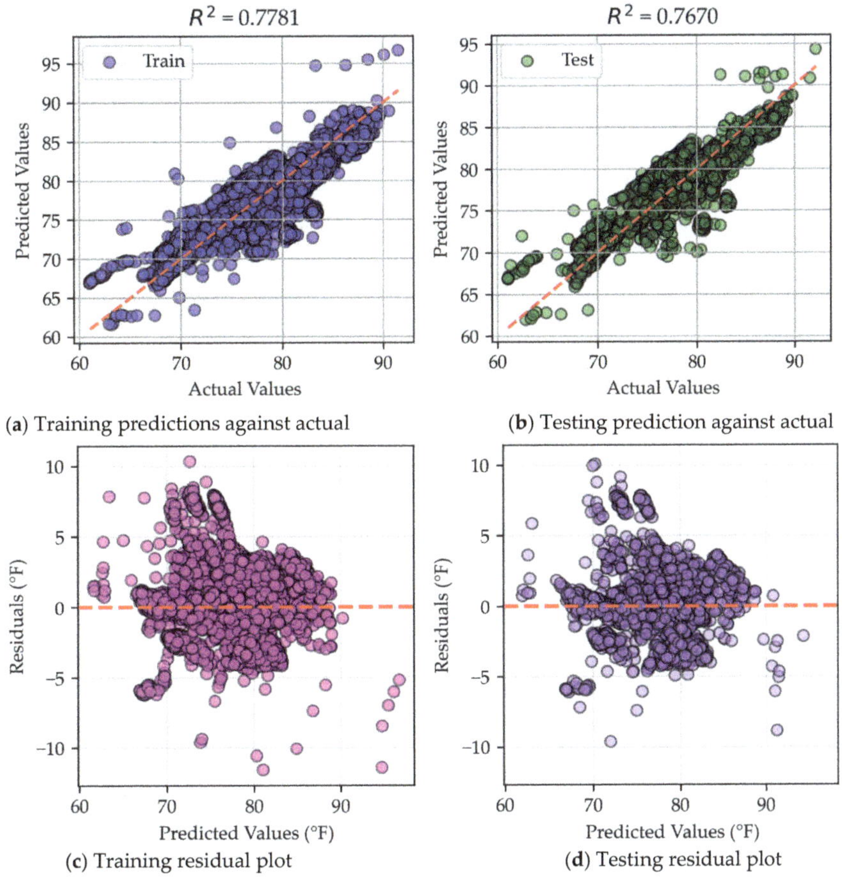

Figure 11. MLR residual and scatter plots.

4.3.2. Neural Network

Neural networks were selected for their ability to model complex relationships with varying levels of complexity. The network architecture primarily consisted of single-layer or right-triangle configurations. Key parameters, such as the number of nodes and layers, were adjusted to identify an optimal predictive model, while bias calculation methods and learning rates remained constant [62]. Since larger networks require more iterations to fine-tune their weights effectively, structural modifications were necessary to balance accuracy and computational efficiency.

The best performing model in this study was a single layer neural network with 100 nodes using a ReLU activation function. While further refinements, such as adjusting the number of nodes or layers, could enhance accuracy, the model's high performance made additional optimization unnecessary. Figure 12 illustrates a simplified neural network with only 10 nodes and their associated connection weights. Although this study may not have identified the most optimal neural network, the tested models demonstrated strong predictive capabilities. More advanced deep learning models could potentially outperform all tested configurations, but within the scope of this research, neural networks provided reliable predictions with some variability across different setups [63].

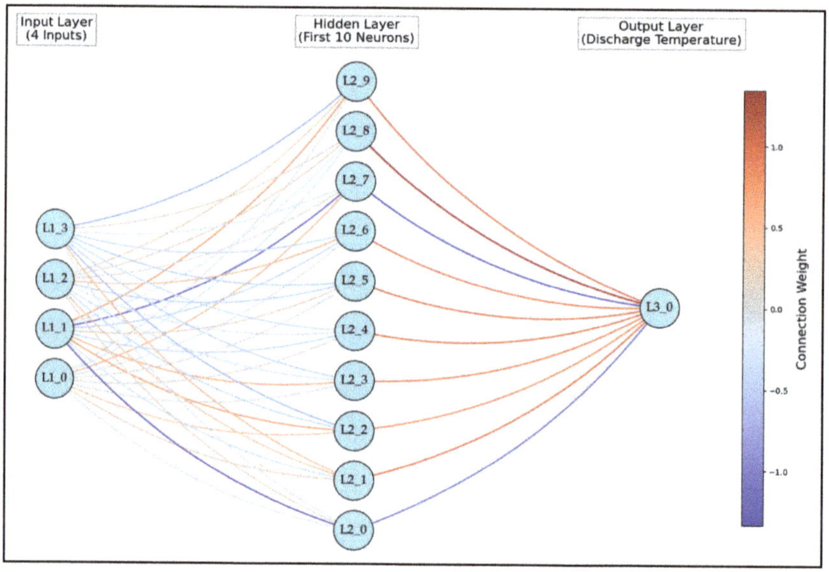

Figure 12. Neural network connection weights.

An analysis of the neural network residuals reveals an interesting trend in that its training and testing performance were nearly identical. By comparing Figure 13a,b, as well as Figure 13c,d, it was evident that the error values were consistent across both phases. Notably, the R^2 value for the training dataset was slightly lower than that for the testing dataset, which appears counterintuitive at first glance. In other models, training typically outperforms testing, but this suggests that the neural network is operating near a local minimum. The model successfully identified patterns in the data and generalizes well to unseen inputs, but it has not necessarily reached a global minimum, as some other models marginally outperformed it. Even still, the neural network strikes a balance between generalization and overfitting. Nearly identical data distribution in both the training and testing scatterplots further supports this claim.

Unlike MLR, which relies on a generalized linear formula, neural networks can capture more complex patterns while maintaining consistency between training and testing results. Appendix A.3 presents an alternative visualization of residual data, reinforcing these observations.

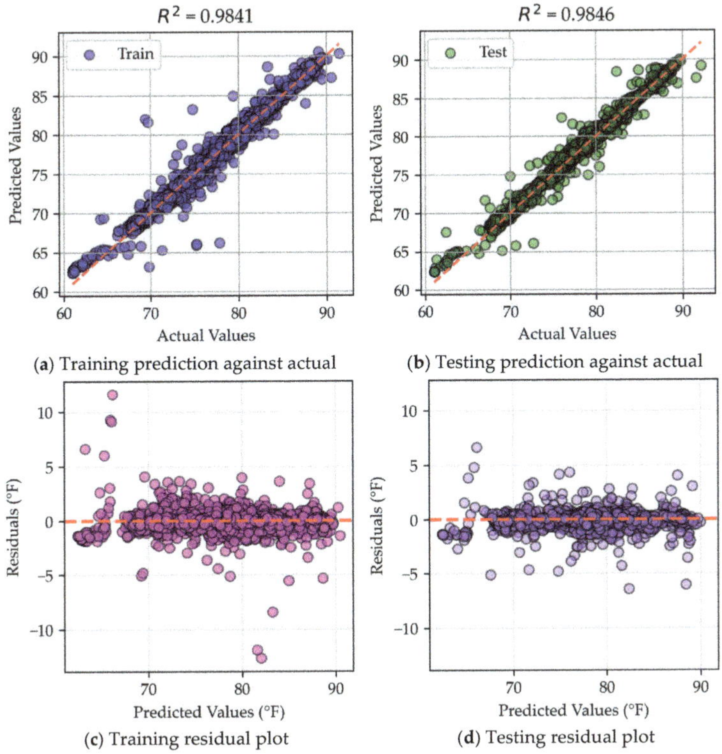

(**a**) Training prediction against actual (**b**) Testing prediction against actual

(**c**) Training residual plot (**d**) Testing residual plot

Figure 13. Neural network residual and scatter plots.

4.3.3. Gradient Boosting

Gradient boosting was selected for its ability to handle complex datasets effectively, particularly in real-world scenarios with outliers or missing values. Additionally, gradient boosting integrates categorical and numerical data within a single framework, making it adaptable to various environments. For this analysis, extreme gradient boosting (XGBoost) was utilized over scikit-learn's gradient boosting technique due to its superior performance. The key distinction between these implementations lies in their data-splitting techniques and regression tree construction [64]. XGBoost is a sophisticated model implemented through an open-source library aimed at being both efficient and effective, making it a preferred method for numerous applications [65].

Like all models, gradient boosting makes assumptions regarding the dataset structure. Figure 14 illustrates the model's predictive process as additional trees are introduced. Sample 9 was accurately predicted after approximately 35 trees, while other data points, such as sample 424, required additional iterations to improve accuracy. However, as more trees were added to improve certain predictions, slight sacrifices in accuracy were made for previously well-predicted data points. This tradeoff is a common characteristic of boosting algorithms, where the model aims to minimize the overall average error rather than improving individual predictions.

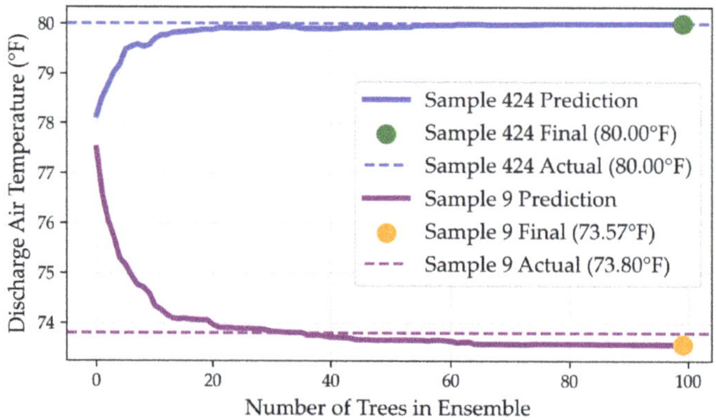

Figure 14. Predicted accumulation of samples for XGBoost.

When compared to the baseline MLR and neural network results, XGBoost consistently produced predictions that are closer to the true values. Figure 15 demonstrates this by showing that most data points are clustered near the central prediction line, indicating strong predictive ability. Although, unlike neural networks, which can capture more intricate data patterns, gradient boosting does not generalize as effectively. The model achieves acceptable error values but exhibits a noticeable performance gap between training and testing. Additionally, the dispersion of data points in testing does not always align with the patterns observed in training. For instance, in Figure 15c, data points around predicted temperature of 85 °F (29.4 °C) show minimal dispersion; while in Figure 15d, multiple outliers appear. Appendix A.4 provides an alternative visual of these outliers.

The discrepancy suggests that XGBoost, while effective, may exhibit some degree of overfitting, preventing it from identifying a fully generalized pattern. While it outperforms simpler models like MLR, it does not consistently achieve the same level of generalization as the neural network.

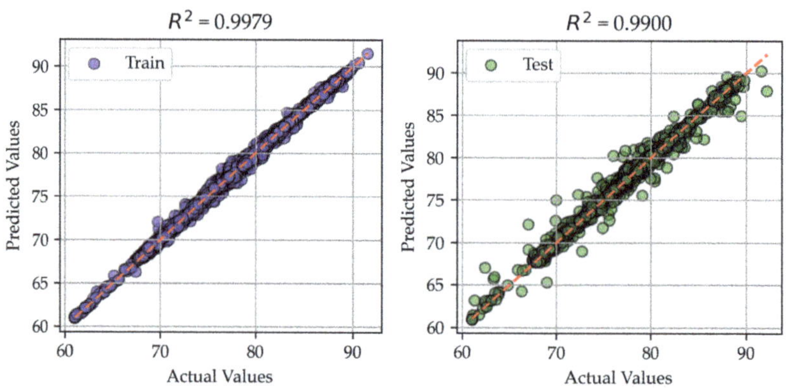

(**a**) Training prediction against actual (**b**) Testing prediction against actual

Figure 15. *Cont.*

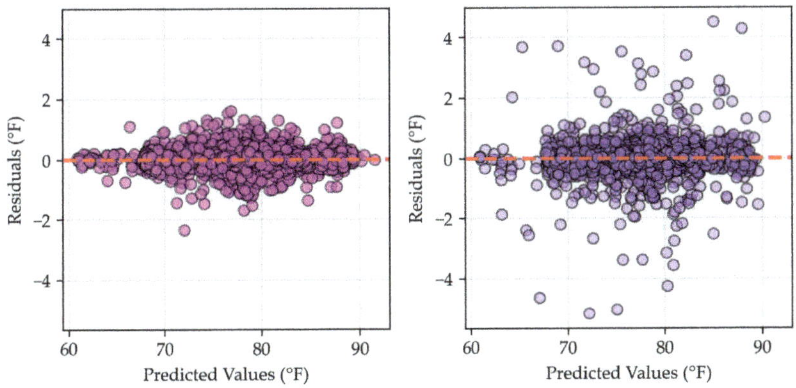

(c) Training residual plot (d) Testing residual plot

Figure 15. XGBoost residual and scatter plots.

4.3.4. Bagging

Bagging was selected for its strong performance, particularly in reducing variance and mitigating the impact of incorrect data points [66]. However further improvements could be achieved by optimizing the number of subsets used. The best performing decision tree with a depth of three is listed in Figure 16, providing insight into how the model prioritizes inputs and makes predictions. These specific characteristic makes bagging highly suitable for HVAC applications, where occasional data inaccuracies may occur [47].

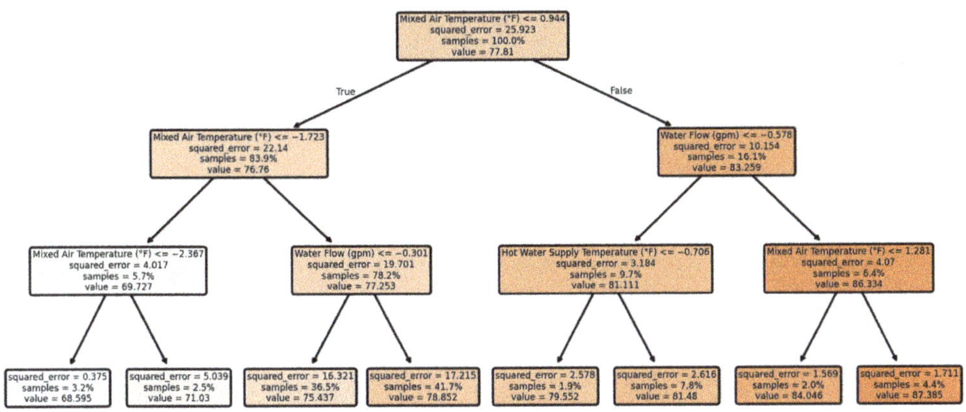

Figure 16. Bagging tree with max depth 3.

A comparison of residual and scatter plots highlights key differences between bagging and other models. In Figure 15, XGBoost displays a wider spread of data points compared to bagging in Figure 17. This suggests that XGBoost is more tolerant of errors but has greater dispersion, whereas bagging brings most predictions closer to expected values but struggles with outliers, leading to a wider spread of extreme residuals. This trade-off implies that bagging is better at capturing the general pattern of the dataset, whereas XGBoost maintains more flexibility in handling variations.

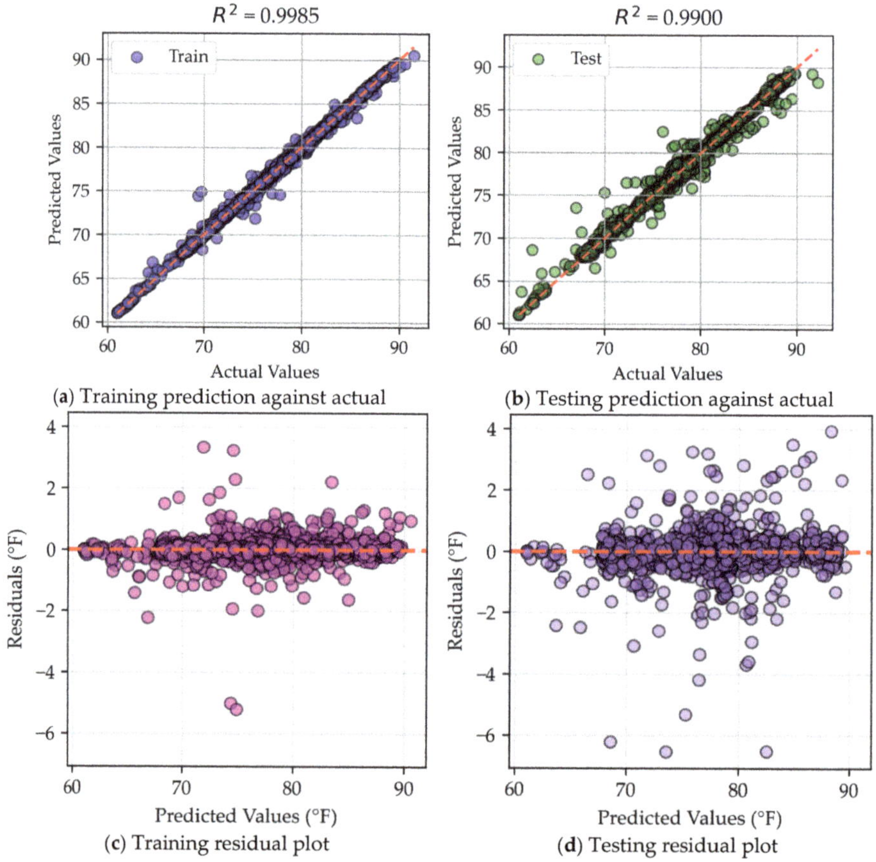

Figure 17. Bagging residual and scatter plots.

Additionally, Figure 17a,b show a similar dispersion pattern, suggesting that bagging is close to an optimal generalization of the data structure. However, differences in error values between training and testing indicate that the model does not fully capture the underlying pattern. Since the neural network successfully identified this pattern, it suggests a clear structure exists within the data that bagging does not fully exploit. While effective, bagging struggles with larger residuals compared to XGBoost, which sacrifices some generalization for better adaptability. Further pattern analysis for bagging is available in Appendix A.5.

4.3.5. Analysis of Model Specialties

Residual plot analysis reveals distinct differences in how each model processes the data. Neural networks excel at identifying patterns, XGBoost is more effective at handling outliers, and bagging strikes the best balance between these approaches. While their overall error values appear similar at a glance, a deeper evaluation is necessary to determine the most suitable model for practical applications.

In this study, bagging emerges as the most effective model, as supported by its error values. Its strength lies in capturing the underlying pattern while distinguishing true trends from noise, allowing it to make more reliable predictions. However, for further advancements, refining model parameters, particularly in neural networks, could lead to even better performance. Since the neural network demonstrated a strong ability to identify general patterns, optimizing its structure and parameters may yield superior results under more complex conditions.

5. Practical Application

5.1. Input Significance

5.1.1. Model Input Weights

Each model assigned specific weights to the inputs, as illustrated in the radar chart below (Figure 18) and detailed in Table 2. The interpretation of these weights varies across models. MLR directly uses coefficients to quantify input significance, while XGBoost and bagging assess feature importance based on their contribution to predictive accuracy. Neural networks rely on connection weights between inputs and hidden layers, requiring node-level analysis for deeper insights.

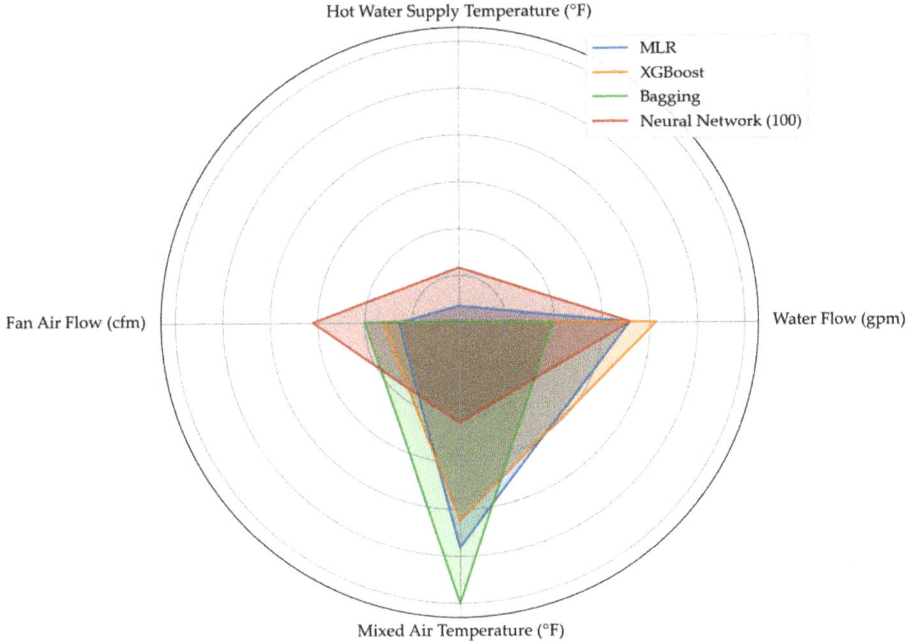

Figure 18. Relative input weights for the models.

Table 2. Input weights for models.

Input Values	Models			
	MLR	Neural Network (100)	Bagging	XGBoost
Hot Water Supply Temperature (°F)	0.035	0.117	0.003	0.003
Water Flow (gpm)	0.357	0.358	0.197	0.414
Mixed Air Temperature (°F)	0.481	0.215	0.600	0.422
Supply Fan Air Flow (cfm)	0.126	0.310	0.199	0.161

A clearer comparison of how each model handles input weighting is provided in Appendix B, with specific distributions for MLR (Appendix B.1), neural networks (Appendix B.2), XGBoost (Appendix B.3), and bagging (Appendix B.4). These appendices offer a more detailed breakdown of model specific weighting approach, reinforcing these findings discussed in this section.

Despite differences in methodology, all models reached a similar conclusion regarding input significance. Hot water supply temperature was consistently identified as the least influential variable, while mixed air temperature emerged as the most critical factor.

5.1.2. Input Weight Considerations

Figures 19 and 20 further examine the relationship between input variables for XGBoost and bagging. Notably, one visualization suggests that hot water temperature has subtle significance, seemingly contradicting previous assessments. This can be explained by the interaction between supply airflow rates and heating coil exposure time. When airflow decreases (i.e., fan speed is reduced), air remains in contact with the heating coil for a longer duration. As a result, lower hot water temperatures can achieve the same supply air temperatures compared to higher fan speeds.

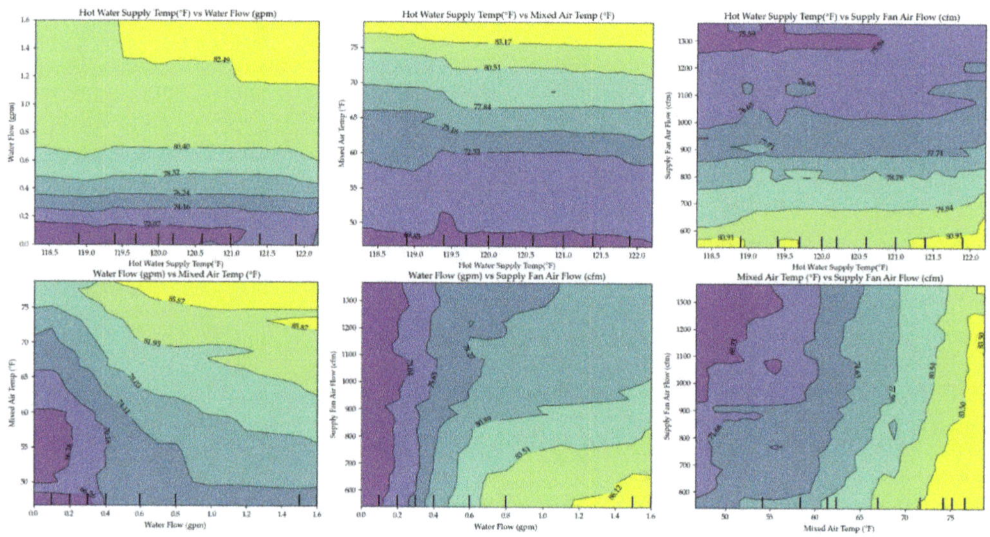

Figure 19. XGBoost input relationships.

Additionally, both models identified a negative correlation between supply airflow rate and supply air temperature. Higher airflow reduces heating time, leading to lower discharge air temperatures, while lower airflow extends exposure time, increasing heat absorption. However, in real-world applications, airflow rates cannot be freely adjusted due to ventilation and thermal comfort requirements. Excessive reductions in airflow may compromise air distribution and ventilation effectiveness, while excessive increases could cause pressure imbalances within the duct system. Understanding these constraints is crucial for applying model-driven optimizations to HVAC control strategies.

5.2. Model Applications

The developed heating coil models can be implemented in real-world building systems to monitor HVAC performance, detect anomalies, and enable proactive responses.

One critical application is cybersecurity. These models can identify unauthorized control attempts or cyber-attacks on the HVAC system, ensuring that only authorized facility managers regulate building conditions. Such safeguards can help maintain a safe and healthy indoor environment for occupants.

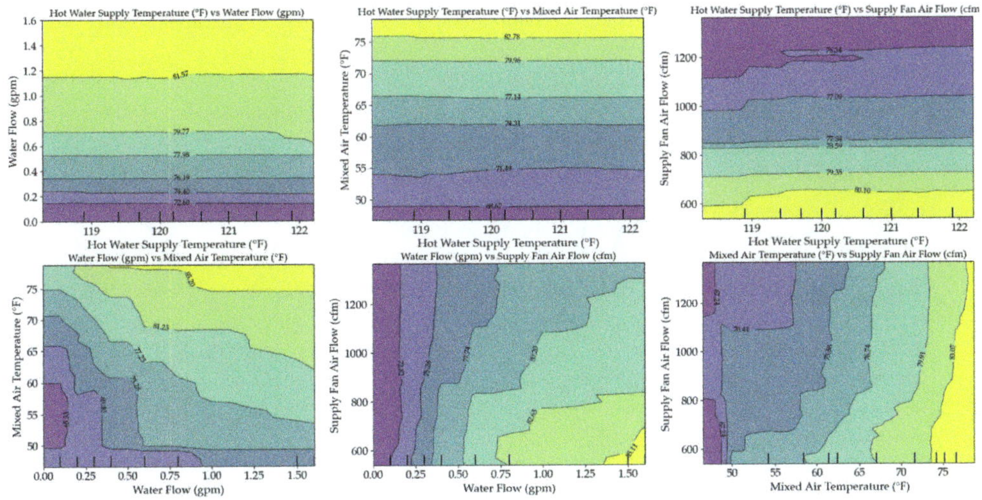

Figure 20. Bagging input relationships.

Additionally, these models facilitate continuous monitoring of heating coil performance, allowing for real-time adjustments to optimize efficiency. By integrating them into energy management systems, facility managers can gain a comprehensive view of actual building performance, enabling data-driven decision-making. This integration supports the implementation of energy efficiency measures and helps identify operational inefficiencies, such as excessive energy use or suboptimal equipment settings.

Another key application is predictive maintenance. The models can forecast potential system failures, minimizing downtime and extending the lifespan of HVAC components by scheduling timely maintenance. During the design phase of new buildings, simulation-based models can guide the selection of appropriate system components and configurations. For existing buildings, these models support retrofit strategies aimed at enhancing performance and reducing energy consumption.

Given the broad range of applications, further refinement and expansion of these foundational models are paramount. Future research should focus on developing advanced control algorithms and extending modeling efforts to other critical HVAC components, such as reheating coils and chilled water coils, to enhance overall system performance and efficiency.

6. Conclusions

This study developed and evaluated predictive models for heating coil performance in HVAC systems, demonstrating that relatively simple models can achieve error thresholds sufficient for laboratory-scale applications. Among the tested models, bagging exhibited strong predictive accuracy, consistently performing well across analyzed error metrics, including R^2, RMSE, MAE, MSE, and CV. It also demonstrated an ability to handle errors effectively and generalize patterns within the data. However, neural networks appeared to show greater potential for future advancements due to their strong performance in error metrics, capacity for developing more complex models, and ability to capture underlying patterns.

While these results highlight promising pathways for improving energy efficiency in HVAC systems, several limitations must be acknowledged. The experimental setup involved a small-scale unit operating under near-ideal conditions, which may not fully capture the complexities of real-world residential or commercial environments. Additionally, the study did not account for reheat and cooling coil loads or internal heat gains.

To enhance model robustness, future work should incorporate datasets from the entire HVAC system to find patterns all coils and operating parameters. This would enable the development of context-specific models, further optimizing energy efficiency across various implementation scenarios. Such advancements could significantly reduce global HVAC energy consumption, aligning with this study's broader objectives of economic savings, ecological sustainability, and energy conservation.

While the current models perform well in controlled settings, their real-world applicability may be limited by unaccounted disturbances and inconsistencies in full-scale systems. Traditional machine learning models rely on mathematical formulations making them less effective at handling unstructured or illogical data without extensive human intervention for tuning and validation. In contrast, deep learning offers a transformative opportunity for HVAC predictive analytics. These models can be designed to simulate logical processes, allowing them to autonomously adapt to dynamic conditions, to identify logical inconsistencies, and to self-correct during operation. Future research should focus on optimizing deep learning parameters to develop more resilient, self-optimizing HVAC models capable of handling real-world variability with minimal manual oversight.

Author Contributions: Conceptualization, N.N. and P.D.; methodology, A.N. and P.D.; software, A.N.; validation, P.D., A.N. and N.N.; formal analysis, A.N.; investigation, A.N.; resources, A.N. and P.D.; data curation, P.D. and A.N.; writing—original draft preparation, A.N. and P.D.; writing—review and editing, P.D. and A.N.; visualization, A.N.; supervision, N.N. All authors have read and agreed to the published version of the manuscript.

Funding: This research received no external funding.

Data Availability Statement: Data used in this study are available upon request.

Conflicts of Interest: The authors declare no conflicts of interest.

Abbreviations

The following abbreviations are used in this manuscript:

AHU	Air Handling Unit
HVAC	Heating Ventilation and Air Conditioning
BEAST	Building Energy Assessments, Solutions, and Technologies
VAV	Variable Air Volume
BAS	Building Automation System
ASHRAE	American Society of Heating Refrigerating and Air-Conditioning Engineers
MLR	Multiple Linear Regression
XGBoost	Extreme Gradient Boosting
R^2	Coefficient of determination
CV	Coefficient of Variance
RMSE	Root Mean Square Error
MSE	Mean Squared Error

MAE	Mean Absolute Error
SHAP	Shapley Additive Explanations
ReLU	Rectified Linear Unit

Appendix A. R^2 and Residual Errors

Appendix A.1. R^2 Values

The following graph shows the R^2 values for the tested models. Blue lines represent testing R^2, and orange lines shows training R^2.

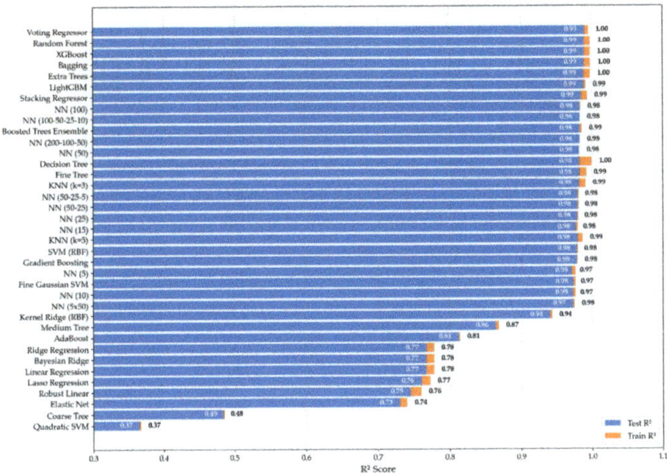

Figure A1. Testing vs. training R^2 values of analyzed models.

Appendix A.2. MLR Predicted vs. Actual Data

The line plot presents a comparison between the predicted and actual values for MLR over samples 750–1250. The orange line represents the actual values, while the blue line represents the predicted values. Areas where the lines overlap indicate accurate predictions. Conversely, greater visibility of the blue line suggests a weaker model performance as it indicates a deviation between predicted and actual values.

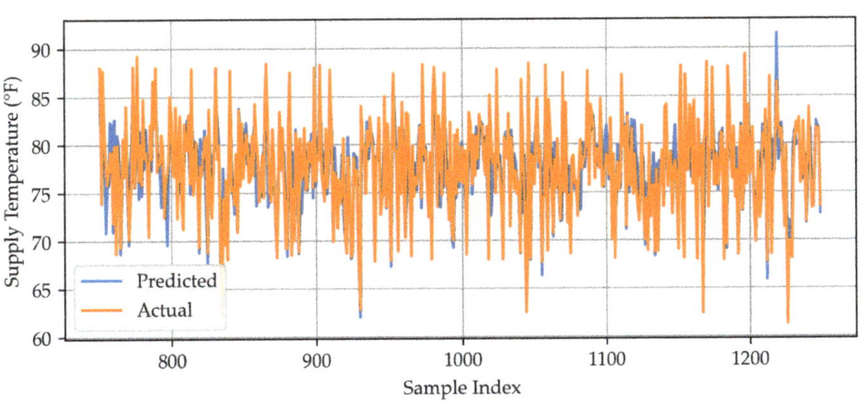

Figure A2. MLR: actual vs. predictions.

Appendix A.3. Neural Network Predicted vs. Actual Data

The line plot shows a line graph connecting the values for the predicted data and actual data of the best-performing neural network for sample 750–1250.

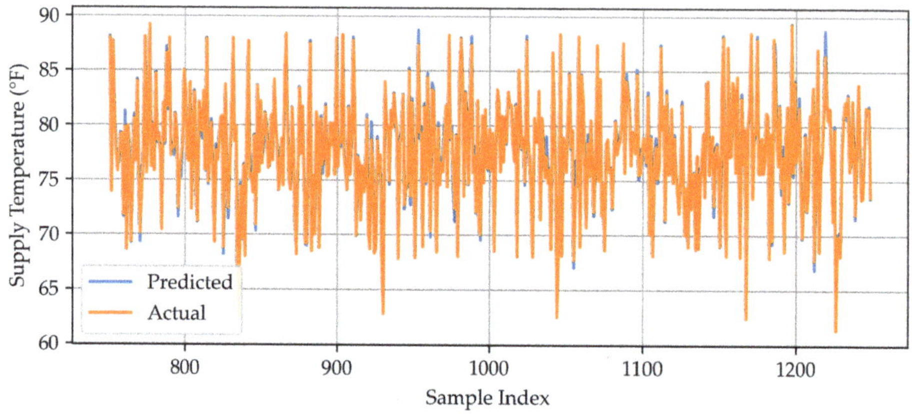

Figure A3. Neural network: actual vs. predictions.

Appendix A.4. XGBoost Predicted vs. Actual Data

The line plot shows a line graph connecting the values for the predicted data and actual data of XGBoost for sample 750–1250.

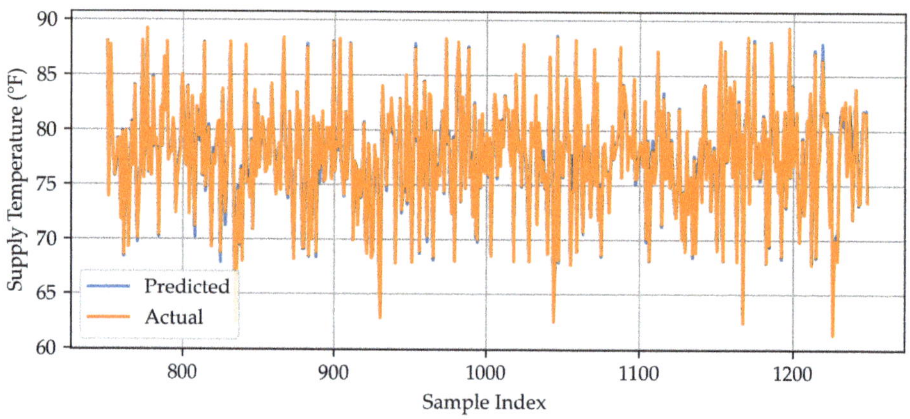

Figure A4. XGBoost: actual vs. predictions.

Appendix A.5. Bagging Predicted vs. Actual Data

The line plot shows a line graph connecting the values for the predicted data and actual data of bagging for sample 750–1250.

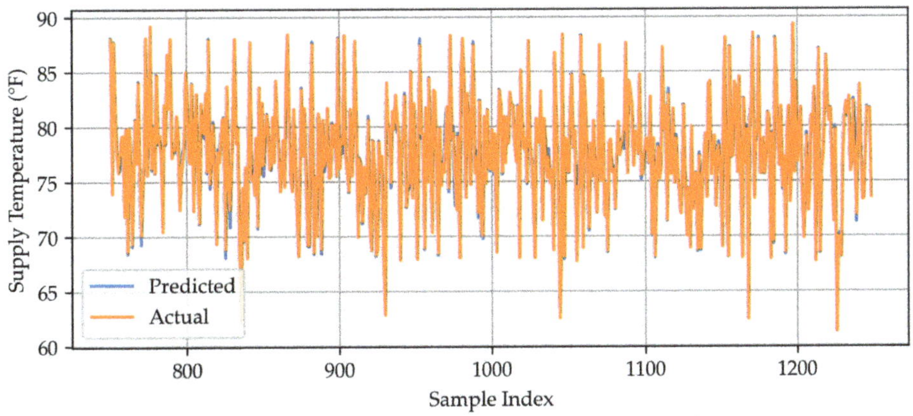

Figure A5. Bagging: actual vs. predictions.

Appendix B. Shapley Additive Explanations (SHAP) Values

Appendix B.1. MLR SHAP Chart

The SHAP chart for MLR visualizes the impact of each input variable on the model's output. The y-axis represents the input features, while the x-axis shows their influence on predictions. Color gradients indicate the relative magnitude of each feature's value across the dataset. A wider spread of data points suggests a stronger impact on the output, highlighting the most influential variables in the model.

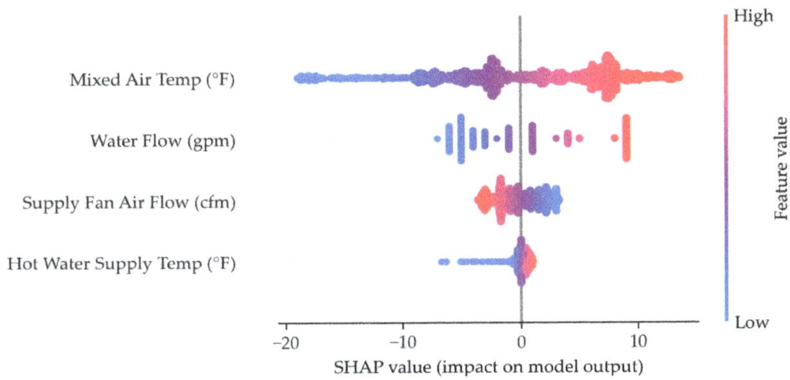

Figure A6. MLR SHAP chart.

Appendix B.2. Neural Network SHAP Chart

SHAP chart for the best performing neural network.

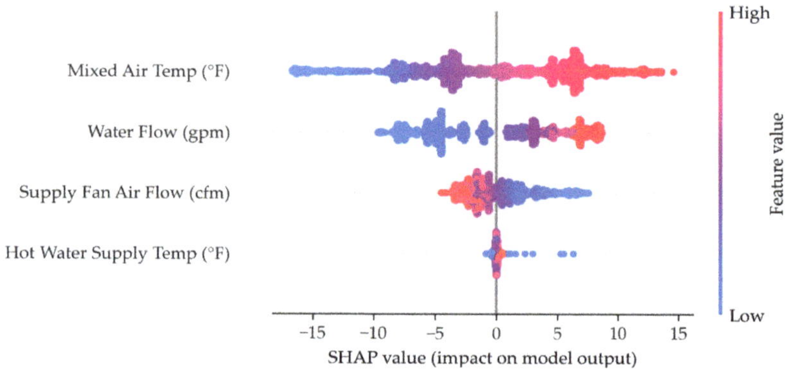

Figure A7. Neural network SHAP chart.

Appendix B.3. XGBoost SHAPSHAP Chart

SHAP chart for XGBoost.

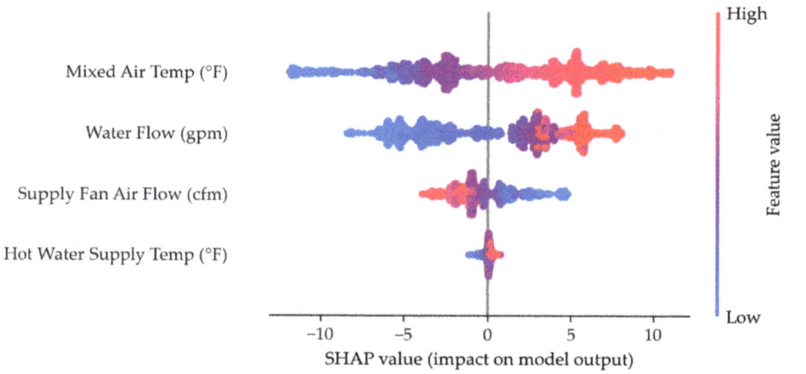

Figure A8. XGBoost SHAP chart.

Appendix B.4. Bagging SHAP Chart

SHAP chart for bagging.

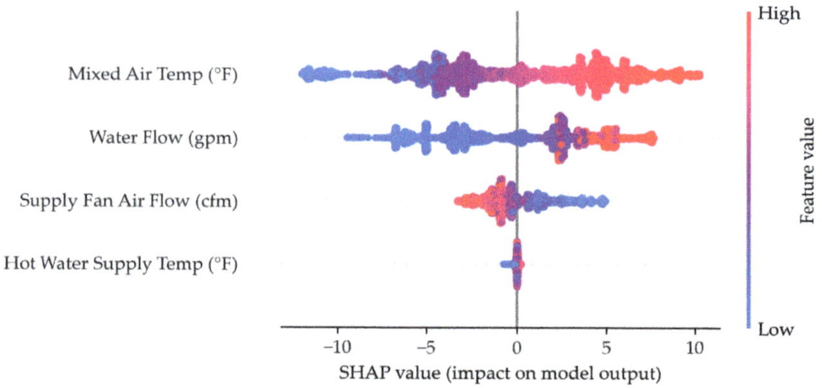

Figure A9. Bagging SHAP chart.

References

1. Paul, W.L.; Taylor, P.A. A Comparison of Occupant Comfort and Satisfaction between a Green Building and a Conventional Building. *Build. Environ.* **2008**, *43*, 1858–1870. [CrossRef]
2. Dharmasena, P.M.; Meddage, D.P.P.; Mendis, A.S.M. Investigating Applicability of Sawdust and Retro-Reflective Materials as External Wall Insulation under Tropical Climatic Conditions. *Asian J. Civ. Eng.* **2022**, *23*, 531–549. [CrossRef]
3. Solano, J.C.; Caamaño-Martín, E.; Olivieri, L.; Almeida-Galárraga, D. HVAC Systems and Thermal Comfort in Buildings Climate Control: An Experimental Case Study. *Energy Rep.* **2021**, *7*, 269–277. [CrossRef]
4. Dharmasena, P.; Nassif, N. Development and Optimization of a Novel Damper Control Strategy Integrating DCV and Duct Static Pressure Setpoint Reset for Energy-Efficient VAV Systems. *Buildings* **2025**, *15*, 518. [CrossRef]
5. Westphalen, D.; Koszalinski, S. *Energy Consumption Characteristics of Commercial Building HVAC Systems Volume I: Chillers, Refrigerant Compressors, and Heating Systems*; Arthur D. Little, Inc.: Boston, MA, USA, 2001.
6. American Socitey of Heating Ventilation and Air-Conditioning Systems and Equipment. *2020—ASHRAE Handbook—HVAC Systems and Equipment*; ASHRAE: Peachtree Corners, GA, USA, 2020; ISBN 978-1-947192-52-2.
7. U.S. Energy Information Administration (EIA). Use of Energy in Commercial Buildings. Available online: https://www.eia.gov/energyexplained/use-of-energy/commercial-buildings.php (accessed on 7 March 2025).
8. Peters, G.P.; Andrew, R.M.; Canadell, J.G.; Friedlingstein, P.; Jackson, R.B.; Korsbakken, J.I.; Le Quéré, C.; Peregon, A. Carbon Dioxide Emissions Continue to Grow amidst Slowly Emerging Climate Policies. *Nat. Clim. Change* **2020**, *10*, 3–6. [CrossRef]
9. Wang, Y.-W.; Cai, W.-J.; Soh, Y.-C.; Li, S.-J.; Lu, L.; Xie, L. A Simplified Modeling of Cooling Coils for Control and Optimization of HVAC Systems. *Energy Convers. Manag.* **2004**, *45*, 2915–2930. [CrossRef]
10. Ebrahimi, P.; Ridwana, I.; Nassif, N. Solutions to Achieve High-Efficient and Clean Building HVAC Systems. *Buildings* **2023**, *13*, 1211. [CrossRef]
11. Yu, B.; Kim, D.; Koh, J.; Kim, J.; Cho, H. Electrification and Decarbonization Using Heat Recovery Heat Pump Technology for Building Space and Water Heating. In Proceedings of the ASME 2023 17th International Conference on Energy Sustainability Collocated with the ASME 2023 Heat Transfer Summer Conference, Washington, DC, USA, 10–12 July 2023.
12. Deason, J.; Borgeson, M. Electrification of Buildings: Potential, Challenges, and Outlook. *Curr. Sustain. Renew. Energy Rep.* **2019**, *6*, 131–139. [CrossRef]
13. Yayla, A.; Świerczewska, K.S.; Kaya, M.; Karaca, B.; Arayici, Y.; Ayözen, Y.E.; Tokdemir, O.B. Artificial Intelligence (AI)-Based Occupant-Centric Heating Ventilation and Air Conditioning (HVAC) Control System for Multi-Zone Commercial Buildings. *Sustainability* **2022**, *14*, 16107. [CrossRef]
14. Zhao, Y.; Li, T.; Zhang, X.; Zhang, C. Artificial Intelligence-Based Fault Detection and Diagnosis Methods for Building Energy Systems: Advantages, Challenges and the Future. *Renew. Sustain. Energy Rev.* **2019**, *109*, 85–101. [CrossRef]
15. Babadi Soultanzadeh, M.; Ouf, M.M.; Nik-Bakht, M.; Paquette, P.; Lupien, S. Fault Detection and Diagnosis in Light Commercial Buildings' HVAC Systems: A Comprehensive Framework, Application, and Performance Evaluation. *Energy Build.* **2024**, *316*, 114341. [CrossRef]
16. *ANSI/ASHRAE/IES Standard 90.1-2022*; Energy Standard for Sites and Buildings Except Low-Rise Residential Buildings. American Socitey of Heating Ventilation and Air-Conditioning Systems and Equipment (ASHRAE): Peachtree Corners, GA, USA, 2022.
17. International Code Council, Inc. (ICC). *2021 International Energy Conservation Code (IECC)*; ICC: Country Club Hills, IL, USA, 2021.
18. The Paris Agreement | UNFCCC. Available online: https://unfccc.int/process-and-meetings/the-paris-agreement (accessed on 12 March 2025).
19. Barney, G.C.; Florez, J. Temperature Prediction Models and Their Application to the Control of Heating Systems. *IFAC Proc. Vol.* **1985**, *18*, 1847–1852. [CrossRef]
20. Asamoah, P.B.; Shittu, E. Evaluating the Performance of Machine Learning Models for Energy Load Prediction in Residential HVAC Systems. *Energy Build.* **2025**, *334*, 115517. [CrossRef]
21. Lu, S.; Zhou, S.; Ding, Y.; Kim, M.K.; Yang, B.; Tian, Z.; Liu, J. Exploring the Comprehensive Integration of Artificial Intelligence in Optimizing HVAC System Operations: A Review and Future Outlook. *Results Eng.* **2025**, *25*, 103765. [CrossRef]
22. Gao, Z.; Yu, J.; Zhao, A.; Hu, Q.; Yang, S. A Hybrid Method of Cooling Load Forecasting for Large Commercial Building Based on Extreme Learning Machine. *Energy* **2022**, *238*, 122073. [CrossRef]
23. Norouzi, P.; Maalej, S.; Mora, R. Applicability of Deep Learning Algorithms for Predicting Indoor Temperatures: Towards the Development of Digital Twin HVAC Systems. *Buildings* **2023**, *13*, 1542. [CrossRef]
24. Maddalena, E.T.; Lian, Y.; Jones, C.N. Data-Driven Methods for Building Control—A Review and Promising Future Directions. *Control Eng. Pract.* **2020**, *95*, 104211. [CrossRef]

25. Halhoul Merabet, G.; Essaaidi, M.; Ben Haddou, M.; Qolomany, B.; Qadir, J.; Anan, M.; Al-Fuqaha, A.; Abid, M.R.; Benhaddou, D. Intelligent Building Control Systems for Thermal Comfort and Energy-Efficiency: A Systematic Review of Artificial Intelligence-Assisted Techniques. *Renew. Sustain. Energy Rev.* **2021**, *144*, 110969. [CrossRef]
26. Platt, G.; Li, J.; Li, R.; Poulton, G.; James, G.; Wall, J. Adaptive HVAC Zone Modeling for Sustainable Buildings. *Energy Build.* **2010**, *42*, 412–421. [CrossRef]
27. Zhang, C.; Zhao, Y.; Fan, C.; Li, T.; Zhang, X.; Li, J. A Generic Prediction Interval Estimation Method for Quantifying the Uncertainties in Ultra-Short-Term Building Cooling Load Prediction. *Appl. Therm. Eng.* **2020**, *173*, 115261. [CrossRef]
28. Ding, Y.; Zhang, Q.; Wang, Z.; Liu, M.; He, Q. A Simplified Model of Dynamic Interior Cooling Load Evaluation for Office Buildings. *Appl. Therm. Eng.* **2016**, *108*, 1190–1199. [CrossRef]
29. Lin, X.; Tian, Z.; Lu, Y.; Zhang, H.; Niu, J. Short-Term Forecast Model of Cooling Load Using Load Component Disaggregation. *Appl. Therm. Eng.* **2019**, *157*, 113630. [CrossRef]
30. Dharmasena, P.; Nassif, N. Testing, Validation, and Simulation of a Novel Economizer Damper Control Strategy to Enhance HVAC System Efficiency. *Buildings* **2024**, *14*, 2937. [CrossRef]
31. Wahba, N.; Rismanchi, B.; Pu, Y.; Aye, L. Efficient HVAC System Identification Using Koopman Operator and Machine Learning for Thermal Comfort Optimisation. *Build. Environ.* **2023**, *242*, 110567. [CrossRef]
32. Babadi Soultanzadeh, M.; Nik-Bakht, M.; Ouf, M.M.; Paquette, P.; Lupien, S. Unsupervised Automated Fault Detection and Diagnosis for Light Commercial Buildings' HVAC Systems. *Build. Environ.* **2025**, *267*, 112312. [CrossRef]
33. Chen, Z.; Xu, P.; Feng, F.; Qiao, Y.; Luo, W. Data Mining Algorithm and Framework for Identifying HVAC Control Strategies in Large Commercial Buildings. *Build. Simul.* **2021**, *14*, 63–74. [CrossRef]
34. Khanuja, A.; Webb, A.L. What We Talk about When We Talk about EEMs: Using Text Mining and Topic Modeling to Understand Building Energy Efficiency Measures (1836-RP). *Sci. Technol. Built Environ.* **2023**, *29*, 4–18. [CrossRef]
35. Ajifowowe, I.; Chang, H.; Lee, C.S.; Chang, S. Prospects and Challenges of Reinforcement Learning-Based HVAC Control. *J. Build. Eng.* **2024**, *98*, 111080. [CrossRef]
36. Jiang, M.L.; Wu, J.Y.; Xu, Y.X.; Wang, R.Z. Transient Characteristics and Performance Analysis of a Vapor Compression Air Conditioning System with Condensing Heat Recovery. *Energy Build.* **2010**, *42*, 2251–2257. [CrossRef]
37. Wang, S.; Ma, Z. Supervisory and Optimal Control of Building HVAC Systems: A Review. *Hvac&R Res.* **2008**, *14*, 3–32. [CrossRef]
38. Belany, P.; Hrabovsky, P.; Sedivy, S.; Cajova Kantova, N.; Florkova, Z. A Comparative Analysis of Polynomial Regression and Artificial Neural Networks for Prediction of Lighting Consumption. *Buildings* **2024**, *14*, 1712. [CrossRef]
39. Fumo, N.; Rafe Biswas, M.A. Regression Analysis for Prediction of Residential Energy Consumption. *Renew. Sustain. Energy Rev.* **2015**, *47*, 332–343. [CrossRef]
40. 6.1—MLR Model Assumptions | STAT 462. Available online: https://online.stat.psu.edu/stat462/node/145/ (accessed on 19 March 2025).
41. Lee, C.-W.; Fu, M.-W.; Wang, C.-C.; Azis, M.I. Evaluating Machine Learning Algorithms for Financial Fraud Detection: Insights from Indonesia. *Mathematics* **2025**, *13*, 600. [CrossRef]
42. Abiodun, O.I.; Jantan, A.; Omolara, A.E.; Dada, K.V.; Mohamed, N.A.; Arshad, H. State-of-the-Art in Artificial Neural Network Applications: A Survey. *Heliyon* **2018**, *4*, e00938. [CrossRef]
43. Han, S.-H.; Kim, K.W.; Kim, S.; Youn, Y.C. Artificial Neural Network: Understanding the Basic Concepts Without Mathematics. *Dement. Neurocogn Disord.* **2018**, *17*, 83–89. [CrossRef]
44. 1.17.Neural Network Models (Supervised). Available online: https://scikit-learn.org/stable/modules/neural_networks_supervised.html (accessed on 19 March 2025).
45. Abdolrasol, M.G.M.; Hussain, S.M.S.; Ustun, T.S.; Sarker, M.R.; Hannan, M.A.; Mohamed, R.; Ali, J.A.; Mekhilef, S.; Milad, A. Artificial Neural Networks Based Optimization Techniques: A Review. *Electronics* **2021**, *10*, 2689. [CrossRef]
46. Engelbrecht, A.P.; Cloete, I.; Zurada, J.M. Determining the Significance of Input Parameters Using Sensitivity Analysis. In Proceedings of the From Natural to Artificial Neural Computation, Torremolinos, Spain, 7–9 June 1995; Mira, J., Sandoval, F., Eds.; Springer: Berlin/Heidelberg, Germany, 1995; pp. 382–388.
47. Breiman, L. Bagging Predictors. *Mach. Learn.* **1996**, *24*, 123–140. [CrossRef]
48. Dietterich, T.G. Machine-Learning Research. *AI Mag.* **1997**, *18*, 97–136. [CrossRef]
49. Friedman, J.H. Greedy Function Approximation: A Gradient Boosting Machine. *Ann. Stat.* **2001**, *29*, 1189–1232. [CrossRef]
50. Nogales, A.G. On Consistency of the Bayes Estimator of the Density. *Mathematics* **2022**, *10*, 636. [CrossRef]
51. Gu, M.; Kang, S.; Xu, Z.; Lin, L.; Zhang, Z. AE-XGBoost: A Novel Approach for Machine Tool Machining Size Prediction Combining XGBoost, AE and SHAP. *Mathematics* **2025**, *13*, 835. [CrossRef]
52. Scikit-Learn: Machine Learning in Python—Scikit-Learn 1.6.1 Documentation. Available online: https://scikit-learn.org/stable/ (accessed on 12 March 2025).

53. Pandas—Python Data Analysis Library. Available online: https://pandas.pydata.org/ (accessed on 12 March 2025).
54. NumPy. Available online: https://numpy.org/ (accessed on 12 March 2025).
55. Li, X.; Shen, Y.; Meng, Q.; Xing, M.; Zhang, Q.; Yang, H. Single-Model Self-Recovering Fringe Projection Profilometry Absolute Phase Recovery Method Based on Deep Learning. *Sensors* **2025**, *25*, 1532. [CrossRef]
56. Jakubec, M.; Cingel, M.; Lieskovská, E.; Drliciak, M. Integrating Neural Networks for Automated Video Analysis of Traffic Flow Routing and Composition at Intersections. *Sustainability* **2025**, *17*, 2150. [CrossRef]
57. James, G.; Witten, D.; Hastie, T.; Tibshirani, R. *An Introduction to Statistical Learning: With Applications in R*; Springer Texts in Statistics; Springer: New York, NY, USA, 2021; ISBN 978-1-0716-1417-4.
58. Qavidelfardi, Z.; Tahsildoost, M.; Zomorodian, Z.S. Using an Ensemble Learning Framework to Predict Residential Energy Consumption in the Hot and Humid Climate of Iran. *Energy Rep.* **2022**, *8*, 12327–12347. [CrossRef]
59. Wang, J.; Yuan, Z.; He, Z.; Zhou, F.; Wu, Z. Critical Factors Affecting Team Work Efficiency in BIM-Based Collaborative Design: An Empirical Study in China. *Buildings* **2021**, *11*, 486. [CrossRef]
60. Neter, J.; Wasserman, W.; Kutner, M.H. *Applied Linear Regression Models*; Richard D. Irwin, Inc.: Homewood, IL, USA, 1983; ISBN 0-256-02547-9.
61. Lian, J.; Jiang, J.; Dong, X.; Wang, H.; Zhou, H.; Wang, P. Coupled Motion Characteristics of Offshore Wind Turbines During the Integrated Transportation Process. *Energies* **2019**, *12*, 2023. [CrossRef]
62. Liu, J.; Wu, Q.; Sui, X.; Chen, Q.; Gu, G.; Wang, L.; Li, S. Research Progress in Optical Neural Networks: Theory, Applications and Developments. *PhotoniX* **2021**, *2*, 5. [CrossRef]
63. Babatunde, D.E.; Anozie, A.; Omoleye, J. Artificial Neural Network and Its Applications in the Energy Sector—An Overview. *Int. J. Energy Econ. Policy* **2020**, *10*, 250–264. [CrossRef]
64. XGBoost Documentation—Xgboost 3.0.0 Documentation. Available online: https://xgboost.readthedocs.io/en/stable/ (accessed on 25 March 2025).
65. Chen, T.; Guestrin, C. XGBoost: Reliable Large-Scale Tree Boosting System. In Proceedings of the 22nd SIGKDD Conference on Knowledge Discovery and Data Mining, San Francisco, CA, USA, 13–17 August 2016.
66. Sehrawat, N.; Vashisht, S.; Singh, A. Solar Irradiance Forecasting Models Using Machine Learning Techniques and Digital Twin: A Case Study with Comparison. *Int. J. Intell. Netw.* **2023**, *4*, 90–102. [CrossRef]

energies

MDPI

Article

BIM-Based Machine Learning Application for Parametric Assessment of Building Energy Performance

Panagiotis Tsikas, Athanasios Chassiakos *, Vasileios Papadimitropoulos and Antonios Papamanolis

Department of Civil Engineering, University of Patras, 26500 Patras, Greece; panostsikas@upatras.gr (P.T.); up1056827@upatras.gr (V.P.)
* Correspondence: a.chassiakos@upatras.gr

Abstract: The energy performance of buildings has become a main concern globally in response to increased energy demand, the environmental impacts of energy production, and the reality of energy poverty. To improve energy efficiency, proper building design should be secured at the early design phase. Digital tools are currently available for performing energy assessment analyses and can efficiently handle complex and technically demanding buildings. However, alternative designs should be checked individually, and this makes the process time-consuming and prone to errors. Machine learning techniques can provide valuable assistance in developing decision support tools. In this paper, typical residential buildings are considered along with eleven factors that highly affect energy performance. A dataset of 337 instances of such parameters is developed. For each dataset, the building energy performance is estimated based on BIM analysis. Next, statistical and machine learning techniques are implemented to provide artificial models of energy performance. They include statistical regression modeling (SRM), decision trees (DTs), random forests (RFs), and artificial neural networks (ANNs). The analysis reveals the contribution of each factor and highlights the ANN as the best performing model. An easy-to-use interface tool has been developed for the instantaneous calculation of the energy performance based on the independent parameter values.

Keywords: building; energy; building information modeling; building energy modeling; machine learning; random forest; artificial neural network

1. Introduction

The increase in population and human activities, followed by the expansion of the built environment, have intensified sustainability problems, among which is the need for significant amounts of energy in contemporary buildings. According to the Global Status Report for Buildings and Construction [1], buildings consume about 36% of annual global energy resources and are responsible for 60% of global electricity production, whilst they contribute 39% of global annual CO_2 emissions.

To facilitate residents' well-being, modern building design requires increased energy consumption. In addition, existing building structures are highly energy-intensive and environmentally unfriendly [2]. Previous studies and construction field experience have demonstrated that energy consumption in buildings can be effectively reduced through simple design solutions, such as the selection of appropriate building shapes and orientations. Thus, one of the challenges of the construction industry is the sustainable design of new buildings and adaptation of existing ones. In this direction, it is imperative to reassess the conventional design and evaluation procedures of structures, particularly through an integrated parametric approach that considers interdependences among leverage factors. In current practice, energy analyses are often conducted separately, as the planning and dimensioning processes are not aligned with the energy assessment process. In fact, most existing buildings were designed in the past with outdated regulations or without any sort of provisions concerning their energy efficiency [3]. Additionally, structure complexity, multiple influential parameters and interdependencies, uncertainties, and limited data often result in assumptions and simplifications, which underestimate or distort the influence of these parameters, leading to inaccurate results and inadequate designs.

The energy efficiency of buildings is affected by several factors related to a building's structure, location, and usage. These factors can be classified into two main classes: physical and human-influenced ones [4]. The first class encompasses technical and physical factors, such as the building envelope, local climate, heating, ventilation, and air conditioning systems. The second category includes human-related factors, such as occupant behavior and energy usage patterns, which exhibit significant variability and uncertainty in energy consumption. Finally, there are social factors, such as energy prices, that have social consequences on energy efficiency.

An effective way to estimate building energy efficiency is the utilization of building energy modeling (BEM) software [5]. However, partially using BEM software without synchronizing it with the global project design and construction process can result in ineffective decisions, which are partly due to discontinuous information flow among digital design models. In addition, the process is time-consuming, tedious, and susceptible to errors, as energy efficiency data must be entered into BEM software manually. The introduction of BIM technology allows for options that can overcome the limitations of conventional energy efficiency assessment processes through BEM, as currently, BEM can act as a subsystem of BIM. The use of BIM can result in accurate and systematic calculations and provide the ability to monitor and control energy use during the operation phase of buildings [6]. The application of BIM technology combined with simulations from the initial stages of project design allow for a wide exploration of alternative design solutions [7]. In fact, this technology provides the ground for understanding and quantifying the actual conditions of a building as well as simulating real-world user behavior, thus improving energy efficiency analyses.

While BIM represents the foundation for performing several types of analyses in building design, these analyses are usually multifactorial and require comparative evaluation of alternative solutions, with user involvement in setting these solutions up and comparing the results. The partial automation of such analyses results in increased time and effort requirements. Instead, one can employ a parametric analysis and use machine learning techniques to easily capture the influence of energy consumption factors and comparatively evaluate alternative building configurations in terms of energy efficiency. The present study first analyzes several factors that predominantly affect the energy efficiency of a typical residential building using BIM technologies. Based on this analysis, a dataset relating the factors influencing energy to their corresponding energy consumption is developed. Machine learning techniques are then developed and evaluated as an instrument for instantaneous energy assessment analyses that can assist in decisions on initial building designs or retrofitting.

2. Background

Existing research has dealt with several aspects of building energy analysis. In the direction of sustainable design, key architectural design variables (ADVs), which influence sustainability characteristics during early building design and integrate computational simulations and stakeholder insights, were examined [8]. This research underscores the importance of including stakeholder perspectives in developing design optimization tools, aligned with real-world needs for effective sustainable design.

The role of BIM in facilitating the LEED® certification process was explored in [9]. The findings underscore BIM's potential to save time and resources, thus enhancing the feasibility of sustainable project management. Further, the implementation of national and international energy standards in low-rise residential buildings in Jordan utilizing BIM models was examined in [10]. The use of BIM tools and regression models to optimize energy performance in early-stage building design was demonstrated in [11]. By simulating various prototype building shapes, the study identified energy-efficient geometries, noting that triangular designs consumed the least energy. A framework combining BIM with a life cycle cost (LCC) analysis and orthogonal testing methods to identify optimal design parameters for energy efficiency in green buildings was introduced in [12]. The study revealed significant annual energy reductions, emphasizing enhancements in building envelope features such as insulation. A BIM-driven approach aiming at net-zero energy in tall buildings within the Malaysian construction sector was the focus of [13]. The results demonstrate significant correlations between BIM usage and factors such as early design integration and predictive energy performance.

Moving on to machine learning applications for energy consumption prediction, the effectiveness of deep learning (DL) methods for short-term cooling load prediction in buildings was explored in [14], emphasizing their potential to outperform traditional physic-based models. A comprehensive review of methods for predicting building energy demand, with a focus on data-driven techniques, like support vector machines (SVMs) and artificial neural networks (ANNs), was presented in [15]. The study highlights the strengths of these methods, including achieving a high predictive accuracy. It further notes that their effectiveness can be hindered by limited data availability. Another review of data-driven models for predicting building energy consumption, emphasizing their role in energy management, conservation, and planning, was performed in [16]. The study recommended focusing on big data analytics and incorporating occupant behavior insights to enhance model accuracy and reliability, paving the way for more tailored prediction models. An analysis of building energy prediction models was performed in [17], categorizing them into white-box, black-box, and grey-box types and highlighted that black-box models, which include machine learning techniques, like SVMs and ANNs, can handle complex data relationships. However, they face challenges related to human behavior and climate variability. A hybrid predictive model that combines empirical mode decomposition (CEEMDAN), deep learning, and ARIMA to enhance short-term heating load forecasts was introduced in [18]. The study demonstrated the potential for hybrid models to address complex data and improve forecast reliability, suggesting future research on dynamic weighting strategies to enhance adaptability. Deep transfer learning (DTL) strategies were evaluated in [19] to overcome data scarcity in cross-building energy prediction. The study concluded that this approach is effective for improving prediction accuracy across various building types and highlights the need for further explorations into diverse application contexts. The use of ANNs for national-level hourly heat demand prediction, focusing on simplifying input requirements while maintaining accuracy, was explored in [20]. The study identifies temperature and radiation as critical factors, allowing for reduced input complexity.

Focusing more closely on optimized and hybrid ML approaches, a comprehensive comparison of various machine learning (ML) techniques to predict annual energy consumption in residential buildings was provided in [21], with a particular focus on aiding early design decisions to enhance energy efficiency. The authors advocate for the use of DNN models for early-design-stage energy consumption forecasting and recommend further studies incorporating larger datasets and exploring ensemble algorithms to enhance predictive reliability. A hybrid machine learning model designed to enhance the accuracy of heating energy consumption predictions in residential buildings was presented in [22]. The study analyzed the support vector regression (SVR) and six meta-heuristic algorithms, including battle royale optimization (BRO) and particle swarm optimization (PSO). The authors recommend extending this method to various climates and building types to verify its broader utility. Energy demand forecasting across seven sectors in Iran was explored in [23], employing machine learning algorithms (ANN, LSTM, and ARIMA models), enhanced with optimization techniques (PSO and Grey Wolf Optimizer). The work concludes that this framework provides a foundation for strategic energy planning and management. An AutoML framework for predicting residential heating and cooling loads was introduced in [24], which automates the feature engineering and model optimization process. The study concludes that the use of SHapley Additive exPlanations (SHAP) values in model explainability enhances the framework's practical utility for early-stage building design. Finally, the integration of evolutionary algorithms with conventional neural networks for residential energy consumption forecasting was proposed in [25], leading to the recommendation of using hybrid techniques for future optimization.

To relate energy analyses and socio-economic factors, artificial neural networks (ANNs) have been utilized as tools for policymakers to identify and assist vulnerable populations, as part of an energy poverty study in Greece, with targeted energy efficiency initiatives and financial aid [26]. The analysis incorporates various socio-economic and geographical factors, including house age and ownership status, to predict seven energy poverty indicators.

The existing literature has dealt with several aspects of the energy efficiency of buildings. However, the analyses mainly focus on fragments of the whole process while the outcomes remain within a rather scientific framework without the appropriate connection to engineering practice and design tools. There are also several research works using ML methods for assessing energy consumption. According to [16], 84% of such studies focus on short-term energy consumption prediction because of its direct relation to the day-to-day operations of buildings, while only 12% of the studies focus on long-term (yearly) energy consumption prediction. The first class refers to energy efficiency improvements, in terms of temperature or lighting, following certain interventions, such as cell insulation upgrades. An indicative study of this kind [27] focuses on an energy simulation of an academic building and aims at evaluating the BIM's potential to improve the building's energy efficiency through the simulation and optimization of design parameters, such as building orientation or insulation characteristics, individually or collectively. A long-term energy efficiency goal refers to a multi-parametric evaluation of energy-efficient building design in a structure's life cycle [3]. Within the latter class, the present work combines BIM modeling and machine learning techniques to evaluate the influencing factors of building energy performance and provides a practical and easy to use tool for energy efficiency analyses of typical residential buildings. The adoption of BIM provides an unlimited capability to simulate the structure and develop energy models for any type and size of building, design characteristics, environmental factors, etc.

3. Methodology

The traditional methods of building energy performance assessment, via analytical calculations, are quite tedious and of low accuracy. This is because they rely on simplified assumptions and limited data that do not fully reflect the linkage of various factors with energy consumption. To address these limitations, a more comprehensive and integrated approach is needed, which considers all relevant factors with their interrelationships. The advancement of building information modeling (BIM) can significantly contribute to energy performance assessment upgrades, the optimal design of new buildings, and the retrofitting of existing ones.

While BIM tools can effectively carry out the desired analysis, the process of comparatively assessing alternative building designs requires the development of each individual BIM design. There are many of these designs and they can be quite different in terms of their characteristics, making the whole process arduous and time-consuming. On the other hand, the existence of a ready-to-use digital solution that could instantly provide an estimate of the building energy performance, based on some prevailing characteristics, without the need to develop each BIM model, could provide a great decision support in designing high-energy-performance buildings. To obtain such a solution, a two-step process is required, which includes the identification of the prevailing parameters and their interrelationships, as well as the energy performance assessment based on those parameters.

In the present study, a set of parameters that influence the building energy performance are developed based on research findings and practical experience. Then, an experiment is performed with alternative sets of these parameters, and the corresponding energy performance is assessed via a BIM application. The input–output set of this process is used to develop artificial prediction models, employing alternative methods of machine learning. A typical group of residential buildings is examined. The energy performance of buildings depends on several parameters with varying levels of contributions. In this work, the ones that have been highlighted in the literature and in practice as having a significant impact on the energy performance of structures are considered. These factors are generally related to the architectural design, the characteristics of the building shell, and the climatic conditions in the area and are in agreement with previous studies [28–30]. The analysis involves eleven main parameters that considerably influence the energy performance of such buildings. These parameters are codified in Table 1 and schematically illustrated in Figure 1. The location parameter corresponds to four different climatic zones, in terms of the average temperature, as is further described in the case study section.

Table 1. Description of design factors and corresponding values.

	Description	Parameter Values	Model Values
X1	Masonry thickness	15, 20, 25 cm	15, 20, 25
X2	Wall insulation thickness	2, 3, 5 cm	2, 3, 5
X3	Slab thickness	15, 20 cm	15, 20
X4	Slab insulation thickness	2, 3, 5 cm	2, 3, 5
X5	Window-to-wall ratio	15%, 20%, 30%	15, 20, 30
X6	Window glass type	Single, double	1, 2
X7	Climate zone at building location	Zone 1, 2, 3, 4	1, 2, 3, 4
X8	Building orientation	South, North, East, West	1, 2, 3, 4
X9	Roof existence	Yes, No	1, 2
X10	Basement existence	Yes, No	1, 2
X11	Building size	1 floor, 2 floors	1, 2

Energy Efficiency Analysis of Buildings Utilizing BIM

This section describes the methodology to assess the energy efficiency of buildings using building information modeling (BIM) technology. This assessment is based on simulated data that have come from full building modeling. The energy efficiency assessment is performed via the quantitative parameter of energy consumption. The methodology can be used in the early design process of buildings for providing decision support for selecting appropriate features for a design.

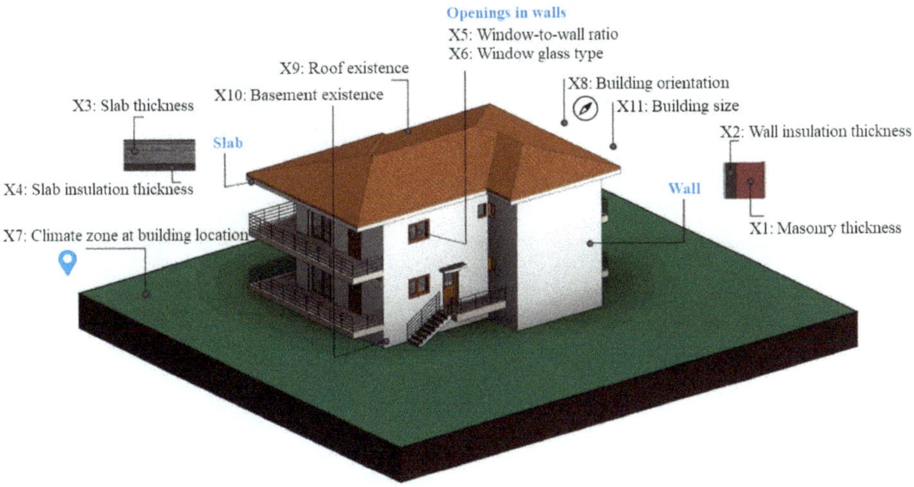

Figure 1. Schematic representation of design parameters.

The BIM design of a building starts with architectural modeling, in which the building elements with their characteristics (geometric dimensions, materials, supplier info, etc.) are entered in the 3D digital model. Then, the building structural characteristics are entered in the model and the building/shell design is created. Finally, the HVAC (heating, ventilation, and air conditioning) design is integrated. The whole building design process is dynamic and subject to continuous revisions and improvements. The BIM model, due to its high interoperability, is continually updated and provides a powerful model and a vivid picture of the alternative designs to all stakeholders. Figure 2 illustrates the BIM modules and their involvement in building energy modeling. Specifically, 3D-BIM is used for modeling the design characteristics of the building (Figure 2a). The solar path and the sunlight radiance components, which are shown in Figures 2b and 2c, respectively, are tools that determine the sun orbit and the intensity and distribution of natural lighting at the building's location as well as their effect on energy consumption. Finally, the energy model is a BIM tool that calculates the energy consumption of a building with specific characteristics (Figure 2d).

BIM technology provides significant benefits in interpreting and quantifying the actual condition of a building, while it further allows for the simulation of user behavior. By incorporating energy simulations into the BIM model, alternative designs can be explored, and energy-efficient solutions can be identified. In this direction, BIM-based building energy assessment is crucial in promoting sustainable construction.

In the following analysis, different pieces of terminology are used in terms of the aim and the output variable of the model. "Energy efficiency" refers broadly to the capability of the building to provide comfortable temperature conditions for inhabitants at relatively low energy consumption rates. "Energy consumption" represents the quantitative assessment of the required energy to attain acceptable levels of comfort and represents the model output variable. "Energy estimation or prediction" highlights the process of calculating the energy consumption level of an existing building or forecasting the energy consumption of a newly designed or reconstructed building.

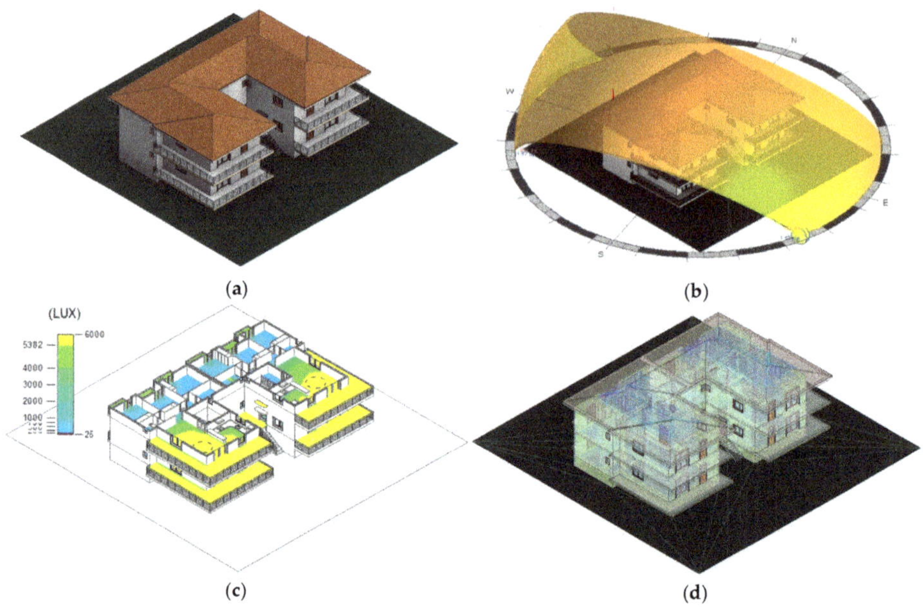

Figure 2. Digital 3D/BIM and energy models of a two-story residential building. (**a**) 3D/BIM model; (**b**) Solar path; (**c**) Sunlight radiance; (**d**) Energy model.

4. Model Development

The energy consumption of a building is affected by several parameters associated with the characteristics of the building and its surroundings. To obtain a generic energy assessment model, the actual building design and energy performance should be analyzed on a parametric basis. This objective can be fulfilled if a representative set of input parameters are developed, and the corresponding energy consumption outputs are computed. Next, a machine learning technique can be employed to develop the corresponding artificial model.

Two general approaches for developing the artificial energy prediction model are examined in this paper. The first includes statistical methods in the form of curve fitting by means of statistical regression analysis (SRM model). The second class includes machine learning techniques, namely decision trees (DT model), random forest (RF model) and artificial neural networks (ANN model). All these methods are particularized as alternatives to energy performance assessments based on BIM modeling and aimed to involve low-effort calculations and provide wide applications without restrictions (Figure 3).

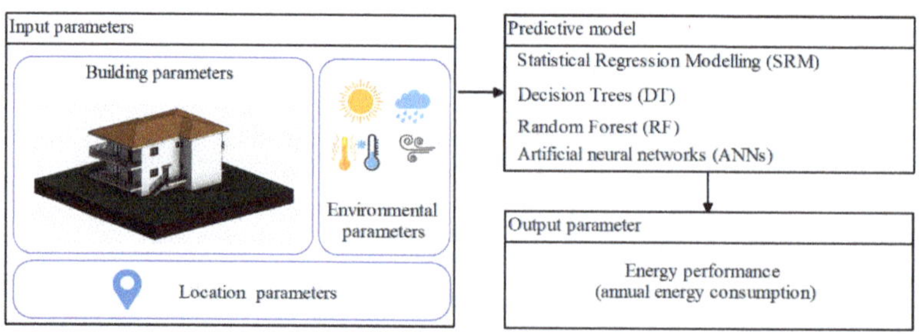

Figure 3. Flowchart of machine-learning-based energy performance assessment.

4.1. Multiple Nonlinear Regression Model

The approximation of the relationship between energy consumption and the factors that leverage it on a statistical ground can be elaborated through regression. The use of multiple nonlinear regression requires the selection (or assumption) of the nonlinear relationship type between input and output data. In many cases, however, the use of multiple nonlinear regression is not sufficient (for strong nonlinear problems), as the relationship type is confined within the available software libraries. Stata software allows for switching among different parameter combinations and exponents (Equation (1)), thus facilitating the determination of functional relationships. Equation (1) describes the general form of the SRM model [31], where Y is the output parameter (e.g., in kWh/m^2 or EUR/m^2 per year), X is the vector of the independent variables, and B_i and P_i are the model coefficients and exponents.

$$Y = B_0 + B_1 \cdot X^{p_1} + B_2 \cdot X^{p_2} + \ldots \quad (1)$$

4.2. Machine Learning Model

4.2.1. Decision Trees (DTs)

In terms of machine learning techniques, the first method employs decision trees (Figure 4). Decision trees (DTs) provide a graphical representation of problem solving through gradual decisions. They can be used in problems with numerical and classification data, being more effective in resolving problems of the latter type. A decision tree algorithm is a way of creating a tree-like model which makes decisions based on the features of the data. The tree starts with a single node known as the root node, which represents the entire dataset. The root node is then split into two or more child nodes based on the values of one of the features in the dataset. This process is repeated for each child node retrospectively, creating a hierarchical structure of nodes that represents the decision process. The algorithm selects the feature that best separates the data into subsets with similar target values to split the data at each node. This is accomplished by gauging the information gain of each feature. The information gain is a measure of how well a feature separates the data into subsets with similar target values. Eventually, the feature with the highest information gain is chosen to split the data at that node. The process of splitting the data continues recursively until it reaches a stopping criterion, i.e., a set of rules that determine when the algorithm should stop creating new nodes. This could be the maximum tree depth, the minimum number of samples per leaf, or the minimum information gain. Once the tree is built, it can be used to make predictions. Each leaf node of the tree represents a prediction for the target variable. The structure of the tree allows one to represent the decision process and the decision boundary in a readable format.

In the current application, the parameters related to the architectural design, the building frame characteristics, and the climate parameters are set up in a question form and the energy efficiency is set up as a decision output (assessment). These components are combined through a learning process to create a decision tree that can be used to estimate energy costs. However, in many cases, if the decision-making process consists of several steps, the decision trees can grow extensively. Still, due to the ease and supervision they offer, DTs are popular and practical tools in decision-making processes.

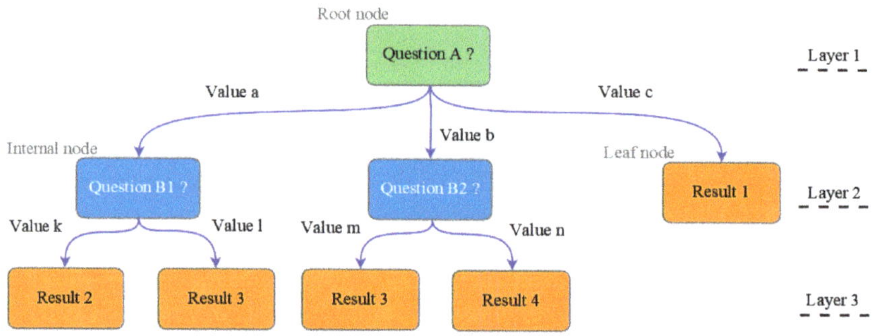

Figure 4. Schematic illustration of a decision tree algorithm.

4.2.2. Random Forests (RFs)

A combination of DT algorithms is often developed to form a random forest algorithm (RF). Random forests (RFs) use a series of different decision trees combining the individual predictions to enhance the accuracy of the output. Figure 5 indicates such a structure with N decision trees. The RF algorithm is powerful and accurate for both problem classes, i.e., regression and classification.

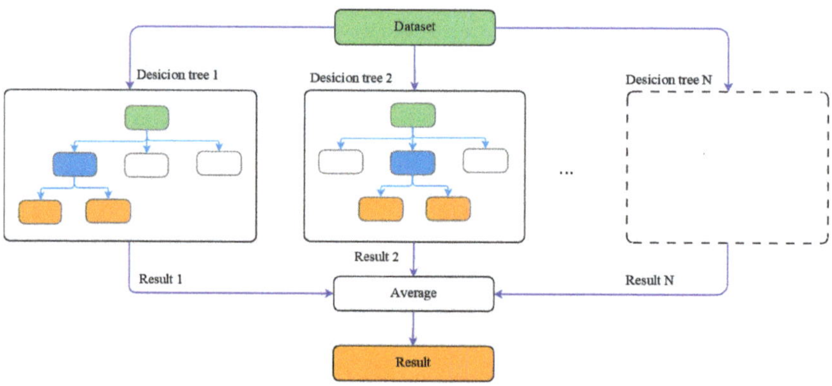

Figure 5. Schematic illustration of random forest algorithm.

An RF algorithm builds multiple decision trees (also known as "forests") and combines their individual predictions to a final decision. A random sample is selected from the dataset for training each decision tree. A random subset is selected for each decision tree for splitting the data at each node. Then each decision tree is trained on its sample of data and selected features. The predictions from all decision trees are finally combined by taking, for instance, the majority of votes for classification problems, or averaging the predictions for regression problems. An advantage of random forests compared to decision trees is the reduced overfitting outcome.

4.2.3. Artificial Neural Network (ANN)

Artificial neural network (ANN) methods are particularly useful in cases in which the cost of implementing an analytical computational process is high or if the way that influencing factors interact is not well known. The ANN structure consists of several input nodes (which refer to the input parameters) and intermediate neurons arranged in layers to process the input data, as well as the output node (Figure 6). The efficiency of an artificial neural network is determined by its architecture and the learning process. The selection of the number of neurons, their layers, and the way that they are connected depend on the problem's nature and characteristics. ANN learning is the process of acquiring knowledge through training examples, which is then recalled, providing an output in similar problems. Neural network operation is based on the transmission of a signal from the initial neurons (input layer) to the final neuron (output layer). Depending on the neuron connection characteristics, the signal is attenuated or amplified during its propagation, so that the physical quantity in the last neuron is calculated.

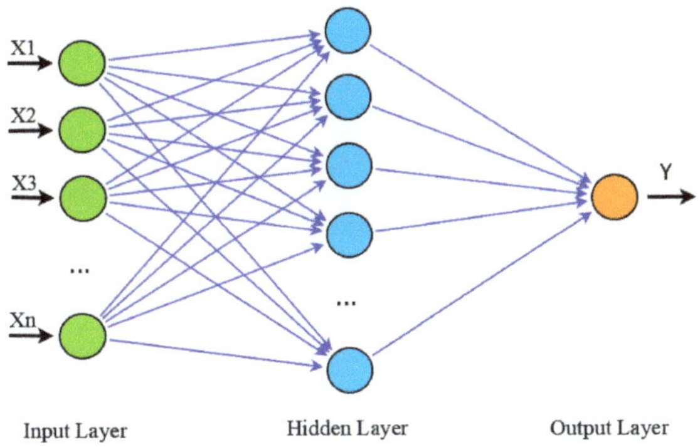

Figure 6. Schematic representation of neural network algorithm.

When developing machine learning models, a balance between model and actual problem complexities should be retained. In this sense, it is not always effective to use large-scale (several input parameters) or complex (in architecture) models. The employment of such a model does not necessarily guarantee more precise predictions. This is because the increase in size and complexity makes the model hypersensitive [32] to such an extent that it cannot work for out-of-training data due to overfitting. On the contrary, a simple model may exhibit high deviation between simulated and actual output data due to underfitting. The above inefficiencies can be regulated by the two complementary error types, variance and bias, which indicate overfitting and underfitting conditions, respectively. An effective model should retain balance between the two errors and be obtained by a trial-and-error development process.

In summary, four compact models are developed based on non-linear multivariate regression, decision trees, random forests, and artificial neural networks. More specifically, the statistical regression modeling (SRM) approach is employed, utilizing Stata 18 software to develop a non-linear and multivariable regression model. The decision tree model consists of 7 levels and 24 terminal nodes and has been trained using the C4.5 algorithm (https://en.wikipedia.org/wiki/C4.5_algorithm (accessed on 10 December 2024)). The random forest model includes five trees (with the same form as that in DT model) and has also been trained using the C4.5 algorithm. The root mean square error (RMSE) statistic is used as the objective (loss) function. The number of epochs in both models (DT and RF) is in the order of 1000.

A multi-layer perceptron (MLP) architecture is implemented for neural network development. The MLP structure consists of 11 input neurons corresponding to the input variables, a hidden layer of 11 neurons, and an output neuron. An activation function of a sigmoid form is employed for the hidden and the output layers. The specific design efficiently serves as a trade-off between model accuracy and simplicity, following experimentation on alternative designs to prevent overfitting and underfitting issues. The training process has been carried out using the Levenberg–Marquardt algorithm, which is a variant of the back-propagation method. The number of epochs is in the order of 5000.

5. Data Management and Development Software

The model's effectiveness highly depends on the size and quality of the available data. In this work, the necessary data have been developed from simulation and parametric analysis of buildings through BIM application. The input parameter selection was made in a way that data are rather representative of all eleven design parameters in Table 1 within a reasonable range of values. The generated sample includes 337 arrays of input–output data. The dataset is randomly divided into training, validation, and testing subsets in proportions of 70%, 15%, and 15%, respectively.

According to [33], the training sample size in such type of analysis should be above an approximate level of fifty plus eight times the number of independent parameters. Further, the work in [34] proposes that 10 to 15 observations per parameter are required to avoid overfitting in statistical and machine learning models. In the present work, the training sample size is roughly twice as high as these levels. As such, it can assure that the model complexity (overfitting) is controllable, and the estimates are statistically reliable and representative of the population, thus avoiding underfitting.

The energy consumption assessment is performed using Autodesk Revit and Green Building Studio (GBS) software (https://gbs.autodesk.com/gbs). Revit has been used to develop the building model while GBS provides energy analysis and consumption optimization. The framework for the evaluation of energy assessment and decision making is illustrated in Figure 7.

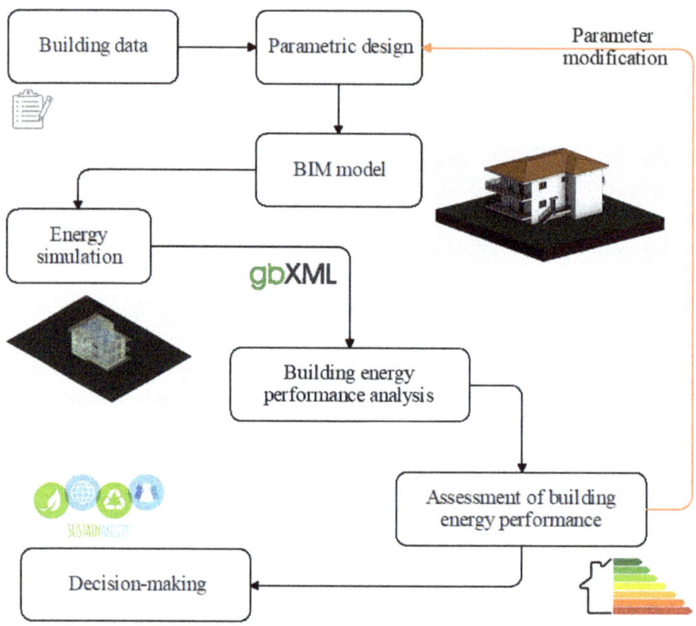

Figure 7. Process flow diagram for data generation.

6. Case Study

6.1. Input Data

A typical residential building is analyzed as part of the case study. The structure consists of two or four apartments, as per the architectural form in Figure 8, over one or two floors. Building type A includes two apartments each with an area of 212 m^2 that are placed either side by side (single floor) or on top of each other (two-floor configuration). Building type B includes four apartments each with an area of 212 m^2 that are placed either side by side (single floor) or on top of each other (two-floor arrangement). The exact building formations are shown in Figure 9.

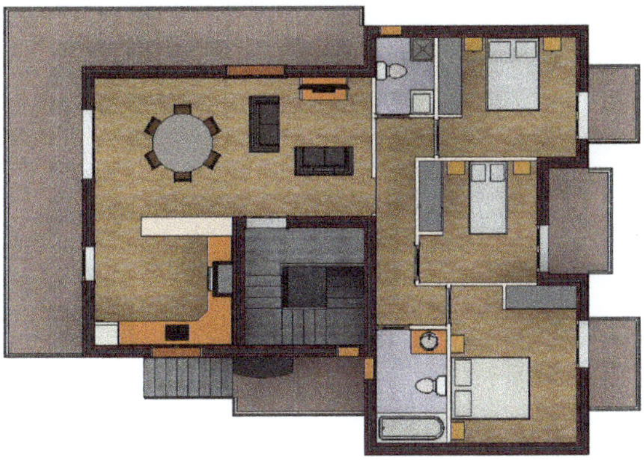

Figure 8. BIM architectural model of a single apartment.

(**a**)

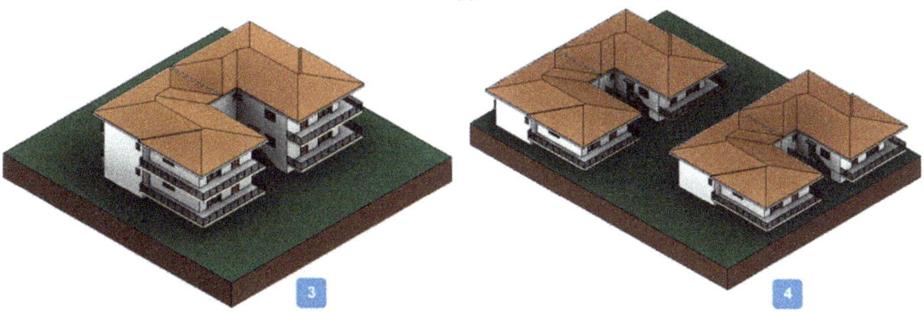

(**b**)

Figure 9. (**a**) Digital 3D/BIM model of type A buildings with one or two floors. (**b**) Digital 3D/BIM model of type B buildings with one or two floors.

Each apartment is a typical four-person family flat. Rooms are cooled and heated via an HVAC system. Rooms are classified into three common typical thermal zones. The first includes the kitchen and the living room, the second comprises the bedrooms, and the third the corridors and bathrooms. Two thermostats—one for heating and one for cooling—are considered. The set points for heating and cooling are at 24 °C and at 21 °C, respectively. The electric loads (for lighting and equipment) are determined based on the number of residents and defined as Watt (W) per square meter (m^2).

Regarding the location parameter (X7), four different climatic zones are considered, based on ambient conditions in Greece and represented by four cities, as shown in Figure 10. These cities are Chania, Patras, Ioannina, and Florina with average annual temperatures of 17.2 °C, 15.5 °C, 12.8 °C, and 10.4 °C, respectively (https://en.climate-data.org).

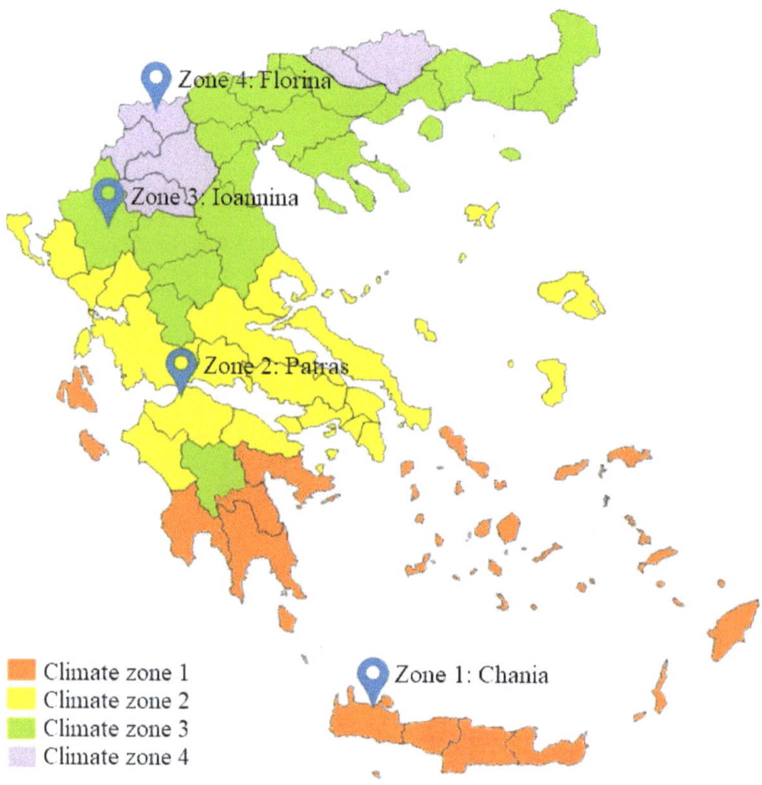

Figure 10. Climatic zones of representative cities in Greece.

To determine the impact of each influencing factor, a default parameter setting is considered. Each factor is then altered to another value, and the effect on energy consumption cost is computed. A bar graph that represents the effect of each factor on the annual energy cost is estimated for each building type (Figure 11). The graph displays the percentage change in energy consumption per each factor when shifting from the default to a new value. The default parameter values are shown in the graph legend. Positive values indicate an increase and negative values indicate a reduction in energy consumption. The annual cost estimate is based on a unit energy cost of 0.12 EUR/KWh.

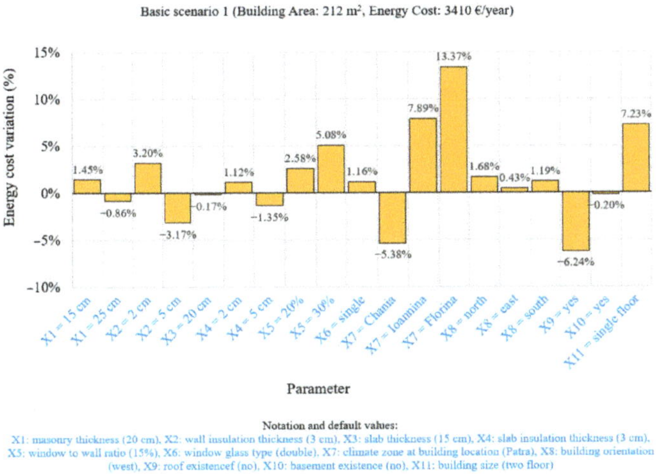

(a)

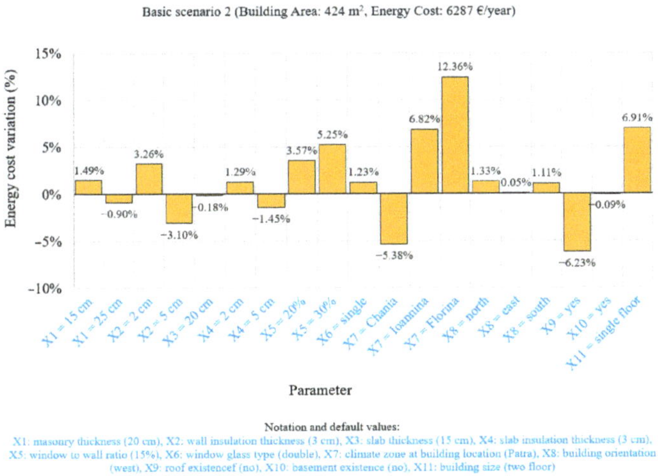

(b)

Figure 11. (**a**) Effect of design parameters for building type A. (**b**) Effect of design parameters for building type B.

The main observations from Figure 11 are as follows:

- The overall view of the two diagrams indicates that individual factors play a similar role in terms of energy consumption or savings in both building types. A rather flat proportion of 1.85 to 1.00 in terms of total energy consumption is observed in these two cases. This is as expected, as type B buildings are more environment-proof, being more condensed than type A buildings. This outcome may highlight an energy performance pattern for buildings of larger sizes.
- The highest influence is observed in terms of the location and the corresponding climatic conditions (parameter X7). For instance, moving from zone 2 (Patras) to zone 4 (Florina), the energy consumption is increased by 13.37% and 12.36% for building types A and B, respectively.

- The presence of adjacent buildings has a positive effect on energy efficiency. This is observed by comparing building configurations 2 and 4 in Figure 9, with the energy savings in the latter case calculated to be 8.80%.
- The energy consumption is increased in one-story compared to two-story structures (parameter X11) by 7.23% and 6.91% for building types A and B, respectively.
- The existence of a roof (parameter X9) within the basic scenario results in energy savings of 6.24% and 6.23% for building types A and B, respectively.
- The window-to-wall ratio (parameter X5) increases from 15% (in the basic scenario) to 30%, leading to increases in energy consumption of 5.08% and 5.25% for building types A and B, respectively.
- Likewise, the thickness variation in the external wall insulation (parameter X2) from one level to another has an influence of about 3% in all cases. The wall thickness itself plays a marginal role in all cases.
- All other parameters have a contribution of less than 2% to energy consumption.

6.2. Models of Energy Performance Prediction

Equation (2) provides the best-fitting relationship for estimating annual energy consumption as a function of the eleven independent factors, as derived from the fractional polynomials in the statistical regression model (SRM).

$$\begin{aligned} y = &-0.12\ X1 - 0.52\ X2 + 0.05\ X3 - 0.16\ X4 + 3.72\ X5 - 0.31\ X6 \\ &+1.06\ X7 - 0.99\ X8 + 0.98\ X9 - 0.10\ X10 - 1.10\ X11 + 21.30 \end{aligned} \quad (2)$$

Figure 12 presents an illustrative part (due to size limitations) of the developed decision tree (DT) model in a self-explanatory graphical form. The random forest (RF) and the neural network (NN) cannot be easily rendered in a graphical way due to their size. The RF consists of multiple decision trees, which makes it difficult to visualize the entire model. The NN structure is rather standardized and globally known. Nevertheless, typical structures of these models are presented earlier in Figures 5 and 6, respectively. The DT model includes 7 levels and 24 end nodes, with part of it being displayed in Figure 12. The RF model consists of 5 individual decision trees. The ANN model is structured upon 11 input nodes, 11 intermediate nodes, and 168 variables in total.

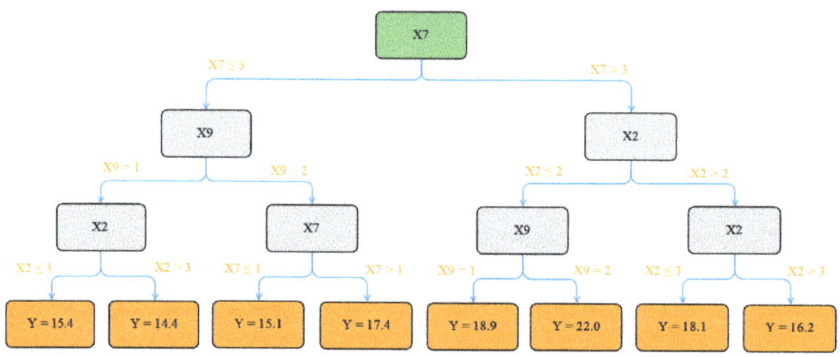

Figure 12. Partial graphic illustration of the DT model.

Table 2 presents the energy consumption estimates, which are derived from the four models for indicative datasets, and the corresponding reference values, which are derived from the BIM model. The results indicate that the model output values do not greatly deviate from each other or from the corresponding reference values. To obtain some quantitative measures, the deviations from the associated reference values are in the ranges of 0.51% to 10.6%, 0.56% to 2.15%, 0.21% to 4.88%, and 0.14% to 1.65% for the SRM, DT, RF, and NN models, respectively. The results highlight the ML model as the best-performing one.

Table 2. Comparison of results of alternative methods.

	Annual Energy Consumption (EUR/m^2)				
	Model-SRM	**Model-DT**	**Model-RF**	**Model-ANN**	**BIM**
1	19.00	18.52	18.39	18.12	18.15
2	15.02	14.49	14.30	14.31	14.33
3	17.45	17.26	17.84	17.14	17.01
4	16.57	16.12	15.92	16.04	15.78
5	17.84	17.39	17.28	17.52	17.75
6	16.77	16.12	15.88	15.92	16.03
7	20.50	22.78	22.19	23.30	22.93
8	17.75	17.37	17.32	17.48	17.27
9	13.26	13.72	14.02	13.37	13.44
10	19.20	18.52	18.71	18.76	18.79

6.3. Model Performance Assesment

Statistical indicators (RMSE, PMRE, and correlation coefficient) are used to evaluate the dependencies between machine learning model output and actual observations. Further, the F-test is a statistical criterion to determine whether model results fit the observed ones. The corresponding mathematical relationships are given in Equations (3)–(5).

$$\text{RMSE} = \sqrt{\frac{1}{n} \cdot \sum_{1}^{n} (x_i - y_i)^2} \tag{3}$$

$$\text{PMRE} = \frac{100}{n} \cdot \sum_{1}^{n} \left| \frac{x_i - y_i}{x_i} \right| \tag{4}$$

$$\text{R}^2 = \left(\frac{\sum_{1}^{n} (x_i - x_{mean}) \cdot (y_i - y_{mean})}{\sqrt{\sum_{1}^{n} (x_i - x_{mean})^2} \cdot \sqrt{\sum_{1}^{n} (y_i - y_{mean})^2}} \right)^2 \tag{5}$$

where x_i is the observed value, y_i is the predicted value, x_{mean} and y_{mean} are the corresponding mean values, and n is the sample size.

For in-depth performance appraisal of the developed models (SRM, DT, RF, and ANN), appropriate statistical analysis indexes are computed (Table 3). They include the root mean square error (RMSE), the percent mean relative error (PMRE), and the correlation coefficient (R^2). The F-test is further used for contrasting the goodness of fit between the observed and the estimated values. The results indicate that the decision tree (DT), random forest (RF), and artificial neural network (ANN) models exhibit improved performance compared to the statistical regression model (SRM), as indicated by the error indices (RMSE and PMRE) and the correlation coefficient (R2). Among pure ML methods, the ANN model presents the best performance and satisfies the F-test in every dataset. The other two methods (DT and RF) partially satisfy the F-test only in the validation and testing datasets.

More specifically, the RMSE of the ANN model output ranges from 0.1278 to 0.2759 for individual data subsets and is 0.1840 for the full dataset. On the other hand, the SRM model presents much higher RMSE values in the order of 0.7514 and 0.8525 in the data subsets and 0.8329 in the full dataset. Similarly, the PMRE values for the ANN model are between 0.54% and 1.23%, while for the SRM, they are from 3.63% to 3.78%. Finally, the F-test is fully satisfied in all tests of the ANN model, partly satisfied in the cases of the DT and RF models (in validation and training subsets), and not satisfied in any dataset of the SRM model. Interestingly, the performance in the validation and testing subsets occasionally appears to be higher than that resulting from the training subset. This effect may be attributed to the randomness of data segmentation in the three subsets. The observed peculiarity is rather insignificant as the model characteristics are not considerably affected. Additionally, this is an indication that the model does not overfit. The superior performance of ML models is due to their inherent ability to model nonlinear relationships and complex interactions among input parameters. Further, ML models hold the ability to adapt to data of different types and sources, which makes them suitable for use in problems with high complexity.

Table 3. Model performance indicators.

Model	Root Mean Square Error (RMSE)	Percent Mean Relative Error (PMRE)	Correlation Coefficient (R2)	F-Test
	All dataset			$F_{2,335,0.05} = 3.03$
SRM	0.8329	3.74%	0.8810	171.55
DT	0.4930	2.17%	0.9460	7.06
RF	0.2722	1.15%	0.9836	5.46
ANN	0.1840	0.75%	0.9925	2.71
	Training			$F_{2,233,0.05} = 3.04$
SRM	0.8525	3.76%	0.8797	119.56
DT	0.4822	2.02%	0.9498	6.70
RF	0.2302	0.96%	0.9888	5.51
ANN	0.1278	0.54%	0.9965	1.60
	Validation			$F_{2,49,0.05} = 3.19$
SRM	0.8190	3.78%	0.8190	23.43
DT	0.5455	2.63%	0.9341	1.15
RF	0.3744	1.74%	0.9688	0.20
ANN	0.2759	1.19%	0.9820	1.01
	Testing			$F_{2,49,0.05} = 3.19$
SRM	0.7514	3.63%	0.7514	28.22
DT	0.4871	2.40%	0.9400	2.34
RF	0.3241	1.43%	0.9727	1.54
ANN	0.2692	1.23%	0.9820	0.14

The model performance comparison can be further visualized in graphical form as in Figures 13–16. These figures illustrate the comparison of the model output with the reference (target) values that have been produced by BIM simulation. The main diagrams show the range and dispersion of the dataset output and target points in relation to the diagram diagonal line, which indicates full convergence. The histograms above and to the right show the distribution of target and model values within their respective ranges. In an ideal convergence case, the two histograms would retain the same distribution. The multi-faceted presentation facilitates a better understanding and interpretation of the results, highlighting the accuracy and effectiveness of the proposed models. These diagrams reveal that, in general, there is an acceptable level of convergence between the observed and predicted values with all models. In a more detailed assessment, the ANN (mainly) and the RF present a stronger relationship between the two set values. Finally, the error distributions of the four prediction methods are shown in Figure 17. The ANN model presents the best performance with zero average values and low standard deviation. Instead, other method distributions shift to non-zero average values and present higher standard deviation values.

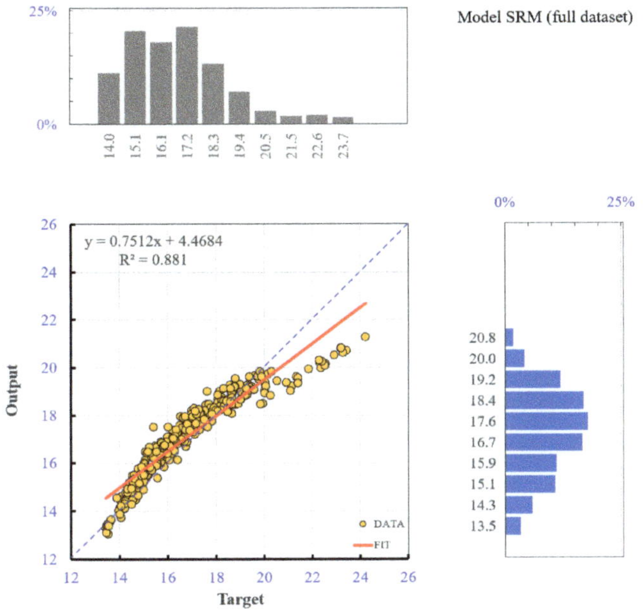

Figure 13. Convergence of observed and predicted values by the SRM.

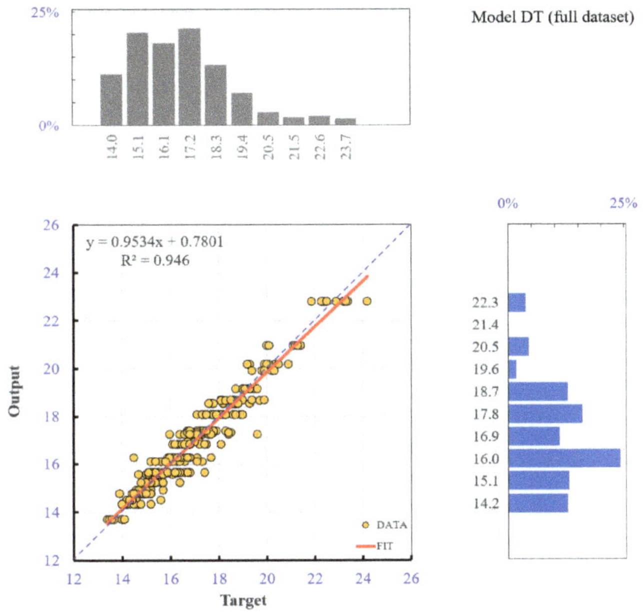

Figure 14. Convergence of observed and predicted values by the DT model.

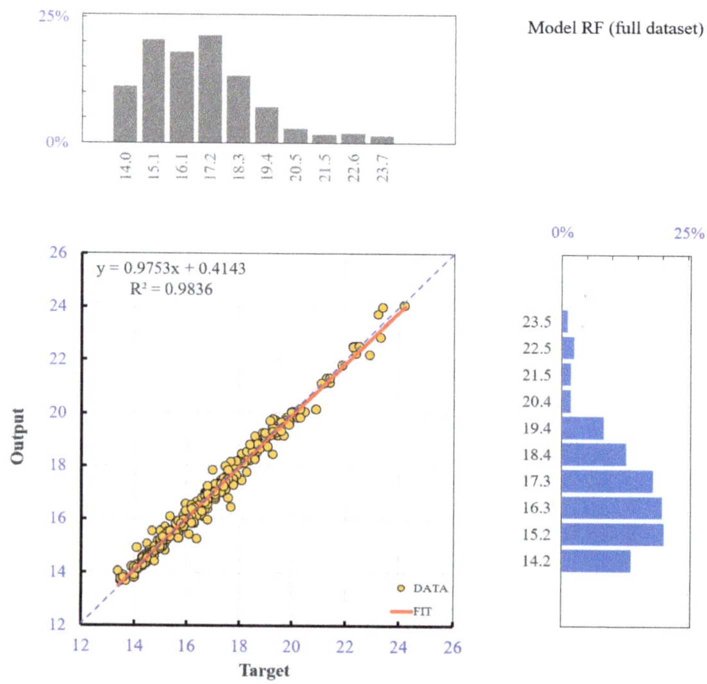

Figure 15. Convergence of observed and predicted values by the RF model.

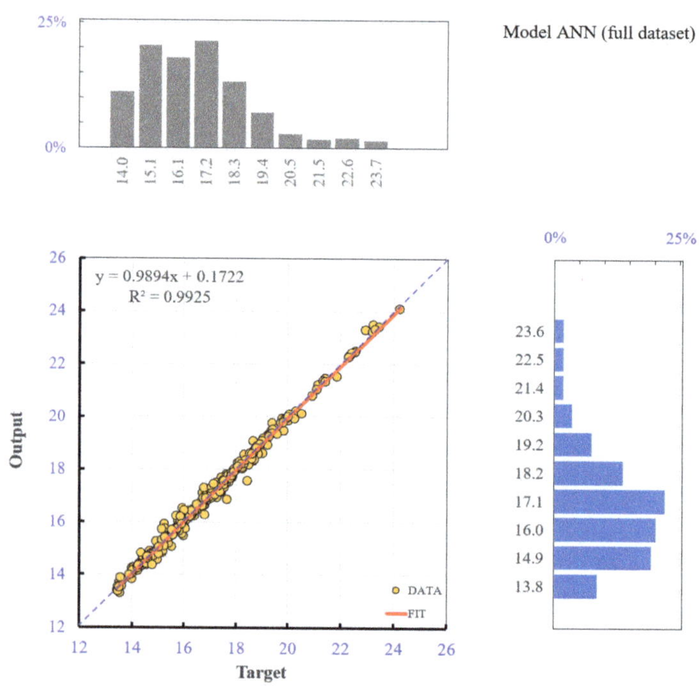

Figure 16. Convergence of observed and predicted values by the ANN model.

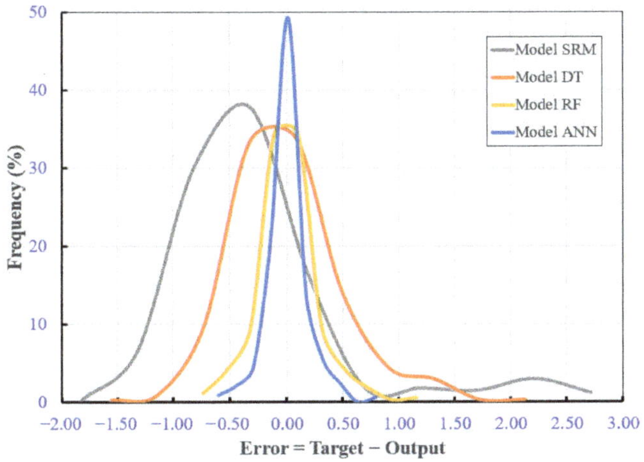

Figure 17. Error distribution between target values from BIM simulation and output from developed models.

The research development and the case study results indicate that the proposed methodology can provide energy cost estimates easily and effectively at the very early design stage of typical residential buildings. Additionally, development and testing for specific building types and characteristics have been performed. The experiment shows that this research, in its current form, can have notable potential for scalability and can be extended to different building structures, places, and environments. More importantly, the adoption of BIM provides unlimited capability to develop energy models for any type and size of buildings, design characteristics, environmental factors, etc. Another important aspect is that building energy performance is highly affected by human-related factors, such as occupant number and behavior in relation to energy usage patterns, or operational and maintenance practices, which exhibit significant variability and uncertainty in energy consumption. Although such parameters may be marginally known at the early design phase, they may be incorporated in the modeling in a parametric or stochastic way. The above issues can be investigated as part of future research.

As a final note, the results of this work are qualitatively consistent with those in the literature. However, a comparative quantitative analysis that would enable meaningful comparisons with and evaluations using previous models and research works is not feasible, as the individual studies' data structures are not fully comparable. Nevertheless, the evaluation of ML model performance is universally performed through the training, testing, and validation processes, which also depend on the size and representativeness of the data used.

In summary, the proposed methodology integrates a BIM simulation, parametric evaluation, and automated estimation of building energy consumption with a simple and effective GUI. The outcome of this effort is the development of ML models that can assess the energy consumption of buildings with distinct characteristics with various levels of performance. These models can effectively assist building designers and researchers in making decisions. The proposed methodology has the advantage of developing energy performance models without the need for the existence or collection of actual data. Instead, it relies on simulated data that can be extracted by consistent BIM-based simulations. Although the current development is confined to a (common) class of residential buildings, the above capability can highly support the repeatability, scalability, and extensibility aims of model development within a practical range of application. In the absence of such models, each individual building design should undergo an energy analysis of several potential solutions based on a multi-parameter evaluation. Another contribution is the development of a simple-to-use GUI (which is described in the next section) for obtaining energy consumption estimates at the early design stage, without the need for comprehensive building modeling and wide energy analyses of alternative building design configurations.

7. Graphical User Interface

The study has developed a method for estimating the long-term energy performance of buildings. BIM technology has been elaborated for the simulation of energy consumption analysis and machine learning techniques to predict building energy efficiency. A drawback of machine learning techniques is their unsuitability for real-life problems. To overcome this obstacle, a user-interface tool has been developed to facilitate the model application (Figure 18). In this application, the user simply inserts the appropriate design parameter values and presses the "Calculation" button. An instant energy consumption calculation is provided based on the selected parameters without any other manual input or calculation.

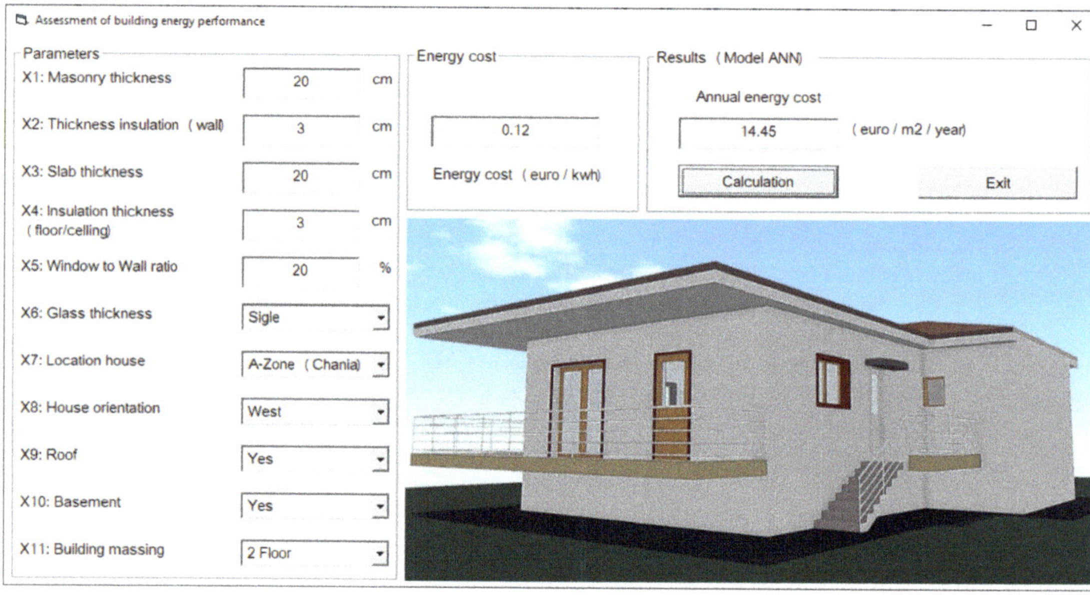

Figure 18. Graphical user interface for energy consumption estimation.

The integration of the trained artificial neural network (ANN) into the graphical user interface (GUI) has been performed using the synaptic weights of the developed model. The ANN model includes 168 parameters, representing the network weight and bias factors. The model has been implemented in Visual Basic programming language, which is suitable for interface development.

8. Conclusions

The energy performance of buildings has become a main concern globally in view of increased energy demand, the environmental impacts of energy production, and the reality of energy poverty in many areas. To increase building energy efficiency, proper building design should be implemented starting at the early design phase or when retrofitting decisions regarding existing buildings need to be made. Digital tools are currently available for performing energy assessment analyses in the form of building information modeling (BIM) and building energy modeling (BEM), and they can efficiently handle complex and technically demanding buildings. However, alternative designs should be subject to individual analyses, and this makes the process time-consuming and prone to errors. Machine learning techniques, on the other hand, can ameliorate such deficiencies and provide effective and easy-to-use tools for such analyses.

In this paper, representative residential buildings are considered along with eleven factors that affect energy performance. A dataset of 337 instances of such parameters is developed. For each dataset, the building energy performance is estimated based on a BIM simulation. Next, statistical and machine learning techniques are implemented to provide artificial models of energy performance. They include statistical regression modeling (SRM), decision trees (DTs), random forests (RFs), and artificial neural networks (ANNs). The above models are appropriately trained, tested, and validated using the dataset.

The case study results reveal that building energy consumption is mainly affected by the local ambient conditions, the number of building floors (one or two), the existence of adjacent buildings, the presence of a roof, and the window-to-wall ratio. Regarding the prediction models, the evaluation results from the case study indicate that all artificial models present satisfactory (for practical use at the early design phase) levels of convergence to the observed values. The comparative evaluation highlights the ANN as the best-performing model, with high statistical performance indicator values. Calculating the modeling energy efficiency of buildings via machine learning techniques appears to be an attractive approach to overcome the computational cost of creating energy efficiency simulations. Further, to enhance the model's applicability, a user interface tool has been developed into which the design parameters of the eleven influencing factors are inserted and the energy cost is instantly calculated.

The experiment in the case study indicates that the proposed models can be directly extensible and scalable, at least within a reasonable parameter extrapolation range. For out-of-range cases, the process of developing the required dataset and training machine learning models can be presumably extended to any type and size of building, along with the rest of the influencing parameters. In a similar path, the development can incorporate human-related factors, which highly affect energy consumption.

The proposed methodology integrates BIM simulations, parametric evaluations, and automated estimations of building energy consumption with a simple and effective GUI. In particular, it provides a useful and handy tool for energy consumption estimations of buildings with varying characteristics without the need for acquiring and using actual data, as it develops these data via a BIM simulation. Another contribution is the development of a simple and easy-to-use GUI for calculating the energy consumption and comparatively assessing the energy performance of alternative building design configurations.

Author Contributions: Conceptualization, P.T. and A.C.; methodology, P.T. and A.C.; software, P.T.; validation, P.T., V.P. and A.P.; formal analysis, P.T., A.C., V.P. and A.P.; investigation, P.T., V.P. and A.P.; resources, A.C.; data curation, P.T., V.P. and A.P.; writing—original draft preparation, P.T., V.P. and A.P.; writing—review and editing, P.T., A.C., V.P. and A.P.; visualization, P.T., A.C., V.P. and A.P.; supervision, A.C.; project administration, P.T. and A.C. All authors have read and agreed to the published version of the manuscript.

Funding: This research received no external funding.

Data Availability Statement: The data that were developed and used in this study will be made available by the authors upon request.

Conflicts of Interest: The authors declare no conflicts of interest.

References

1. International Energy Agency. *2019 Global Status Report for Buildings and Construction*; International Energy Agency: Paris, France, 2019; ISBN 9789280736861.
2. Mauro, G.M.; Menna, C.; Vitiello, U.; Asprone, D.; Ascione, F.; Bianco, N.; Prota, A.; Vanoli, G.P. A Multi-Step Approach to Assess the Lifecycle Economic Impact of Seismic Risk on Optimal Energy Retrofit. *Sustainability* **2017**, *9*, 989. [CrossRef]
3. Pohoryles, D.A.; Maduta, C.; Bournas, D.A.; Kouris, L.A. Energy Performance of Existing Residential Buildings in Europe: A Novel Approach Combining Energy with Seismic Retrofitting. *Energy Build.* **2020**, *223*, 110024. [CrossRef]
4. Yoshino, H.; Hong, T.; Nord, N. IEA EBC Annex 53: Total Energy Use in Buildings—Analysis and Evaluation Methods. *Energy Build.* **2017**, *152*, 124–136. [CrossRef]
5. Gao, H.; Koch, C.; Wu, Y. Building Information Modelling Based Building Energy Modelling: A Review. *Appl. Energy* **2019**, *238*, 320–343. [CrossRef]
6. Reeves, T.; Olbina, S.; Issa, R.R.A. Guidelines for Using Building Information Modeling for Energy Analysis of Buildings. *Buildings* **2015**, *5*, 1361–1388. [CrossRef]
7. Habibi, S. The Promise of BIM for Improving Building Performance. *Energy Build.* **2017**, *153*, 525–548. [CrossRef]
8. Wang, X.; Teigland, R.; Hollberg, A. Identifying Influential Architectural Design Variables for Early-Stage Building Sustainability Optimization. *Build. Environ.* **2024**, *252*, 111295. [CrossRef]
9. Azhar, S.; Carlton, W.A.; Olsen, D.; Ahmad, I. Building Information Modeling for Sustainable Design and LEED® Rating Analysis. *Autom. Constr.* **2011**, *20*, 217–224. [CrossRef]
10. Albdour, M.S.; Shalby, M.; Salah, A.A.; Alhomaidat, F. Evaluating and Enhancing the Energy Efficiency of Representative Residential Buildings by Applying National and International Standards Using BIM. *Energies* **2022**, *15*, 7763. [CrossRef]
11. Tahmasebinia, F.; Jiang, R.; Sepasgozar, S.; Wei, J.; Ding, Y.; Ma, H. Using Regression Model to Develop Green Building Energy Simulation by BIM Tools. *Sustainability* **2022**, *14*, 6262. [CrossRef]
12. Li, X.; Lin, M.; Jiang, M.; Jim, C.Y.; Liu, K.; Tserng, H. Bim and Orthogonal Test Methods To Optimize the Energy Consumption of Green Buildings. *J. Civ. Eng. Manag.* **2024**, *30*, 670–690. [CrossRef]
13. Sajjad, M.; Hu, A.; Alshehri, A.M.; Waqar, A.; Khan, A.M.; Bageis, A.S.; Elaraki, Y.G.; Shohan, A.A.A.; Benjeddou, O. BIM-Driven Energy Simulation and Optimization for Net-Zero Tall Buildings: Sustainable Construction Management. *Front. Built Environ.* **2024**, *10*, 1296817. [CrossRef]
14. Fan, C.; Xiao, F.; Zhao, Y. A Short-Term Building Cooling Load Prediction Method Using Deep Learning Algorithms. *Appl. Energy* **2017**, *195*, 222–233. [CrossRef]
15. Ahmad, T.; Chen, H.; Guo, Y.; Wang, J. A Comprehensive Overview on the Data Driven and Large Scale Based Approaches for Forecasting of Building Energy Demand: A Review. *Energy Build.* **2018**, *165*, 301–320. [CrossRef]
16. Amasyali, K.; El-Gohary, N.M. A Review of Data-Driven Building Energy Consumption Prediction Studies. *Renew. Sustain. Energy Rev.* **2018**, *81*, 1192–1205. [CrossRef]
17. Yu, J.; Chang, W.S.; Dong, Y. Building Energy Prediction Models and Related Uncertainties: A Review. *Buildings* **2022**, *12*, 1284. [CrossRef]
18. Zhou, Y.; Wang, L.; Qian, J. Application of Combined Models Based on Empirical Mode Decomposition, Deep Learning, and Autoregressive Integrated Moving Average Model for Short-Term Heating Load Predictions. *Sustainability* **2022**, *14*, 7349. [CrossRef]
19. Li, G.; Wu, Y.; Liu, J.; Fang, X.; Wang, Z. Performance Evaluation of Short-Term Cross-Building Energy Predictions Using Deep Transfer Learning Strategies. *Energy Build.* **2022**, *275*, 112461. [CrossRef]
20. Zhang, M.; Millar, M.-A.; Chen, S.; Ren, Y.; Yu, Z.; Yu, J. Enhancing Hourly Heat Demand Prediction through Artificial Neural Networks: A National Level Case Study. *Energy AI* **2024**, *15*, 100315. [CrossRef]
21. Olu-Ajayi, R.; Alaka, H. Building Energy Consumption Prediction Using Deep Learning. *Environ. Des. Manag. Conf.* **2021**, *202*, 1.
22. Khajavi, H.; Rastgoo, A. Improving the Prediction of Heating Energy Consumed at Residential Buildings Using a Combination of Support Vector Regression and Meta-Heuristic Algorithms. *Energy* **2023**, *272*, 127069. [CrossRef]
23. Javanmard, M.E.; Ghaderi, S.F. Energy Demand Forecasting in Seven Sectors by an Optimization Model Based on Machine Learning Algorithms. *Sustain. Cities Soc.* **2023**, *95*, 104623. [CrossRef]
24. Lu, C.; Li, S.; Penaka, S.R.; Olofsson, T. Automated Machine Learning-Based Framework of Heating and Cooling Load Prediction for Quick Residential Building Design. *Energy* **2023**, *274*, 127334. [CrossRef]

25. Wang, G.; Mukhtar, A.; Moayedi, H.; Khalilpoor, N.; Tt, Q. Application and Evaluation of the Evolutionary Algorithms Combined with Conventional Neural Network to Determine the Building Energy Consumption of the Residential Sector. *Energy* **2024**, *298*, 131312. [CrossRef]
26. Papada, L.; Kaliampakos, D. Artificial Neural Networks as a Tool to Understand Complex Energy Poverty Relationships: The Case of Greece. *Energies* **2024**, *17*, 3163. [CrossRef]
27. Maglad, A.M.; Houda, M.; Alrowais, R.; Khan, A.M.; Jameel, M.; Rehman, S.K.U.; Khan, H.; Javed, M.F.; Rehman, M.F. Bim-Based Energy Analysis and Optimization Using Insight 360 (Case Study). *Case Stud. Constr. Mater.* **2023**, *18*, e01755. [CrossRef]
28. Tsanas, A.; Xifara, A. Accurate quantitative estimation of energy performance of residential buildings using statistical machine learning tools. *Energy Build.* **2012**, *49*, 560–567. [CrossRef]
29. Ibrahim, M.; Ghaddar, N.; Ghali, K. Optimal Location and Thickness of Insulation Layers for Minimizing Building Energy Consumption. *J. Build. Perform. Simul.* **2012**, *5*, 384–398. [CrossRef]
30. Banihashemi, S.; Ding, G.; Wang, J. Developing a Hybrid Model of Prediction and Classification Algorithms for Building Energy Consumption. *Energy Procedia* **2017**, *110*, 371–376. [CrossRef]
31. StataCorp. *Stata User Guide*; Muthén & Muthén: Los Angeles, CA, USA, 2021; ISBN 9781597181709.
32. Huang, W.; Martin, P.; Zhuang, H.L. Machine-Learning Phase Prediction of High-Entropy Alloys. *Acta Mater.* **2019**, *169*, 225–236. [CrossRef]
33. Green, S.B. How Many Subjects Does It Take To Do A Regression Analysis. *Multivar. Behav. Res.* **1991**, *26*, 499–510. [CrossRef] [PubMed]
34. Babyak, M.A. What You See May Not Be What You Get: A Brief, Nontechnical Introduction to Overfitting in Regression-Type Models. *Psychosom. Med.* **2004**, *66*, 411–421.

energies

MDPI

Article

CFD and Statistical Analysis of the Impact of Surface Physical Parameters on the Thermal Resistance of Layered Partitions in ETICS Systems

Arkadiusz Urzędowski [1,*], Andrzej Sachajdak [2], Arkadiusz Syta [1] and Jacek Zaburko [1]

[1] Faculty of Mathematics and Information Technology, Lublin University of Technology, Nadbystrzycka 38, 20-618 Lublin, Poland; a.syta@pollub.pl (A.S.)

[2] Department of Thermal Technology, Silesian University of Technology, Konarskiego 22, 44-100 Gliwice, Poland; andrzej.sachajdak@polsl.pl

* Correspondence: a.urzedowski@pollub.pl; Tel.: +48-690006777

Abstract: In the article, the authors attempted to analyze the impact of such materials factors as surface emissivity, surface roughness, air gap thickness, and type of concrete on heat transport in the microstructure of vertical multilayer building walls. The surface analysis conducted using three-dimensional modeling tools provided information about the formation of its microstructure before and after the application of a reflection-smoothing coating, which has a direct impact on the emissivity of the surface and was reduced from 0.93 to 0.29. Thermal analyses demonstrated that after applying the reflective coating, thermal resistance increased significantly in the air gap, by approximately 86%, which resulted in a 28% improvement of the evaluated walls samples. The studies have shown that increasing the gap thickness between concrete and thermal insulation results in a thermal resistance increase. It is feasible to enhance the thermal insulation of walls while simultaneously reducing their thickness, a development that holds significant potential for application in the production of prefabricated sandwich panels. The statistical analyzes performed showed significant differences between the analyzed configurations.

Keywords: heat transfer in walls; CFD; surface emissivity

1. Introduction

A key issue in improving the thermal insulation of buildings is the reduction of heat loss through the building envelope, which accounts for up to 25% for walls and 30% for roofs [1,2]. The mechanisms of heat transfer by conduction in single- and multi-layer walls are well-studied, investigated, and described [3,4]. However, the complex mechanism of heat transfer between rough surfaces, involving simultaneous conduction, convection, and radiation, requires further analysis [5]. Understanding heat transfer phenomena in building structures is a critical aspect from the perspective of energy efficiency and durability. Analyzing these phenomena enables the optimization of construction and insulation methods for new buildings and the improvement of retrofitting techniques for existing structures [6].

The issue of heat transfer in multilayer partitions is further complicated by their porous and heterogeneous structure. The heat transfer mechanisms in civil engineering structures described to date have not accounted for the influence of microstructure in gap regions on the contact thermal resistance of layered constructions [7–11].

This study addresses an issue not yet explored in the existing scientific literature, specifically involving the determination of the heat transfer coefficient, with a particular focus on calculating thermal contact resistance values with varying surface preparations on the side of air gaps in walls. Conducting a geometric and thermal analysis of the considered materials and their contact area provided insights into the actual thermal conditions and their impact on the insulation parameters of the building partition. In order to carry out research, authors used tools for numerical fluid mechanics, which was preceded by the construction of three-dimensional models reflecting the multi-layer building walls working conditions. Reverse engineering tools were used to reproduce the surface shape, before and after the application of reflective smoothing coating. The research focused on change in heat transport by radiation and thermal resistance in air gaps between materials and was carried out according to the chart presented in Figure 1.

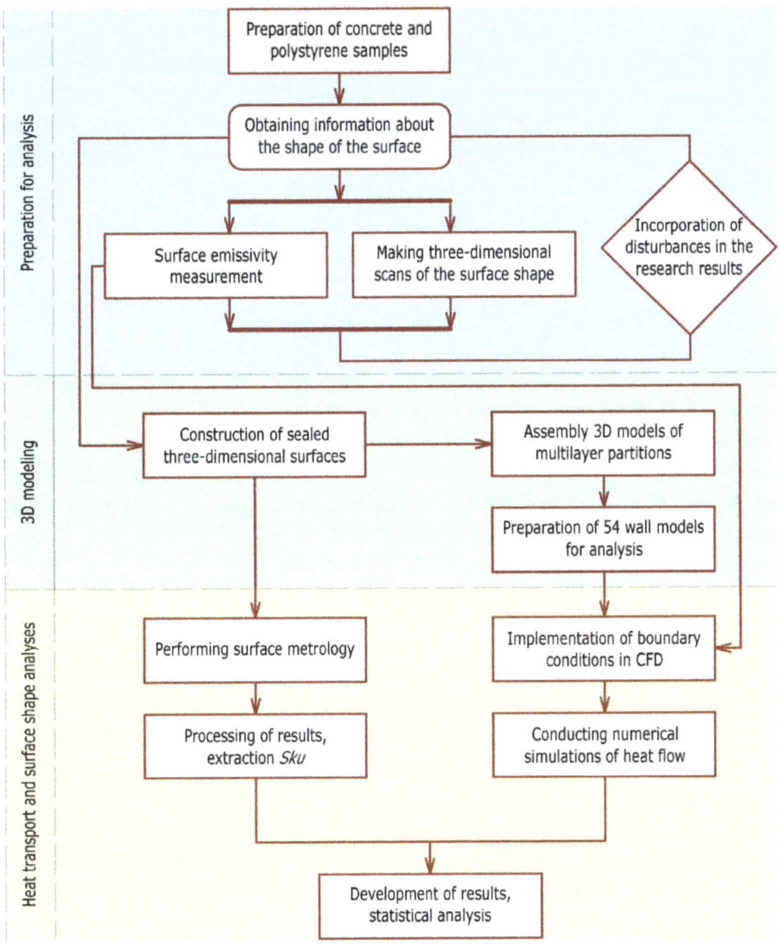

Figure 1. Research flow chart.

1.1. Thermophysical Phenomena Occurring in Multilayer Walls

In the case of the temperature distribution at specific points in a body determined solely by position T = f(x,y,z), it is referred to as steady-state heat transfer. However, if the temperature also depends on time T = f(x,y,z,τ), transient heat transfer occurs. In practice, unsteady heat transfer typically occurs in building partitions. Nevertheless, due to relatively slow in time changes of thermal and moisture conditions in typical building partitions, steady-state conduction is most often assumed. The heat flux flows exclusively in an environment with a temperature difference, from areas of higher temperature to areas of lower temperature, until equilibrium is reached [12].

Heat transfer, depending on the type of medium, can occur through [13]:

- Conduction—characteristic of solids, it is described by Fourier's law, which relates the heat flux density at a specific point in a body to the temperature gradient at that point (1):

$$q = -\lambda \nabla T \qquad \left[\mathrm{W/m^2}\right] \tag{1}$$

- Convection—characteristic of liquids and gases, the solution for fluid region bases on the continuity equation, the momentum conservation equation, and the energy equation, when heat is added to a fluid and the fluid density varies with temperature, a flow can be induced due to the force of gravity acting on the density variations. Such buoyancy-driven flows are termed natural-convection (or mixed-convection) (2)–(4) [14,15].

$$\nabla \cdot (\rho v) = 0 \tag{2}$$

$$\nabla \cdot (\rho v v) = -\nabla p + \nabla \bar{\bar{\tau}} + \rho g \tag{3}$$

$$\nabla \cdot (v \rho h) = \nabla \cdot (\lambda \nabla T + \bar{\bar{\tau}} v) \tag{4}$$

- Radiation—characteristic of solids in a gaseous medium. In this article, a two-layer vertical wall consisting of a structural layer insulated with polystyrene is analyzed, where heat transfer is realized through all three modes of exchange: conduction in the structural layer, thermal insulation and adhesive, as well as convection and radiation in the air gap [12,14], where the amount of heat exchanged by radiation from the surface to the surface (heat flux) can be described by the Formula (5):

$$Q_{1-2} = \varepsilon_{1-2} C_C \varphi_{1-2} F_1 \left[\left(\frac{T_1}{100} \right)^4 - \left(\frac{T_2}{100} \right)^4 \right] \tag{5}$$

The thermal resistance of joints in building walls depends on several factors, including inter alia (the value of the temperature of the bodies in the contact zone, the value of the temperature of the cores of the bodies, the roughness and waviness of the surface, and the size of the joints [16–18]). In the analyzed work, the effect of changing the distance between concrete and thermal insulation was examined, which results from the technology of thermal insulation of buildings in the ETICS system [19,20]. An important role in heat transport in the air gaps is played by the emissivity of the surface, which can be modified by applying reflective and smoothing coatings used in the experiments presented in the article.

1.2. Problem Formulation

The European Union committed itself to reducing the Union's economy-wide net greenhouse gas emissions by at least 55% by 2030 below 1990 levels in the updated nationally determined contribution submitted to the UNFCCC Secretariat on 17 December 2020 [21], whereas by 2050, all buildings should meet zero-energy requirements. In recent years, numerous publications have highlighted challenges related to improving the thermal performance of building envelopes to meet these requirements [22–24].

Thermal studies are being conducted on innovative materials [25], windows [26], heat transfer within multilayer walls [27] and at the joints of building envelopes [28]. Very detailed studies are conducted leading to even slight improvements, among other, research on wall corners insulated with PIR boards with aluminum facing, demonstrated that thermal coupling coefficient (L2D), with PU glue is equal to 0.2834 W/(mK), whereas without PU glue: 0.2948 W/(mK) [29]. Research on phenomena occurring in air gaps is particularly challenging due to the complexity of the processes involved, hence numerical techniques are applied to conduct analyses [30,31]. Air gap studies carried out by Saber H. H., have shown that for single and double airspaces subjected to an upward heat flow, the effective thermal resistance changed significantly with a changing A_R (enclosed region aspect ratio) for the full range of effective emittance of the enclosed airspace [22]. However, previous studies on heat transfer in multilayer walls [32–34] have not accounted for thermal phenomena occurring in air gaps after modifications to the structural layer and changes in its emissivity. Authors of this article conducted research to solve and describe these issues.

To prepare the digital layered wall models, materials were collected that accurately represent the actual heat transfer phenomena occurring in vertical external building. The selected fragment represents a wall made of two layers: structural material in the form of concrete 0.24 m thick and insulating material in the form of polystyrene 0.20 m thick, which is mounted to the wall with adhesive mortar. The study did not take into account the plaster on the outside and inside of the wall, because its share is insignificant in the heat transport, it is difficult to clearly determine its thickness and it often depends on the thickness of the installations that it has to cover.

According to the ETICS (External Thermal Insulation Composite System), the thickness of the adhesive varies between 10 mm and 20 mm, and it creates an air gap in which combined heat transport is carried out by conduction, heat radiation and convection, hence, for the study, three thickness ranges have been adopted: 10, 15 and 20 mm. According to the guidelines [20,35], the adhesive mortar is applied to polystyrene boards using the circumferential point method, as a result of which it fills approx. 35% of the board surface, while the rest is filled with air. For better illustration of the phenomena occurring in the air gaps, a fragment of thickness between 70 and 80 mm (depending on the width of the gap) and an area of 0.1 m^2 was distinguished for further analysis (Figure 2).

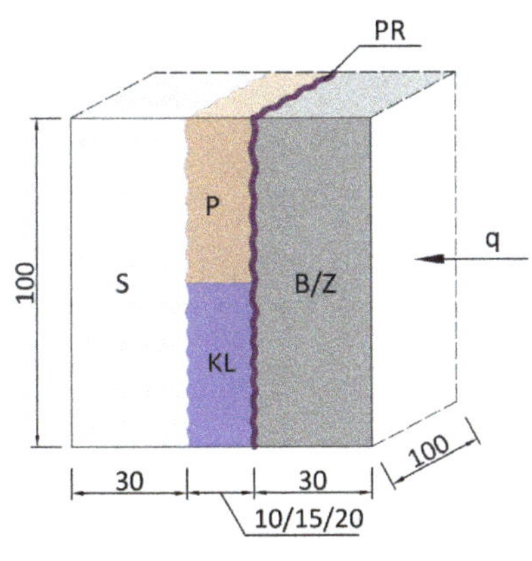

Figure 2. Model of a section of the layered wall subjected to testing.

The following material designations were adopted:

B	concrete R.23;
Z	concrete R.23.1;
S	polystyrene;
KL	adhesive mortar for polystyrene;
P	air;
PR	reflective smoothing coating, acrylic enamel spray.

Samples, as well as the components of the R.23 and R23.1 concrete mixtures, were obtained from the company, which specializes in the production of pre-casting elements for residential, industrial and road construction. The formulation of the analyzed R23 and R23.1 concretes was developed as part of research and development activities within the company, focusing on the production of innovative, architectural self-compacting concrete mixtures. The difference in the composition of the concrete mix R23 and R23.1 consisted of reducing the superplasticizer from 3.1 to 2.9 kg/m^3 and adding the ingredient bonding accelerator in the amount of 2.3 kg/m^3, which resulted in increased workability of the mix.

Thermal analyses were conducted on 54 samples in the simulation studies which preceded categorized by concrete, gap thickness, and filling of the gap between the layers, and were summarized in Table 1. Wall layer systems were divided into three groups, assigned a code to facilitate statistical analyses. Each code included three samples with structural layers made from Z-type concrete and three with B-type concrete. The differences between them resulted from the filling of the space between the thermal insulation and the structural layer, which was denoted by the symbol *d*, with an adopted value of 10, 15 or 20 mm. In the system described by code 0, the closed gap is filled with air where no turbulent fluid motion occurs. In the Code 1 model, the space is filled with adhesive, and heat transfer occurs exclusively through conduction. In contrast, the Code 2 model replicates a configuration with an air-filled gap, where the structural layer is additionally coated with a reflective-smoothing layer. Prior to computer simulations, the CFD environment was configured to incorporate the physical properties of the samples, environmental conditions, and thermal resistances, following thermophysical algorithms derived from laboratory research. Both systems accounted for variations in the gap thickness between the thermal insulation layer and the structural layer within the specified ranges analyzed in this study.

Table 1. Analyzed layered wall systems, classified by code, type of concrete and air gap thickness.

Code	**0**		**1**		**2**	
Concrete Type	**Z1**	**B1**	**Z1**	**B1**	**Z1**	**B1**
d = 0.01 m	Z1+P10+S	B1+P10+S	Z1+KL10+S	B1+KL10+S	Z1+PR+P10+S	B1+PR+P10+S
	Z1+P10+S	B2+P10+S	Z2+KL10+S	B2+KL10+S	Z2+PR+P10+S	B2+PR+P10+S
	Z1+P10+S	B3+P10+S	Z3+KL10+S	B3+KL10+S	Z3+PR+P10+S	B3+PR+P10+S
d = 0.015 m	Z1+P15+S	B1+P15+S	Z1+KL15+S	B1+KL15+S	Z1+PR+P15+S	B1+PR+P15+S
	Z2+P15+S	B2+P15+S	Z2+KL15+S	B2+KL15+S	Z2+PR+P15+S	B2+PR+P15+S
	Z3+P15+S	B3+P15+S	Z3+KL15+S	B3+KL15+S	Z3+PR+P15+S	B3+PR+P15+S
d = 0.02 m	Z1+P20+S	B1+P20+S	Z1+KL20+S	B1+KL20+S	Z1+PR+P20+S	B1+PR+P20+S
	Z2+P20+S	B2+P20+S	Z2+KL20+S	B2+KL20+S	Z2+PR+P20+S	B2+PR+P20+S
	Z3+P20+S	B3+P20+S	Z3+KL20+S	B3+KL20+S	Z3+PR+P20+S	B3+PR+P20+S

1.3. Modification of Heat Transfer by Radiation Through the Reduction of Surface Irregularities

It is a well-known phenomenon that the emissivity of a surface depends strongly on its roughness. Agabov [36,37] developed a simple yet effective method for determining the emissivity of rough surfaces with uniform thermal and reflective properties. As shown in Figure 3, he modeled local surface roughness as a depression in area A_r and examined the total radiation leaving the ideally smooth reference surface A_s, taking into account the effects of reflections from the actual surface A_r [36].

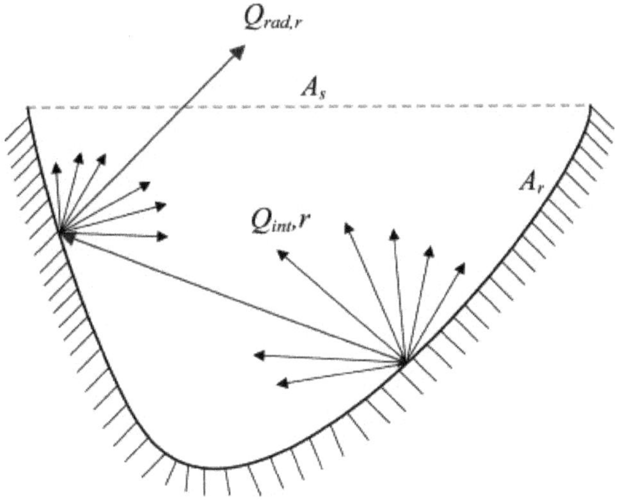

Figure 3. Modeling the behavior of incident flux on surface irregularities [36].

Agababov described the relationship between the emissivity of a gray body's surface and its geometry using the Formula (6):

$$\varepsilon_r = \left[1 + \left(\frac{1}{\varepsilon_s} - 1\right) R_{rs}\right]^{-1} \tag{6}$$

where the roughness coefficient of the gray body R_{rs} is (7):

$$R_{rs} = \frac{A_s}{A_r} \tag{7}$$

For a rectangular cavity with side length l, where the surface area A_s is l^2, and the surface A_r is discretized into M-micro-surfaces, the roughness of the body is expressed by the Formula (8):

$$R_{rs} = \frac{l^2}{\sum_{i=1}^{M} A_{ri}} \tag{8}$$

Based on these relationships and working with three-dimensional models, Zezhan [37] proposed a method for determining surface emissivity that accounts for its roughness in three steps:

1. measuring the root mean square roughness S_q,
2. determining A_r and calculating the coefficient R_{rs} from Formula (8),
3. calculating ε_s of the reference surface.

1.4. Solving Thermal Problems of Heat Transfer in CFD

Numerical fluid mechanics, or computational fluid dynamics (CFD), is a method used to solve equations describing fluid flow, system behaviors, heat transfer, mass transfer, and other similar physical phenomena [38–40]. Its application provides essential information on the distribution of velocity fields, pressure fields, heat movement, temperature fields, and other associated phenomena [28,41–43]. CFD enables the analysis of issues without the need for time-consuming and costly experimental studies [44–46]. However, simulation studies should be preceded by verification and validation of the model under real or laboratory conditions. CFD methods are applied in simulations of pressure drops during fluid flow, lift forces on aircraft wings, rotor thrust, airflow in air conditioning systems, temperature distribution in rooms, mixing processes, and more [28,47–49].

The Ansys Fluent program was chosen for thermal calculations due to its advanced calculation engine, module that allows it to import 3D geometry from reverse engineering programs and module for determining the results of calculations taking into account heat transfer paths. The phenomena occurring in the partition were calculated using the finite volume method (FVM), which is a numerical technique that converts volume integrals with divergence terms into surface integrals using the divergence theorem [25,50]. This allows for partial differential equations to be expressed and solved as algebraic equations within the FVM framework.

2. Materials and Methods

2.1. Collection of Geometric Data on the Concrete Surface Before and After the Application of the Reflective Smoothing Coating

Three-dimensional surface measurements of the samples (Figure 4a) were performed using the ATOS Core 3D optical scanner, (LENSO SP. Z O.O., Poznań, Poland) with a single scan field of 0.2 × 0.15 m, camera resolution of 5 MP, and sensor dimensions of 0.21 × 0.21 × 0.06 m. Data collection (points in space) was carried out using triangulation and recorded in GOM Scan software (2019, GOM GmbHa ZEISS, Braunschweig, Germany), where the point cloud was processed into a triangular mesh (polygonization).

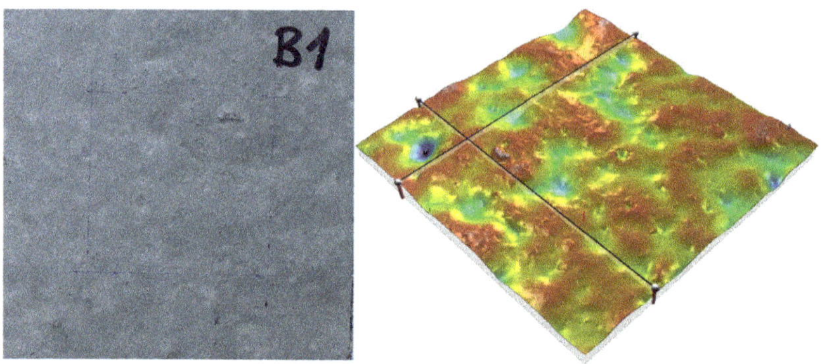

(a) (b)

Figure 4. Sample B1: (**a**) view of B1 sample surface, (**b**) an isometric view of the 3D scanned model of B1 sample surface with a color-coded interpretation of depressions and peaks.

Surface metrology of the scanned samples was conducted based on representative point clouds using MountainsLab Premium software by Digital Surf, version 8.1. Detailed morphometric measurements, geometric profiles, roughness analysis, assessments of surface waviness and spatial locations of these phenomena were performed [51]. Parameters defining surface topography, including height, frequency, hybrid, and functional characteristics were also determined [51–53]. Advanced software options allowed for the visualization of 3D surfaces in 24-bit resolution with real-time manipulation. Rendering with color palettes was applied (Figure 4b) and measurement data were cleaned using an advanced filter to remove anomalies, which facilitated the detection of surface features and potential irregularities.

After scanning and conducting surface structure analysis, sheet piling meshes have been converted into solids from which the models of multilayer walls were assembled in 3D computer-aided design software. Models consisting of three solids representing: concrete, polystyrene and air have been converted to .sat format and implemented to the environment of computational fluid mechanics in order to conduct heat transfer simulations.

The emissivity of the surface was examined using a custom-designed and constructed setup, which consisted of a measurement system: Flir T440bx thermal imaging camera (Teledyne FLIR LLC., Wilsonville, OR, USA) mounted on a tripod, an enclosure to limit ambient emissions and point reflections, a heating system with a thermostat to raise the sample surface temperature by 20 °C, and a multi-channel temperature recorder with probes [53]. Research methodology was based on guidelines provided in the literature, the FLIR device manual, and the instructions from the FLIR Tools/Tools+ software manual [54,55].

2.2. Boundary Conditions Implemented into the Model in the Computer Fluid Dynamics Program

For heat transport calculations using numerical fluid dynamics, the Ansys Fluent program (2020 R1) was used, which conducted the thermal issues considered in the article using the finite volume method. The simulation was preceded by importing the wall model, describing the model, assigning physical properties of materials and describing the model into finite volumes (Figure 5a). The volume of geometrical model is divided into 33 million cells using periodic system at the side surfaces for reducing size of the model but with regards to wider area of real multilayer wall at the buildings. A stationary tetrahedral non-structural mesh is used for calculations, as shown in the Figure 5a. That type of mesh allows for a simplified process of mesh creation compared to structural mesh which can be difficult to process at unregular scanned surfaces. The mesh density is mainly due to the irregularity of the scanned surfaces. Because the important role of natural convection is expected inside the air gap, the proper wall layer was created with five layers with the inflation coefficient set to 1.1. Y+ boundary layer thickness takes values below 20. Default parameters of k-w were assumed.

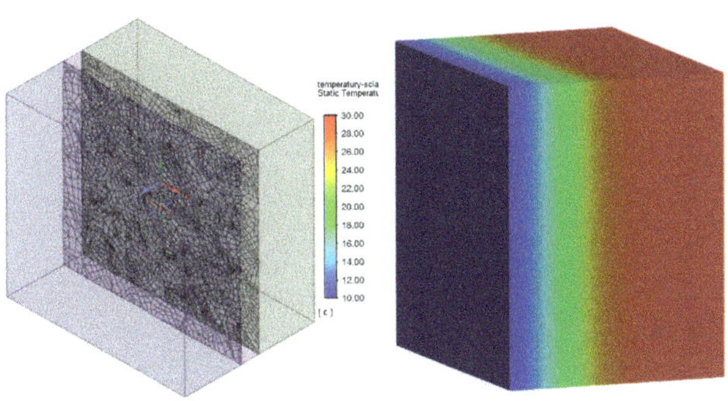

(a) (b)

Figure 5. Model of multilayer partition with representation of the surface shape in the CFD program window: (**a**) during mesh discretization, (**b**) visualization of temperature distribution after performing heat transport simulations.

Boundary conditions are assumed as follows:

- wall temperature at the warm side of the wall t = 30 °C,
- wall temperature at the cold side of the wall t = 10 °C,
- periodic boundary condition at the side surfaces of the air gap,
- heat flux equal to 0 at the side surfaces of the solid regions (pseudo one-dimensional heat transfer in solid regions),
- coupled heat transfer at the inner surfaces of the air gap with the emissivity assumed, respectively.

Results of the simulation were converted into the values of the heat transfer coefficient and thermal resistance of each layer, and after post-processing, they were presented in a graphic form with visualization of temperature distribution (Figure 5b).

For a diffuse surface, the reflectivity is independent of the incidence and reflection directions. In models analyzed in the paper, in which the space between the layers is filled with air, the equation transport (5) for flux leaving the surface k is determined in the Ansys [56–58] from (9):

$$q_{out,s} = \varepsilon_s \sigma T_k^4 + \rho_k + q_{in,k} \quad (9)$$

The turbulent flow model has been adopted with a model $k - \omega$ SST, which combines the advantages of the model $k - \varepsilon$ and model $k - \omega$, and introduces an additional component limiting the overproduction of kinetic energy of turbulence in areas of strong positive pressure gradients (dam points, areas of detachment of the boundary layer) [59,60]. The assumed model of the real gas-air is an incompressible ideal gas, in which the density is determined from the Clapeyron ideal gas equation [61,62]. The radiative heat flow for the air void was determined using the S2S (surface-to-surface) model [63]. The S2S radiation model in Ansys assumes that surfaces are diffusible. The emissivity and absorption of the grey surface is independent of the wavelength, and according to Kirchoff's law, they are equal ($\varepsilon = \alpha$) [10,64,65].

2.3. *Collection and Processing of Results from the CFD Analyses*

For the purpose of determining the heat transfer coefficient of the analyzed wall models, formulas for U (10) were derived taking into account the total heat flux $\dot{Q}$, which is obtained from the numerical analyses conducted.

$$U = \frac{\dot{Q}}{A} / (T_1 - T_2) \quad (10)$$

To determine the U-value, it was also necessary to obtain information on the interior and exterior surface temperatures of the component and implement the physical properties of the models, which values for B1+P20+S model was recorded and presented in Table 2. Using computational fluid dynamics (CFD) tools, it was also possible to identify the heat transport paths and determine the value of the heat flux transferred through radiation $\dot{Q}_{RAD}$.

For the accurate interpretation of the impact of changes in the emissivity of the construction layer's surface on thermal phenomena occurring in the analyzed walls, it was necessary to determine the thermal resistances in the air gaps between the concrete and the thermal insulation using Equation (11):

$$R_S = \frac{(A \cdot \Delta T)}{\dot{Q}} \quad (11)$$

where

$$\Delta T = T_{B\_SRC} - T_{S\_TRG}$$

Table 2. Summary of results and parameters implemented into the algorithms in order to determine and compare the U of partitions, for the model in the B1+P20+S configuration.

$\dot{Q}$	W	0.1778
$\dot{Q}_{RAD}$	W	0.1221
A	m2	0.009
d	m	0.08
T_1	°C	30
T_2	°C	10
U	W/m^{2}K	0.9852
λ_z	W/mK	0.079

For each of the 54 analyzed cases, calculations were performed according to Formulas (6) and (7), which were compiled in spreadsheets presented for a selected model constructed of B1 concrete without a coating and with a 20 mm air gap between the structural layer and the polystyrene (Table 3).

Table 3. Summary of the results and parameters implemented in the algorithms to determine the thermal resistances of gaps, for model in the B+P20+S configuration.

$\dot{Q}$	W	0.1778
T_{B_SRC}	°C	29.3
T_{S_TRG}	°C	26.3
ΔT	°C	3.0
A	m^2	0.009
R_S	(m^{2}K)/W	0.1513

2.4. Determination of Samples Roughness Parameters

Among parameters describing surface roughness, four groups can be distinguished: hybrids, volumes, functional and heights in which the most important for the course of thermal phenomena considered in the work in terms of emissions are: *Sq* and *Sku*. *Sku* is the kurtosis of the 3D surface texture, respectively [3,66]. Figuratively, a histogram of the heights of all measured points is established and the symmetry and deviation form an ideal normal (i.e., bell curve) distribution, while mathematically, it is evaluated as follows (12) and (13):

$$S_{ku} = \frac{1}{S_q^4}\sqrt{\iint_a \left(Z(x,y)^4\right)dxdy} \tag{12}$$

where:

$$S_q = \sqrt{\iint_a \left(Z(x,y)^2\right)dxdy} \tag{13}$$

Sku indicates the presence of inordinately high peaks/deep valleys (*Sku* > 3.00) or lack thereof (*Sku* < 3.00) making up the texture. If the surface heights are normally distributed (i.e., bell curve), then *Ssk* is 0.00 and *Sku* is 3.00. Surfaces described as gradually varying, free of extreme peaks or valley features, will tend to have *Sku* < 3.00.

Sq, root mean square roughness, is evaluated over the complete 3D surface, respectively. It represent an overall measure of the texture comprising the surface and is insensitive in differentiating peaks, valleys and the spacing of the various texture features [67]. Thus, *Sq* may be misleading in that many surfaces with grossly different spatial and height symmetry features (e.g., milled vs. honed) may have the same *Sq*, but function quite differently [68].

The results of the analysis of surface roughness before and after applying the reflective coating indicate a significant, nearly tenfold, reduction in the depressions and peaks on the concrete surface after the application of the coating (Table 4). These results may have a significant impact on heat transfer because the reflection of radiation on smooth surfaces is specular, whereas on matte surfaces it is diffuse and more readily absorbed.

Table 4. Change in the height parameters of the B1 sample surface before and after applying a reflective and smoothing coating.

Parameter			B1		Z1	
				+PR		+PR
Root Mean Square Roughness	S_q	µm	1.88	4.20	2.47	5.52
Kurtosis of the 3D surface texture	S_{ku}		4.47	0.55	5.62	0.69

3. Results

3.1. Development and Presentation of the Overall Research Results

The application of a low-emissivity coating on the concrete surface adjacent to the polystyrene layer significantly improves the thermal insulation properties of the entire partition and reduces the wall's thermal transmittance coefficient.

After applying the reflective and smoothing coating, the U-value of thermal transmittance for models constructed with B-type concrete was reduced by 27.8% for a 10 mm air gap, 38.1% for a 15 mm gap, and 42.1% for a 20 mm gap (Figure 6a). Thus, the average reduction in the U-value of thermal transmittance achieved for the analyzed layered partition models was 36.0%. Similarly, after applying the reflective and smoothing coating, the U-value of thermal transmittance for models constructed with Z-type concrete was reduced by 28.2% for a 10 mm air gap, 38.7% for a 15 mm gap, and 43.0% for a 20 mm gap (Figure 6b). Consequently, the average reduction in the U-value of thermal transmittance achieved for the analyzed layered partition models was 36.6%.

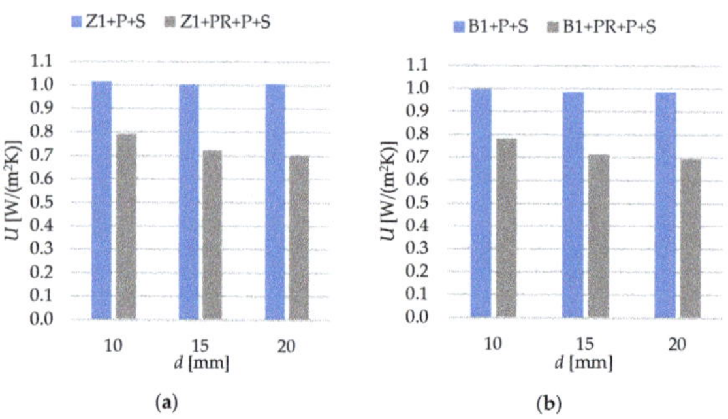

Figure 6. Change in heat transfer coefficient *U* of multilayered wall models after applying a reflective coating: (**a**) to the concrete surface B1, (**b**) to the concrete surface Z1.

For both layered models with B-type and Z-type concrete, the U-value is lowest when the distance between the structural layer and the thermal insulation is 15 mm (Figure 7). A slight correlation was observed between changes in the thickness of the air-filled gap and the U-value. In the B+P+S models, reducing the gap thickness to 10 mm resulted in an increase in the thermal transmittance coefficient by 1.3%, while increasing the gap thickness to 20 mm caused a 1.3% increase compared to models with a 15 mm gap. Similarly, for the Z+P+S models, reducing the gap from 15 mm to 10 mm led to a 1.1% increase in the thermal transmittance coefficient, whereas a gap thickness of 20 mm resulted in a 0.9% increase compared to models with a 15 mm gap.

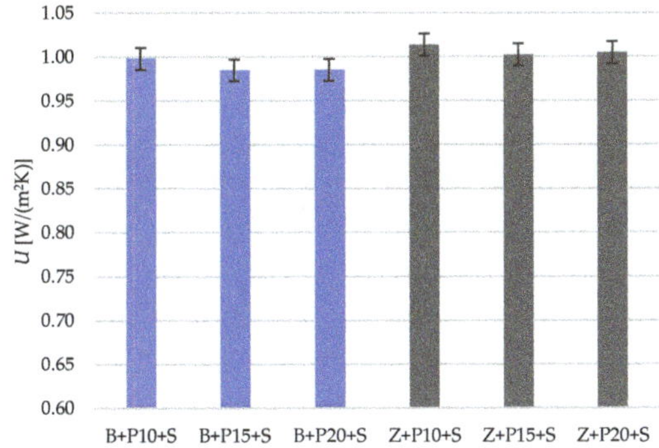

Figure 7. Change in heat transfer coefficient of multilayered wall models depending on the air gap thickness.

To investigate the statistically significant differences in heat transfer across various constructions, the following parameters were used: thermal resistance of the wall, thermal resistance of the air gap, categorize the models based on the gap-filling material, air gap thickness, emissivity of the concrete surface involved in radiative heat exchange, kurtosis of the 3D surface texture (Table 5). For each studied characteristic, a representative number of eighteen samples was used. Identical measurement conditions were ensured for all samples. The analyses considered simulation results conducted in a computational fluid dynamics environment on layered component models in the baseline configuration (Table 1) and the results of the heat transport studies are presented in Table 5. The measurements exhibit a certain degree of variability.

Table 5. Comparison of parameters of surface type B1 before and after applying a reflective and smoothing coating.

No.	Type	Class	d [m]	Rs [(m^{2}K)/W]	Rp [(m^{2}K)/W]	ε	Sku
1	Z1+P10+S	0	10	0.1311	0.9872	0.93	5.616
2	Z2+P10+S	0	10	0.1295	0.9867	0.93	14.079
3	Z3+P10+S	0	10	0.1296	0.9870	0.93	9.802
4	Z1+P15+S	0	15	0.1474	0.9982	0.93	5.616
5	Z2+P15+S	0	15	0.1474	0.9978	0.93	14.079
6	Z3+P15+S	0	15	0.1472	0.9979	0.93	9.802
7	Z1+P20+S	0	20	0.1504	0.9956	0.93	5.616
8	Z2+P20+S	0	20	0.1505	0.9956	0.93	14.079
9	Z3+P20+S	0	20	0.1503	0.9954	0.93	9.802

Table 5. *Cont.*

No.	Type	Class	*d* [m]	*Rs* [(m²K)/W]	*Rp* [(m²K)/W]	ε	*Sku*
10	Z1+KL10+S	1	10	0.0228	0.9921	0.93	5.616
11	Z2+KL10+S	1	10	0.0227	0.9921	0.93	14.079
12	Z3+KL10+S	1	10	0.0228	0.9921	0.93	9.802
13	Z1+KL15+S	1	15	0.0343	1.0063	0.93	5.616
14	Z2+KL15+S	1	15	0.0342	1.0062	0.93	14.079
15	Z3+KL15+S	1	15	0.0342	1.0063	0.93	9.802
16	Z1+KL20+S	1	20	0.0457	1.0204	0.93	5.616
17	Z2+KL20+S	1	20	0.0456	1.0204	0.93	14.079
18	Z3+KL20+S	1	20	0.0457	1.0204	0.93	9.802
19	Z1+PR+P10+S	2	10	0.2440	1.2663	0.29	0.691
20	Z2+PR+P10+S	2	10	0.2425	1.2623	0.29	1.732
21	Z3+PR+P10+S	2	10	0.2445	1.2670	0.29	1.206
22	Z1+PR+P15+S	2	15	0.2900	1.3851	0.29	0.691
23	Z2+PR+P15+S	2	15	0.2889	1.3814	0.29	1.732
24	Z3+PR+P15+S	2	15	0.2904	1.3856	0.29	1.206
25	Z1+PR+P20+S	2	20	0.2940	1.4236	0.29	0.691
26	Z2+PR+P20+S	2	20	0.2933	1.4214	0.29	1.732
27	Z3+PR+P20+S	2	20	0.2942	1.4246	0.29	1.206
28	B1+P10+S	0	10	0.1319	1.0017	0.93	4.470
29	B2+P10+S	0	10	0.1316	1.0018	0.93	9.283
30	B3+P10+S	0	10	0.1322	1.0024	0.93	7.213
31	B1+P15+S	0	15	0.1474	1.0154	0.93	4.470
32	B2+P15+S	0	15	0.1474	1.0155	0.93	9.283
33	B3+P15+S	0	15	0.1475	1.0156	0.93	7.213
34	B1+P20+S	0	20	0.1513	1.0150	0.93	4.470
35	B2+P20+S	0	20	0.1513	1.0154	0.93	9.283
36	B3+P20+S	0	20	0.1514	1.0153	0.93	7.213
37	B1+KL10+S	1	10	0.0229	0.9985	0.93	4.470
38	B2+KL10+S	1	10	0.0228	0.9984	0.93	9.283
39	B3+KL10+S	1	10	0.0228	0.9984	0.93	7.213
40	B1+KL15+S	1	15	0.0343	1.0126	0.93	4.470
41	B2+KL15+S	1	15	0.0342	1.0126	0.93	9.283
42	B3+KL15+S	1	15	0.0343	1.0126	0.93	7.213
43	B1+KL20+S	1	20	0.0457	1.0268	0.93	4.470
44	B2+KL20+S	1	20	0.0457	1.0268	0.93	9.283
45	B3+KL20+S	1	20	0.0457	1.0268	0.93	7.213
46	B1+PR+P10+S	2	10	0.2454	1.2812	0.29	0.551
47	B2+PR+P10+S	2	10	0.2444	1.2790	0.29	1.142
48	B3+PR+P10+S	2	10	0.2453	1.2812	0.29	0.887
49	B1+PR+P15+S	2	15	0.2908	1.4024	0.29	0.551
50	B2+PR+P15+S	2	15	0.2901	1.4008	0.29	1.142
51	B3+PR+P15+S	2	15	0.2908	1.4026	0.29	0.887
52	B1+PR+P20+S	2	20	0.2950	1.4432	0.29	0.551
53	B2+PR+P20+S	2	20	0.2946	1.4422	0.29	1.142
54	B3+PR+P20+S	2	20	0.2951	1.4434	0.29	0.887

CFD analysis, by its nature, is time-intensive and demands significant computational resources. Direct application of CFD tools to building thermal design or control poses numerous challenges. A potential solution lies in translating CFD results into efficient statistical models. Differentiation between various material configurations (Classes 0, 1, 2) can be achieved using statistical methods and artificial intelligence. In this context, group statistical tests for data distribution and machine learning classification models prove to be particularly valuable.

3.2. Statistical Analysis

Preliminary comparison of variable distributions across individual groups can be conducted using box plots. These plots graphically present central tendencies, variability, and skewness of the data. A box plot illustrates the median (line inside the box), skewness as the position of the median, dispersion as the interquartile range (IQR, length of the boxes), and outliers (points outside the whiskers). The box plots of the *Rs*, *Rp* and *Sku* values summarized in Table 5 are presented in Figure 8.

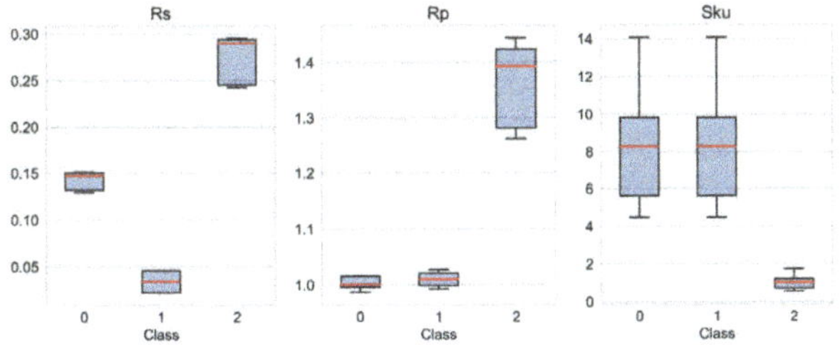

Figure 8. The box plots of the input data set.

The tabular summary of independent variables (Table 5) shows a finite set of values taken by the variables d (three values) and eps (two values) for all values of the dependent variable (Class). This means that they do not introduce additional information to the analysis and can be excluded from it. For the remaining independent variables (Figure 6), the largest statistical differences occur for the *Rs* values, clearly distinguishing all three classes (0, 1, and 2). The *Rp* values also allow for distinguishing the individual classes, but the differences between class 0 and 1 are less noticeable. Meanwhile, the *Sku* values show differences only for two groups (0–1 and 2). Nevertheless, it is worth emphasizing that even simple descriptive statistics, presented graphically in the form of box plots, show significant differences in the case of material configuration with an air gap and an additional reflective-smoothing coating (class 2).

More information regarding the statistical differences between observations can be provided by statistical tests. Here, the non-parametric Kruskal–Wallis test was chosen, which is used to compare at least two independent groups. This test is an alternative to one-way ANOVA but does not assume normality of the data distribution. Using the implementation of the Kruskal–Wallis test from the scipy package (Scipy v1.14.1), we compared the groups of features *Rs*, *Rp*, and *Sku*. At a significance level of $\alpha = 0.01$, the null hypothesis (H0) was formulated as follows: the medians of all groups are equal, indicating no significant differences between the groups in terms of the median. Conversely, the alternative hypothesis (H1) posits that at least one of the groups differs in median from the others, suggesting significant differences between the groups in terms of the median. Given the p-value of (8.96×10^{-27}) at a significance level of $\alpha = 0.01$, we can reject the null hypothesis (H0). This extremely low p-value indicates that there are significant differences between the groups in terms of their medians.

4. Conclusions

Surface metrology has shown that the application of a reflective-smoothing coating significantly affects the configuration of the surface microstructure, particularly the kurtosis of the 3D surface texture, resulting in an almost tenfold reduction in irregularities on average, which have a direct impact on the emissivity of the surface, which was reduced from 0.93 to 0.29. Thermal analyses conducted have shown that, in accordance with transport phenomena theory, there is a significant reduction in the density of heat flux exchanged through radiation in air gaps, by approximately 86%, which resulted in a 28% improvement of the evaluated walls samples. This approach enables the enhancement of wall thermal insulation while simultaneously reducing their thickness, making it particularly applicable in the construction industry, especially for prefabricated sandwich panels.

Heat transfer in the analyzed, closed air gaps between concrete and thermal insulation occurs primarily through radiation. The thickness of the gap affects the heat exchange phenomena. The studies have shown that for the Z+P+S models, increasing the gap thickness from 10 mm to 15 mm results in a thermal resistance increase average of 13%, and from 15 mm to 20 mm of 2%. After applying the reflective and smoothing coating, these values were 18% and 1%, respectively. For models constructed from Z concrete, the proportions were very similar. This leads to the conclusion that increasing the air gap thickness from 10 mm to 15 mm is more advantageous, as it is associated with the occurrence of convective movements in larger gaps, and the coating enhances thermal insulation of walls.

A further research direction is to conduct CFD simulations and explore statistical correlations between changes in the surface emissivity of polystyrene through the application of a reflective coating and variations in thermal resistance within the air gaps of multilayer walls. This approach could be implemented in practice at the stage of production of ETICS system components, increasing the availability of the solution.

Author Contributions: Conceptualization, A.S. (Andrzej Sachajdak); Methodology, A.U. and A.S. (Andrzej Sachajdak); Software, A.U., A.S. (Andrzej Sachajdak) and J.Z.; Validation, A.U.; Formal analysis, A.U., A.S. (Andrzej Sachajdak) and A.S. (Arkadiusz Syta); Investigation, A.U.; Resources, A.U.; Data curation, A.S. (Arkadiusz Syta); Writing—original draft, A.U.; Writing—review & editing, A.S. (Andrzej Sachajdak), A.S. (Arkadiusz Syta) and J.Z.; Visualization, J.Z.; Supervision, A.S. (Andrzej Sachajdak). All authors have read and agreed to the published version of the manuscript.

Funding: The study was funded by research grants from the Polish Ministry of Education and Science in Warsaw, Poland: FD-20/IS-6/050; The study was funded by research grants from the Polish Ministry of Education and Science in Warsaw, Poland: FN-52.

Data Availability Statement: The original contributions presented in the study are included in the article, further inquiries can be directed to the corresponding author.

Conflicts of Interest: The authors declare no conflict of interest.

Nomenclature

A_s	surface area [m^2],
C_C	blackbody radiation coefficient equal to 5.77 [W/(m^{2}K)],
d	air gap thickness [m],
g	gravitational vector,
h	enthalpy [J/kg],
l	length,
p	static pressure [Pa],
q	heat flux density [W/m^2],
$q_{in,k}$	flux of energy incident on the surface from the environment [W/m^2],
$q_{out,k}$	flux of energy leaving the surface [W/m^2],
R_p	thermal resistance of the wall [(m^{2}K)/W],
R_s	thermal resistance of the air gap [(m^{2}K)/W],
R_{rs}	roughness coefficient of the gray body,

S_{ku}	kurtosis of the 3D surface texture,
S_q	root mean square roughness [mm],
T	temperature [K],
T_1	interior surface temperature [K],
T_2	exterior surface temperature [K],
T_{B_SRC}	surface temperature of the concrete on the air gap side [K],
T_{S_TRG}	surface temperature of the polystyrene on the air gap side [K].
Greek symbols	
ε_s	surface emissivity,
ε_{1-2}	equivalent emissivity,
λ	thermal conductivity [W/(mK)],
ρ	density [kg/m^3],
$\bar{\bar{\tau}}$	stress tensor,
v	velocity vector,
φ_{1-2}	configuration ratio.
Acronyms	
CAD	Computer Aided Design,
CFD	Computational Fluid Dynamics,
EPS	Expanded Polystyrene Insulation,
ETICS	External Thermal Insulation Composite System,
FVM	Finite Volume Method,
RBF	Radial Basis Functions,
SST	Shear-Stress Transport.

References

1. Zheng, H.; Gao, G.; Zhong, X.; Zhao, L. Monitoring and Diagnostics of Buildings' Heat Loss Based on 3D IR Model of Multiple Buildings. *Energy Build.* **2022**, *259*, 111889. [CrossRef]
2. Gong, Q.; Kou, F.; Sun, X.; Zou, Y.; Mo, J.; Wang, X. Towards Zero Energy Buildings: A Novel Passive Solar House Integrated with Flat Gravity-Assisted Heat Pipes. *Appl. Energy* **2022**, *306*, 117981. [CrossRef]
3. Chen, J.; Wang, H.; Xie, P.; Najm, H. Analysis of Thermal Conductivity of Porous Concrete Using Laboratory Measurements and Microstructure Models. *Constr. Build. Mater.* **2019**, *218*, 90–98. [CrossRef]
4. Park, H.M.; Lee, J.H.; Kim, K.D. Wall Temperature Prediction at Critical Heat Flux Using a Machine Learning Model. *Ann. Nucl. Energy* **2020**, *141*, 107334. [CrossRef]
5. Xue, X.; Han, S.; Guo, D.; Zhao, Z.; Zhou, B.; Li, F. Study of the Convective Heat Transfer Coefficient of Different Building Envelope Exterior Surfaces. *Buildings* **2022**, *12*, 860. [CrossRef]
6. Clarke, J.A. *Energy Simulation in Building Design*, 2nd ed.; Butterworth-Heinemann: Oxford, UK, 2001; ISBN 0750650826.
7. Thomas, T.R.; Probert, S.D. Thermal Contact Resistance: The Directional Effect and Other Problems. *Int. J. Heat Mass Transf.* **1970**, *13*, 789–807. [CrossRef]
8. Yüncü, H. Thermal Contact Conductance of Nominaly Flat Surfaces. *Heat Mass Transf.* **2006**, *43*, 1–5. [CrossRef]
9. Defraeye, T.; Blocken, B.; Carmeliet, J. Convective Heat Transfer Coefficients for Exterior Building Surfaces: Existing Correlations and CFD Modelling. *Energy Convers. Manag.* **2011**, *52*, 512–522. [CrossRef]
10. Wodołażski, A.; Howaniec, N.; Jura, B.; Bąk, A.; Smoliński, A. Cfd Numerical Modelling of a Pv–Teg Hybrid System Cooled by Air Heat Sink Coupled with a Single-Phase Inverter. *Materials* **2021**, *14*, 5800. [CrossRef]
11. Zeng, H.; Lu, C.; Zhang, L.; Yang, T.; Jin, M.; Ma, Y.; Liu, J. Prediction of Temperature Distribution in Concrete under Variable Environmental Factors through a Three-Dimensional Heat Transfer Model. *Materials* **2022**, *15*, 1510. [CrossRef]
12. Morris Grenfell, D. Heat Transfer by Radiation. In *Building Heat Transfer*; Routledge: Abingdon, UK, 2004; pp. 115–148, ISBN 9780470020555.
13. Rohsenow, W.M.; Hartnett, J.P.; Cho, Y.I. (Eds.) *Handbook of Heat Transfer*, 1st ed.; McGraw-Hill Education: New York, NY, USA, 1998; ISBN 9780070535558.
14. Oosthuizen, P.H.; Naylor, D. Heat Transfer Through a Double-Glazed Window. In *Heat-Transfer Calculations*; Kutz, M., Ed.; McGraw-Hill Education: New York, NY, USA, 2006; ISBN 9780071410410.
15. Ohlsson, K.E.A.; Östin, R.; Grundberg, S.; Olofsson, T. Dynamic Model for Measurement of Convective Heat Transfer Coefficient at External Building Surfaces. *J. Build. Eng.* **2016**, *7*, 239–245. [CrossRef]

16. Barnat-Hunek, D.; Grzegorczyk-Frańczak, M.; Suchorab, Z. Surface Hydrophobisation of Mortars with Waste Aggregate by Nanopolymer Trietoxi-Isobutyl-Silane and Methyl Silicon Resin. *Constr. Build. Mater.* **2020**, *264*, 120175. [CrossRef]
17. Jayamaha, S.E.G.; Wijeysundera, N.E.; Chou, S.K. Measurement of the Heat Transfer Coefficient for Walls. *Build. Environ.* **1996**, *31*, 399–407. [CrossRef]
18. Peng, Z.; Carrigan, S.; Kornadt, O. Investigation of the Influence of Different Room Geometries and Wall Thermal Transmittances on the Heat Transfer in Rooms with Floor Heating. *Energy Build.* **2024**, *317*, 114391. [CrossRef]
19. Viegas, C.A.; Borsoi, G.; Moreira, L.M.; Parracha, J.L.; Nunes, L.; Malanho, S.; Veiga, R.; Flores-Colen, I. Diversity and Distribution of Microbial Communities on the Surface of External Thermal Insulation Composite Systems (ETICS) Facades in Residential Buildings. *Int. Biodeterior. Biodegrad.* **2023**, *184*, 105658. [CrossRef]
20. Xu, H.; Wang, H.; Huo, Q.; Qin, Y.; Zhou, H. Comparative Study of Chinese, European and ISO External Thermal Insulation Composite System (ETICS) Standards and Technical Recommendations. *J. Build. Eng.* **2023**, *68*, 105687. [CrossRef]
21. Directive (EU) 2024/1275 of the European Parliament and of the Council of 24 April 2024 on the Energy Performance of Buildings; 2024; p. 68. Available online: https://eur-lex.europa.eu/eli/dir/2024/1275 (accessed on 1 July 2024).
22. Saber, H.H.; Yarbrough, D.W. Advanced Modeling of Enclosed Airspaces to Determine Thermal Resistance for Building Applications. *Energies* **2021**, *14*, 7772. [CrossRef]
23. Orlik-Kożdoń, B. Polystyrene Waste in Panels for Thermal Retrofitting of Historical Buildings: Experimental Study. *Energies* **2021**, *14*, 1844. [CrossRef]
24. Golder, S.; Narayanan, R.; Hossain, M.R.; Islam, M.R. Experimental and CFD Investigation on the Application for Aerogel Insulation in Buildings. *Energies* **2021**, *14*, 3310. [CrossRef]
25. Rehman, A.U.; Sheikh, S.R.; Kausar, Z.; McCormack, S.J. Numerical Simulation of a Novel Dual Layered Phase Change Material Brick Wall for Human Comfort in Hot and Cold Climatic Conditions. *Energies* **2021**, *14*, 4032. [CrossRef]
26. Basok, B.; Pavlenko, A.; Novikov, V.; Koshlak, H.; Ciosek, A.; Moroz, M. Comprehensive Investigation of the Thermal Performance of an Electrically Heated Double-Glazed Window: A Theoretical and Experimental Approach. *Energies* **2024**, *17*, 4491. [CrossRef]
27. Alaidroos, A. Transient Behavior Analysis of the Infiltration Heat Recovery of Exterior Building Walls. *Energies* **2023**, *16*, 7198. [CrossRef]
28. Urzędowski, A.; Zaburko, J.; Gleń, P.; Łagód, G. Numerical and Experimental Investigation of the Influence of External Wall Moisture on Building Temperature Distributions. *Adv. Sci. Technol. Res. J.* **2021**, *15*, 297–308. [CrossRef]
29. Makaveckas, T.; Bliūdžius, R.; Burlingis, A. The Influence of Different Facings of Polyisocyanurate Boards on Heat Transfer Through the Wall Corners of Insulated Buildings. *Energies* **2020**, *13*, 1991. [CrossRef]
30. Babiarz, B.; Krawczyk, D.A.; Siuta-Olcha, A.; Manuel, C.D.; Jaworski, A.; Barnat, E.; Cholewa, T.; Sadowska, B.; Bocian, M.; Gnieciak, M.; et al. Energy Efficiency in Buildings: Toward Climate Neutrality. *Energies* **2024**, *17*, 4680. [CrossRef]
31. Fan, C.; Xia, X.L.; Du, W.; Sun, C.; Li, Y. Numerical Investigations of the Coupled Conductive-Radiative Heat Transfer in Alumina Ceramics. *Int. Commun. Heat Mass Transf.* **2022**, *135*, 106097. [CrossRef]
32. Urzędowski, A.; Styczeń, J.; Wójcicka-Migasiuk, D. Analysis of thermal properties and heat loss in construction and isothermal materials of multilayer building walls. *Adv. Sci. Technol. Res. J.* **2017**, *11*, 33–37. [CrossRef]
33. Kats, D.; Wang, Z.; Gan, Z.; Liu, W.K.; Wagner, G.J.; Lian, Y. A Physics-Informed Machine Learning Method for Predicting Grain Structure Characteristics in Directed Energy Deposition. *Comput. Mater. Sci.* **2022**, *202*, 110958. [CrossRef]
34. Chen, L.; Zhang, Y.; Zhou, X.; Shi, X.; Yang, L.; Jin, X. A New Method for Measuring Thermal Resistance of Building Walls and Analyses of Influencing Factors. *Constr. Build. Mater.* **2023**, *385*, 131438. [CrossRef]
35. Gonçalves, M.; Simões, N.; Serra, C.; Almeida, J.; Flores-Colen, I.; Vieira de Castro, N.; Duarte, L. Onsite Monitoring of ETICS Comparing Different Exposure Conditions and Insulation Materials. *J. Build. Eng.* **2021**, *42*, 103067. [CrossRef]
36. Agababov, S.G. Effect of the Roughness Factor on Radiation Properties of Solids. *Teplofiz. Vysok. Temp.* **1970**, *8*, 770–773.
37. Zhang, Z.; Chen, M.; Yu, P.; Huang, H.; Li, H.; Yu, F.; Zhang, Z.; Niu, Y.; Gao, S.; Wang, C.; et al. Study of the Roughness Effect on the Normal Spectral Emissivity of GH3044. *Infrared Phys. Technol.* **2023**, *133*, 104831. [CrossRef]
38. Tomkiewicz, D.; Cudzewicz, A.; Górecka-Orzechowska, J. Zastosowanie Programu FLUENT Do Modelowania Zjawisk Termodynamicznych w Komorze Klimatycznej. *Inżynieria Rol.* **2011**, *15*, 239–245.
39. Sachajdak, A.; Słoma, J.; Szczygieł, I. Thermal Model of the Gas Metal Arc Welding Hardfacing Process. *Appl. Therm. Eng.* **2018**, *141*, 378–385. [CrossRef]
40. Korkerd, K.; Zhou, Z.; Zou, R.; Piumsomboon, P.; Chalermsinsuwan, B. Effect of Immersed Tubes Configurations on Mixing and Heat Transfer of Mixed Biomass and Silica Sand in a Bubbling Fluidized Bed Using CFD-DEM and Statistical Experimental Design Analysis. *Powder Technol.* **2024**, *437*, 119542. [CrossRef]
41. Jamińska-Gadomska, P.; Lipecki, T.; Pieńko, M.; Podgórski, J. Wind Velocity Changes Along the Passage Between Two Angled Walls—CFD Simulations and Full-Scale Measurements. *Build. Environ.* **2019**, *157*, 391–401. [CrossRef]

42. Słoma, J.; Szczygieł, I.; Sachajdak, A. Modelling of Thermal Phenomena in Electric Arc During Surfacing. *Arch. Civ. Mech. Eng.* **2011**, *11*, 437–449. [CrossRef]
43. Sachajdak, A.; Lappalainen, J.; Mikkonen, H. Dynamic Simulation in Development of Contemporary Energy Systems—Oxy Combustion Case Study. *Energy* **2019**, *181*, 964–973. [CrossRef]
44. Hoang, M.L.; Verboven, P.; De Baerdemaeker, J.; Nicolaï, B.M. Analysis of the Air Flow in a Cold Store by Means of Computational Fluid Dynamics. *Int. J. Refrig.* **2000**, *23*, 127–140. [CrossRef]
45. Kaleta, A.; Górnicki, K.; Peroń, S.; Szwast, Z.; Kiryjow, J. *Podstawy Techniki Cieplnej w Inżynierii Rolniczej*; Wyd. 2.; Wydawnictwo SGGW: Warsaw, Poland, 2015; ISBN 9788375831054.
46. Norton, T.; Sun, D.W.; Grant, J.; Fallon, R.; Dodd, V. Applications of Computational Fluid Dynamics (CFD) in the Modelling and Design of Ventilation Systems in the Agricultural Industry: A Review. *Bioresour. Technol.* **2007**, *98*, 2386–2414. [CrossRef]
47. Syta, A.; Czarnigowski, J.; Jakliński, P. Detection of Cylinder Misfire in an Aircraft Engine Using Linear and Non-Linear Signal Analysis. *Measurement* **2021**, *174*, 108982. [CrossRef]
48. Jedliński, Ł.; Syta, A.; Gajewski, J.; Jonak, J. Nonlinear Analysis of Cylindrical Gear Dynamics Under Varying Tooth Breakage. *Measurement* **2022**, *190*, 110721. [CrossRef]
49. Barnat-Hunek, D.; Lagod, G.; Klimek, B. Evaluation of the Contact Angle and Frost Resistance of Hydrophobised Heat-Insulating Mortars with Polystyrene. *AIP Conf. Proc.* **2017**, *1866*, 040004.
50. Thongsuk, S.; Songsukthawan, P.; Lertwanitrot, P.; Ananwattanaporn, S.; Yoomak, S.; Pothisarn, C. Impact of Building Envelope Materials on Energy Usage and Performance of Evaporative Cooling System in Residential Building. *Energies* **2024**, *17*, 3748. [CrossRef]
51. Schabowicz, K.; Wójcicka-Migasiuk, D.; Urzędowski, A.; Wróblewski, K. Nondestructive Investigations of Expansion Gap Concrete Roughness. *Measurement* **2021**, *182*, 109603. [CrossRef]
52. Sánchez-Calderón, I.; Merillas, B.; Bernardo, V.; Rodríguez-Pérez, M.Á. Methodology for Measuring the Thermal Conductivity of Insulating Samples with Small Dimensions by Heat Flow Meter Technique. *J. Therm. Anal. Calorim.* **2022**, *147*, 12523–12533. [CrossRef]
53. Urzędowski, A.; Wójcicka-Migasiuk, D.; Buraczyńska, B. Visual Effects of Surface Emissivity in Thermal Imaging. *Adv. Sci. Technol. Res. J.* **2020**, *14*, 215–222. [CrossRef]
54. Nowak, H. The Influence of Environmental Thermal Radiation on the Results of Thermographic Investigation of Buildings. *Bud. Arch.* **2013**, *12*, 059–066. [CrossRef]
55. Shu, S.; Wu, T.; Wang, Z.; Zhang, Y.; Yang, Z.; Liang, H. Research on the Normal Emissivity of Graphite between 150 and 500 °C by an Infrared Camera for Nuclear Fusion Devices. *Nucl. Mater. Energy* **2022**, *31*, 101182. [CrossRef]
56. Węgrzyński, W.; Vigne, G. Experimental and Numerical Evaluation of the Influence of the Soot Yield on the Visibility in Smoke in CFD Analysis. *Fire Saf. J.* **2017**, *91*, 389–398. [CrossRef]
57. Steiner, T.R. High Temperature Steady-State Experiment for Computational Radiative Heat Transfer Validation Using COMSOL and ANSYS. *Results Eng.* **2022**, *13*, 100354. [CrossRef]
58. Sohrabi, N.; Hammoodi, K.A.; Hammoud, A.; Jasim, D.J.; Hashemi Karouei, S.H.; Kheyri, J.; Nabi, H. Using Different Geometries on the Amount of Heat Transfer in a Shell and Tube Heat Exchanger Using the Finite Volume Method. *Case Stud. Therm. Eng.* **2024**, *55*, 104037. [CrossRef]
59. Menter, F.; Kuntz, M.; Langtry, R.B. Ten Years of Industrial Experience with the SST Turbulence Model. *Heat Mass Transf.* **2003**, *4*, 625–632.
60. Menter, F. Two Equation Eddy-Viscosity Turbulence Modeling for Engineering Applications. *AIAA J.* **1994**, *32*, 1598–1605. [CrossRef]
61. Pogorzelski, J.A. Ostrożnie z Niestacjonarnymi Badaniami Przewodności Cieplnej. *Pr. Inst. Tech. Bud.* **2000**, *113*, 38–52.
62. Gawin, D.; Klemm, P.; Konca, P.; Więckowska, A.; Woźniak, J. *Komputerowa Fizyka Budowli: Komputerowa Symulacja Procesów Wymiany Masy i Energii w Budynku: Przykłady Zastosowań*; Wydawnictwo Politechniki Łódzkiej: Łódź, Poland, 1998; ISBN 8387198412.
63. Siegel, A.F.; Wagner, M.R. Chapter 15—ANOVA: Testing for Differences Among Many Samples and Much More. In *Practical Business Statistics*, 8th ed.; Siegel, A.F., Wagner, M.R., Eds.; Academic Press: Cambridge, MA, USA, 2022; pp. 485–510, ISBN 978-0-12-820025-4.
64. Saini, M.C.; Jakhar, O.P. CFD Simulation and Experimental Validation of PCM Thermal Energy Storage System for Micro Trigeneration System Application. *Int. J. Refrig.* **2023**, *149*, 119–134. [CrossRef]
65. Modest, M.F. *Radiative Heat Transfer*; Academic Press Inc.: Cambridge, MA, USA, 2003; ISBN 9780080515632.

66. Andrzejuk, W.; Grzegorczyk-Frańczak, M.; Barnat-Hunek, D.; Franus, M.; Łagód, G. Microstructure, Durability and Surface Free Energy of Lightweight Aggregate Modification of Sanitary Ceramic Wastes and Sewage Sludge. *J. Build. Eng.* **2024**, *93*, 109725. [CrossRef]
67. Chen, J.; Chu, R.; Wang, H.; Xie, P. Experimental Measurement and Microstructure-Based Simulation of Thermal Conductivity of Unbound Aggregates. *Constr. Build. Mater.* **2018**, *189*, 8–18. [CrossRef]
68. Deconinck, L.; Bernardo Quejido, E.; Villa Vidaller, M.T.; Jägle, E.A.; Verbeken, K.; Depover, T. The Mechanism behind the Effect of Building Orientation and Surface Roughness on Hydrogen Embrittlement of Laser Powder Bed Fused Ti-6Al-4V. *Addit. Manuf.* **2023**, *72*, 103613. [CrossRef]

energies

Article

Eddy–Viscosity Reynolds-Averaged Navier–Stokes Modeling of Air Distribution in a Sidewall Jet Supplied into a Room

Maria Hurnik, Piotr Ciuman and Zbigniew Popiolek *

Faculty of Energy and Environmental Engineering, Silesian University of Technology, Konarskiego 20, 44-100 Gliwice, Poland; maria.hurnik@polsl.pl (M.H.); piotr.ciuman@polsl.pl (P.C.)
* Correspondence: zbigniew.popiolek@polsl.pl; Tel.: +48-32-237-11-51

Abstract: Air velocity is one of the key parameters affecting the sensation of thermal comfort. In mixing ventilation, the air is most often supplied above the occupied zone, and the air movement in a room is caused by jets that generate recirculating flows. An effective tool for predicting airflow in a room is CFD numerical modeling. In order to reproduce the air velocity distribution, it is essential to select a proper turbulence model. In this paper, seven *Eddy–Viscosity RANS* turbulence models were used to carry out CFD simulations of a sidewall air jet supplied into a room through a wall diffuser. The goal was to determine which model was the most suitable to adopt in this type of airflow. The CFD results were validated using experimental data by comparing the gross and integral parameters, along with the parameters of the quasi-free jet model. The numerical results obtained for *Std k-ε* and *EVTM* models were most consistent with the measurements. Their error values slightly exceeded 15%. On the contrary, the *k-ω* and *RNG k-ε* models did not reproduce the quasi-free jet parameters correctly. The research findings can prove beneficial for simulating air distribution in supplied air jets during the initial conceptual phases of HVAC system design.

Keywords: ventilation; air distribution; sidewall jet; CFD prediction; validation; eddy viscosity turbulence model

1. Introduction

Properly organized air distribution in rooms should ensure thermal comfort and high air quality with the lowest possible energy consumption. Requirements regarding thermal conditions in rooms, depending on the category of the indoor environment, are specified in various standards [1,2]. The sensation of thermal comfort is significantly influenced by air velocity. This parameter can also improve the circulation and distribution of fresh air, which can reduce the concentration of pollutants and thus affect the air quality in the room, as well as limit the spread of viruses and bacteria. COVID-19 diffusion is a problem linked to indoor air ventilation [3,4]. Poor ventilation can lead to higher concentrations of viral particles, increasing the risk of infection among occupants. Effective indoor air ventilation plays a crucial role in mitigating the spread of COVID by diluting and removing airborne contaminants, thus reducing the likelihood of transmission within indoor environments.

In mixing ventilation, the air is most often supplied above the occupied zone, and the air movement in a room is caused by jets that generate recirculating, i.e., induced secondary flows. The air speed in the occupied zone is very highly correlated with the momentum flux of the supplied jet [5,6]. The terms velocity and speed are used as synonyms in everyday language. In this paper, we use these terms as defined in ANSI/ASHRAE 113-2005 [7]. According to this standard, velocity and speed are different physical quantities. Velocity is a vector quantity, $\vec{U} = i\,U_x + j\,U_y + k\,U_z$, while speed is a scalar quantity, i.e., it is the value of the modulus of the velocity vector, $W = \left|\vec{U}\right| = \left(U_x^2 + U_y^2 + U_z^2\right)^{1/2}$.

A comparison of various advanced air distribution systems, which mainly depend on the type and location of supplied jets, is presented in a prior review paper [8]. The simplest tool for designing air distribution systems is based on engineering models of ventilation air jets. However, these models do not provide complete information about the air distribution in the entire room. An effective tool for predicting airflow, both in the supplied jets and in the occupied zone, is CFD numerical modeling [5].

The quality of CFD modeling depends on many factors, such as the adopted turbulence model, the type of discretization grid and number of cells, the way the boundary conditions are defined, and others. All these factors are sources of uncertainty for the CFD calculation results. Therefore, validation of the CFD codes, which means assessing the uncertainty of simulation results by comparing them with experimental data, is necessary [9–12]. Uncertainty quantification analysis is essential for enhancing the reliability, robustness, and applicability of flow dynamic simulations in various engineering and scientific applications [13,14].

1.1. Turbulence Models

To obtain reliable CFD results, accurate turbulence modeling is necessary. The turbulence model should be selected and adapted to the type of flow. Some experience and knowledge in this field are also needed.

Air distribution in a room can be modeled with the use of various turbulence models that are available in computing software, e.g., Ansys CFX 22.1 [15]. The first group of turbulence models is the *Reynolds-Averaged Navier–Stokes Equation* (*RANS*), which includes turbulence models from *Eddy–Viscosity Models* (*EVM*) and *Reynolds Stress Models* (*RSM*) subgroups. Moreover, the *Unsteady RANS* (*URANS*) and vortex-resolving *Large Eddy Simulation* (*LES*) turbulence models are available. The *LES* technique directly solves the filtered Navier–Stokes equations, thus solving large scales of motion. Smaller scales are modeled with a suitable *sub-grid scale model* (*SGS*). A review and discussion of the prospects of *LES* development are presented in a previous paper [16].

The averaged equations of the *RANS* turbulence models include expressions containing the fluctuating components, the so-called Reynolds stresses, which are considered as the impact of turbulence on the average airflow. Their presence means that the system of the Reynolds equations is not a closed one. Each turbulence model based on time-averaging differs in the way it determines the values of Reynolds stress and, thus, creates the equations that close the system.

In the *EVM* group, the turbulent viscosity coefficient v_t is introduced to determine the values of Reynolds stress describing the local state of turbulence and, unlike the molecular coefficient v_m, it is a variable dependent on location and time. *EVM* models differ in the method of determining this coefficient. The names of the models are related to the number of transport equations based on which this coefficient is calculated. In the zero-equation model, the differential transport equations are not used to determine turbulent viscosity, and only algebraic equations containing empirical coefficients are used. However, the zero-equation model has a weak physical basis; it is not recommended for use in numerical predictions.

The one-equation *Eddy–Viscosity Transport Model* (*EVTM*) includes a single transport equation to determine the turbulent viscosity coefficient. One of the most popular turbulence models is the two-equation *k-ε* model, which consists of two differential transport equations: the turbulence kinetic energy *k* equation and the turbulence energy dissipation rate ε. Its modified form is the *Re-Normalisation Group* (*RNG*) *k-ε* model, in which the transport equations are the same, but the constants of the model are different. It was developed for highly turbulent isotropic flows. One of the advantages of another popular model, the two-equation *k-ω* model (in which *k* is the turbulence kinetic energy and ω is the turbulence vorticity), is that is provides a more accurate method of calculating the boundary-layer flows for low Reynolds numbers compared to the *k-ε* model. The model does not require the modifications that are necessary in this region in the *k-ε* model. It is used in the standard version, known as the *Wilcox k-ω* model, as well as in the *Baseline* (*BSL*) *k-ω* model and the *Shear Stress Transport* (*SST*) model created by combining it with the *k-ε* model. The *BSL* model integrates the *k*-ω model close to solid walls with the standard *k*-ε model presuming distance from them. The *SST* model is also a combination of the *k*-ω model (in the inner boundary layer) and the *k*-ε model (in the outer boundary layer and beyond), but additionally a limitation of the value of shear stress in the area of the reverse pressure gradient is implemented in it. This model is frequently recommended for calculating the airflow of supply jets. The *EVM RANS* turbulence models available in the ANSYS CFX 22.1 software are presented in Table 1.

Table 1. *EVM RANS* turbulence models available in the ANSYS CFX 22.1 software.

Zero Equation Model		
One-Equation Model	*EVTM* model	
Two-Equation Models	*k*-ε model	Standard *k*-ε
		RNG k-ε
	k-ω model	*Wilcox k*-ω
		BSL k-ω
		SST

In the one-equation *Eddy–Viscosity Transport Model* (*EVTM*) the transport equation of turbulent viscosity coefficient υ_t is used:

$$\frac{\partial\rho}{\partial t}\overline{\upsilon}_t+\frac{\partial\rho V_j\overline{\upsilon}_t}{\partial x_j}=c_1\rho\overline{\upsilon}_t S-c_2\rho\left(\frac{\overline{\upsilon}_t}{L_{\nu k}}\right)^2+\left[\left(\mu+\frac{\rho\overline{\upsilon}_t}{\sigma}\right)\frac{\partial\overline{\upsilon}}{\partial x_j}\right] \tag{1}$$

where $\overline{\upsilon}$ is the kinematic viscosity coefficient of the vortex, $\overline{\upsilon}_t$ is the turbulent viscosity coefficient of the vortex, and σ is a constant in the model. The model includes a decay term ${L_{\nu k}}^2$ which expresses the turbulence structure and is based on the Karman length scale.

The two-equation *k*-ε model consists of two differential transport equations:

- Turbulence kinetic energy *k*, which is a measure of the portion of energy flow that arises from velocity fluctuations:

$$\frac{\partial(\rho k)}{\partial t}+\frac{\partial}{\partial x_j}(\rho V_j k)=\frac{\partial}{\partial x_j}\left[\left(\mu+\frac{\mu_t}{\sigma_k}\right)\frac{\partial k}{\partial x_j}\right]+P_k-\rho\varepsilon+P_{kb} \tag{2}$$

- Turbulence energy dissipation rate ε, which is a measure of the conversion of turbulent kinetic energy into heat per unit time:

$$\frac{\partial(\rho\varepsilon)}{\partial t}+\frac{\partial}{\partial x_j}(\rho V_j\varepsilon)=\frac{\partial}{\partial x_j}\left[\left(\mu+\frac{\mu_t}{\sigma_\varepsilon}\right)\frac{\partial\varepsilon}{\partial x_j}\right]+\frac{\varepsilon}{k(C_{\varepsilon1}P_k-C_{\varepsilon2}\rho\varepsilon+C_{\varepsilon1}P_{\varepsilon b})} \tag{3}$$

where $C_{\varepsilon1}$, $C_{\varepsilon2}$, σ_k and σ_ε are constants, P_{kb} and $P_{\varepsilon b}$ represent buoyant forces, and P_k expresses the production of turbulence caused by viscous forces:

$$P_k=\mu_t\left(\frac{\partial V_i}{\partial x_j}+\frac{\partial V_j}{\partial x_i}\right)\frac{\partial V_i}{\partial x_j}-\frac{2}{3}\frac{\partial V_k}{(\partial x_k)\left(3\mu_t\frac{\partial V_k}{\partial x_k}+\rho k\right)} \tag{4}$$

The turbulent viscosity coefficient is related to the model parameters through a correlation:

$$\nu_t=C_\mu\rho\frac{k^2}{\varepsilon} \tag{5}$$

where C_μ is a constant.

The *Re-Normalisation Group* (*RNG*) *k*-ε model is based on the mathematical technique of renormalisation group in reference to the Navier–Stokes equations. The transport equations for turbulence generation and its dissipation are the same as in the standard *k*-ε model, whereas the difference lies in the model's constants, the constant $C_{\varepsilon 1}$ is replaced by the constant $C_{\varepsilon 1RNG}$.

The standard *Wilcox k*-ω model assumes that the turbulent viscosity coefficient is related to the turbulence kinetic energy and the turbulence frequency through the equation:

$$\nu_t = \rho \frac{k}{\omega} \tag{6}$$

This model resolves two transport equations:

- Turbulence kinetic energy k:

$$\frac{\partial(\rho k)}{\partial t} + \frac{\partial}{\partial x_j}(\rho V_j k) = \frac{\partial}{\partial x_j}\left[\left(\mu + \frac{\mu_t}{\sigma_k}\right)\frac{\partial k}{\partial x_j}\right] + P_k - \beta' \rho k \omega + P_{kb} \tag{7}$$

- Turbulence vorticity ω:

$$\frac{\partial(\rho \omega)}{\partial t} + \frac{\partial}{\partial x_j}(\rho V_j \omega) = \frac{\partial}{\partial x_j}\left[\left(\mu + \frac{\mu_t}{\sigma_\omega}\right)\frac{\partial \omega}{\partial x_j}\right] + \alpha \frac{\omega}{k} P_k - \beta \rho \omega^2 + P_{\omega b} \tag{8}$$

where P_k is the turbulence generation rate, P_{kb} and $P_{\omega b}$ are terms accounting for thermal buoyancy. β', β, α, σ_k, σ_ω are constants in the model.

In the *Baseline* (*BSL*) *k*-ω model the Wilcox model equation is multiplied by an appropriately selected transition function F_1, while the transformed *k*-ε model is multiplied by the function 1-F_1. F_1 equals 1 near the surface and decreases to 0 outside the boundary layer (it is a function of distance from the wall). Therefore, at the edge of the boundary layer and outside it, the computational model becomes the standard *k*-ε model.

In the *SST* model, to reduce the shear stress values in separated flow, a limit is imposed on the calculated value of the turbulent viscosity coefficient:

$$v_t = \frac{\alpha_1 k}{\max(\alpha_1 \omega, SF_2)} \tag{9}$$

where F_2 is a blending function similar to F_1, which imposes a limitation only for the wall boundary layer. S is a quantity describing the rate of deformation.

1.2. Impact of Turbulence Model on CFD Results

The scientific literature includes examples of numerous research articles related to the numerical modeling of turbulent airflows in ventilated buildings. For many years, the influence of turbulence models from the *RANS*, *URANS*, and *LES* groups on the accuracy of predicting air distribution in rooms has been tested [17–24]. Various boundary conditions were also considered, i.e., inlet geometry, supply air velocity, Reynolds number, isothermal or non-isothermal flows, as well as inlet turbulence quantities: kinetic energy, intensity, dissipation rate, and scale length. Exemplary results from related studies are presented in refs [25–29]. Selected representative studies conducted in this area in the last few years are presented in Table 2.

Table 2. Overview of CFD validation studies of the air jets at different distribution systems in rooms.

Authors (Year) [Ref.]	Type of Airflow in Room	Method/Turbulence Model	Validation Method	Conclusion/Preferred Method and Turbulence Model
Gurgul and Fornalik-Wajs (2023) [30]	Impinging round jet	*SST k-ω*, *RNG k-ε*, *Intermittency Transition Model (SST k-ω)*, *Transition SST*, v^2-*f*	Comparison of calculated local Nusselt number distribution with literature experimental data and inlet velocity profile with *DNS* simulation	*SST k-ω*, *SST k-ω*, and *Intermittency Transition* models have the best agreement with experimental and numerical data
Chen et al. (2022) [31]	Forced, natural, and mixed convection	*RANS/Data-driven RNG k-ε*, *conventional RNG k-ε* *LES/WALE*, and *Smagorinsky–Lilly subgrid scale*	Artificial neural network was used to determine the coefficient of high-order terms; *RANS* validated with *LES*	Data-driven model is more accurate than conventional *RNG k-ε*
Hurnik et al. (2022) [32]	Sidewall jet, recirculating flow in an occupied zone	*URANS/Standard k-ε*, *wall-modeled LES/S-Omega subgrid-scale model*	Comparison of local, gross, and integral parameters in the jet zone, and cumulative distribution of mean air speed in the occupied zone with LDA [1] and LVTA [2]	*LES* is in better agreement with measurements than *RANS* and *URANS*
Kang and van Hooff (2022) [33]	Non-isothermal side-wall jet	*RANS/Standard k-ε*, *Realizable k-ε*, *RNG k-ε*, *LRN k-ε*, *RSM (SW)*, *RSM (BSL)*, and *SST k-ω*	Comparison of measured with three-hot-wire anemometer and predicted dimensionless velocity magnitude, air temperature, and turbulence kinetic energy	*SST k-ω* is the optimal turbulence model for CFD calculations in a room with a non-isothermal supplied jet
Thysen et al. (2021) [34]	Two opposing plane wall jets in an empty airplane cabin	*RANS/RNG k-ε*, *LRN k-ε*, *SST k-ω* *LES/WALE*, and *kinetic energy subgrid scale*	Comparison of measured with PIV [3] and predicted contour maps, mean decay of dimensionless maximum velocity, and jet growth profiles	*RANS* is in acceptable agreement with measurements; *SST k-ω* performs better than the *k-ε*; *LES* performed much better than *RANS*
Sánchez et al. (2020) [35]	Ventilated façade	*Sparlat–Allmaras*, *Standard k-ε*, *RNG k-ε*, *REA k-ε*, *Standard k-ω*, *SST k-ω*	Comparison of measured with PIV [3] vertical velocity component profiles	*RNG k-ε* model is in the best agreement with measurements
Morozowa et al. (2020) [36]	Differentially heated cavity, mixed convection	Direct numerical simulation (*DNS*) *No-model* *LES/WALE* and *S3PQ* *URANS/Standard k-ε* and *SST-k-ω*	Comparison of calculated global, integral airflow quantities: Nusselt number, stratification, kinetic energy, enstrophy, and average temperature, with reference values obtained in *DNS* simulation	*LES* and no-model predict global, integral airflow quantities with higher accuracy than *URANS*
Khayrullina et al. (2019) [37]	Impinging plane jets	*RANS/Standard k-ε*, *Realizable k-ε*, *RNG k-ε*, *SST k-ω* and *Reynolds stress* model	Comparison of velocity distributions predicted and measured using PIV [3]	The differences in the validation metric are negligibly small. It is impossible to distinguish the best model
Lestinen et al. (2019) [38]	Two plane opposed jets	*URANS/SST k-ω*, *hybrid RANS-LES*—detached eddy simulation (*DES*), hybrid *RANS-LES* stress-blended eddy simulation (*SBES*)/*SST-k-ω RANS* was merged with *LES*	Comparison of velocity distributions predicted and measured using LVTA [2]	There are no final conclusions regarding the preferred turbulence model

Table 2. *Cont.*

Authors (Year) [Ref.]	Type of Airflow in Room	Method/Turbulence Model	Validation Method	Conclusion/Preferred Method and Turbulence Model
Kosutova et al. (2018) [39]	Non-isothermal mixing ventilation in an enclosure with a heated floor	*RANS/RNG k-ε, Low Reynolds number k-ε, SST k-ω, Std k-ω* and *RSM*	Comparison of velocity distributions predicted and measured using LDA [1] and temperature distributions predicted and measured using thermocouples	*Low-Reynolds-number k-ε* performed best in velocity prediction. Temperature was most accurately reproduced by *SST k-ω*
Kobayashi et al. (2017) [40]	Impinging jet	*RANS/Standard k-ε, RNG k-ε, SST k-ω,* and *Low-Re SST k-ω*	Comparison of measured and predicted vertical profiles of velocity, turbulent kinetic energy, and temperature; velocity measured with hot wire and ultrasonic anemometers	*SST k-ω* is optimal for accuracy and computational economy
Moureh and Yataghene (2017) [41]	Air curtain	*RANS/Standard k-ε, LES/Dynamic Smagorinsky subgrid scale*	Comparison of velocity distributions predicted and measured using LDA [1] and PIV [3]	*LES* predicts jet characteristics better than *RANS k-ε*, but *LES* strongly underestimates the jet deviation outwards in comparison with PIV [3]
van Hooff et al. (2017) [42]	Cross ventilation	*RANS/Standard k-ε, RNG k-ε, Realizable k-ε, SST k-ω, RSM* *LES/Dynamic Smagorinsky subgrid scale*	Comparison of measured with constant temperature anemometry system and predicted parameters: mean velocity, turbulent kinetic energy, ventilation flow rate, and spreading width	*RANS* models fail to reproduce turbulent kinetic energy, *LES* better reproduces velocity, turbulence kinetic energy and volume flow rate
Achari and Das (2015) [43]	Impinging plane jet	*RANS/Standard k-ε, Low Reynolds number k-ε* proposed by Launder and Sharma (*LS*) and Yang and Shih (*YS*), *standard k-ω*	Comparison of calculated velocity component profiles with literature experimental data	*Low-Reynolds-number k-ε* Yang and Shih (*YS*) performed best
Hurnik et al. (2015) [44]	Sidewall jet, recirculating flow in the occupied zone	*RANS/Standard k-ε with enhanced wall treatment*	Comparison of predicted local, gross, and integral parameters in the jet and occupancy zones with LDA and LVTA measurements	Reproduction of the jet momentum is necessary for accurate air speed modeling in the occupied zone
Miltner et al. (2015) [45]	Straight and slightly rotating turbulent free jets	*RANS/One-equation, Standard k-ε, RNG k-ε, Realizable k-ε, Standard k-ω, SST* and *RSM*	Comparison of velocity distributions predicted and measured using LDA	The best results of validation in terms of axial and tangential velocity components and turbulence intensity are obtained with *RSM*

[1] LDA—laser Doppler anemometer. [2] LVTA—low-velocity thermal anemometer. [3] PIV—particle image velocimetry.

In the studies presented in Table 2 and within several older papers, the air movement in the room was caused by air jets such as plane jets [41,46], plane wall jets [31,34,36,38,39,46], 3D circular quasi-free sidewall jets [32,33,42,44], impinging jets [37,40,43], slightly swirling free jets [45], and confluent jets [47]. In order to validate the CFD results in the jet zone, the air velocity measurements were performed using LDA [32,40,41,44,45], a hot-wire anemometer [40,42], a three-hot-wire anemometer [33], PIV anemometry [34,37,41], a low-velocity thermal anemometer (LVTA) [38], and an ultrasonic anemometer [40]. In a previous paper [31], *RANS* CFD results were validated using *LES* results. Direct numerical simulation (*DNS*) results were used as reference data for *URANS* and *LES* validation [36]. Most of the studies presented in Table 2 analyzed the usefulness of *RANS* turbulence models; three cases concerned *URANS* models and half of the cases involved *LES* models.

1.3. Methods of CFD Validation

Most of the papers listed in Table 2 include, among other things, a comparison of calculated and measured flow parameters in the form of contour maps and profiles. This comparison is rather qualitative and its usefulness in selecting a turbulence model when providing more accurate results is questionable.

Previous papers [33,37,39,42] presented a quantitative comparison based on determining the validation metrics *FAC*(*H*). These metrics describe the fraction of the data within the range $(1/H; H)$:

$$FAC(H) = \frac{1}{n}\Sigma_{i=1}^{n} B \ for \ B = \begin{Bmatrix} 1 \ for \ 1/H \leq \frac{predicted\ value_i}{measured\ value_i} \leq H \\ 0 \ else \end{Bmatrix} \quad (10)$$

where H = 1.05, 1.1, 1.25, 1.3, 1.5, or 2.

In some cases, the global and integral parameters were used to quantitatively compare the CFD results. In an earlier paper [36], predicted values of the Nusselt number, stratification, average temperature, and average kinetic energy were quantitatively compared with the reference data.

In other previous work [32,44], the maximum mean velocity $\overline{U}_{xm}$, jet width R, and momentum flux M were used to validate the CFD predictions. These gross and integral parameters, that characterize the velocity distribution in the jet at a certain distance from the inlet, are presented in more detail in Section 2.3.

Average Speed in Occupied Zone Versus Jet Momentum Flux

So far, the benchmark test [48] has been used to validate CFD results for three types of turbulence models, i.e., *RANS Std k-ε* model [44], unsteady Reynolds-averaged Navier–Stokes (*URANS*), and vortex-resolving large eddy simulation *LES* [32]. The *RANS Std k-ε* model was used for two cases (denoted A and B) with different boundary conditions. The *LES* results with 16 and 35 million cells were analyzed. In the case of the benchmark test [48], the jet can be considered quasi-free at a distance x/D_e = 10–32 and, in this region, the airflow can be approximated using a point source of momentum model (PSM). Based on the available data, the relationship between the average air speed in the occupied zone, $\overline{W}_{aver}$, and the square root of the jet momentum flux $(M/\rho)^{1/2}$ is determined, see Figure 1.

Figure 1 confirms that the average air speed in the occupied zone, $\overline{W}_{aver}$, is very highly correlated with the square root of the jet momentum flux $(M/\rho)^{1/2}$. Incorrect modeling of the momentum flux may result in the classification of a room's thermal conditions into the wrong category. To avoid this, the modeling uncertainty of the square root of momentum flux $(M/\rho)^{1/2}$ should be less than about 5%.

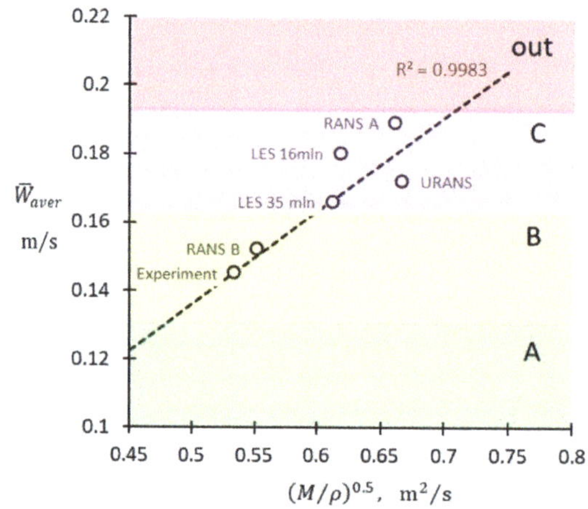

Figure 1. Correlation between the average mean speed in the occupied zone and the square root of momentum flux of the jet. Source of data: Experiment, *URANS*, *LES* 16 and 35 million [32], *RANS* A and B [44]; color scale: category of thermal local discomfort due to draft *DR*; category A—*DR* < 15%; category B—*DR* < 20%; category C—*DR* < 25%; assumed air temperature—20 °C; and turbulence intensity 40%.

1.4. Recommended Turbulence Models

In all cases comparing the *RANS* and *URANS* models with the results of the *LES* approach it was found that the *LES* results better reproduce the tested airflows in the room. In previous work [33,34,39,40], it was indicated that the *RANS/SST k-ω* model better agreed with measurement data compared to other *RANS* models. Earlier studies [37,38] did not identify the best turbulence model among the *RANS* models. Recent studies were mainly based on a statistical approach when estimating the uncertainties arising from the adopted turbulence model, whereas data-driven methods were predominantly used to reduce these uncertainties. In a prior paper [31], a data-driven *RANS* nonlinear model with coefficients of high-order terms determined using an artificial neural network was proposed. Three indoor airflows were selected as a training set, and four other flows were used to verify the model. The results show that this model can better predict anisotropic indoor flows.

Based on a review of the literature, it can be stated that *LES* modeling undeniably provides more accurate and reliable results than *RANS*. However, the *LES* models require higher computational costs and are more time consuming. Therefore, the LES method is rarely used in engineering applications. The author of ref. [49] concluded that *RANS* models are not obsolete because *RANS* is still widely used in engineering research and practice. Although *LES* is superior in its own right, it incurs greater simulation complexity and significantly higher computational costs. In the review conducted in ref. [12], the authors stated that CFD simulations of industrial flows in the coming decades will still mostly be based on the *RANS* turbulence model, and the uncertainties in the *RANS* model will remain a major obstacle to the predictive ability of these simulations. Thus, quantifying uncertainties in *RANS* predictions is essential to achieving the goal of certified CFD simulations.

In the case of the benchmark test [48], the jet can be considered quasi-free at a distance of x/D_e = 10–32. In this region, the airflow in the jet can be approximated and compared using a point source of momentum model (PSM). So far, this benchmark test has been used to validate CFD results for three types of turbulence models, i.e., the *Std k-ε* model available in the Fluent Airpak 3.0.16 commercial code [44], unsteady Reynolds-averaged Navier–Stokes (*URANS*) [32], and vortex-resolving large eddy simulation (*LES*) [32]. The last two are available in ANSYS Fluent.

The aim of the tests presented here is the qualitative and quantitative assessment of the results of the CFD calculation of airflow in a sidewall jet based on benchmark data [48]. The tests were carried out with steady-state conditions using seven Eddy–Viscosity Models (*EVM*) available in the ANSYS CFX 22.1 software. The goal was to find an *EVM* turbulence model that could provide the most similar results to the experiment and CFD results obtained with the *LES* turbulence model. So far, no validation of the CFD calculations for such cases has been performed.

2. Methods

2.1. Benchmark of a Room with a Sidewall Jet

The geometry of the test room with a sidewall jet, which was proposed by Hurnik et al. [48] for CFD validation, is shown in Figure 2. The room's dimensions correspond to a medium-sized office or living room. The air was supplied from a rectangular opening with the dimensions 0.144 m × 0.096 m and the velocity U_0 = 5.16 m/s. The measurements were performed both in the jet region and in the occupied zone. A two-dimensional laser Doppler anemometer was used to measure velocity components in the jet region in two perpendicular planes (Figure 2). In the occupied zone, the air velocity was measured using a low-velocity thermal anemometer with omnidirectional sensors. A detailed description of the benchmark and the full set of measurement results are presented in earlier papers [44,48].

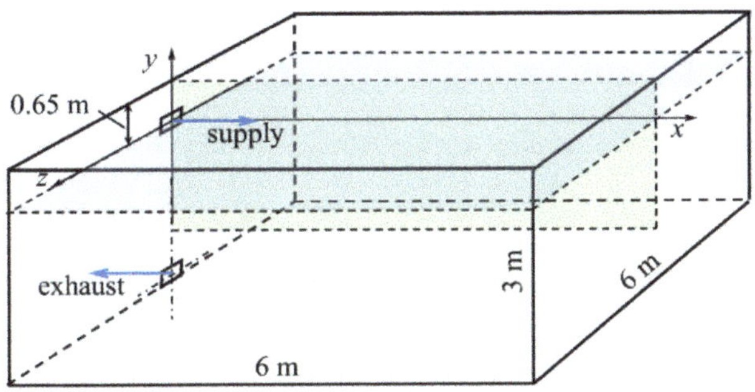

Figure 2. Tested room and the measurement planes in the jet zone.

2.2. Numerical Method

Numerical calculations were carried out with the use of Ansys CFX 22.1 software in steady-state and isothermal conditions for half of the test room due to its symmetry (Figure 2). The dimensions of the modeled half of the air supply opening were 0.072 m × 0.096 m (1/2 width × height). The intensity of the turbulence in the opening was set equal to 5%. The Navier–Stokes differential equations were discretized using the Finite Volume Method. The second-order upwind discretization scheme and Rhie–Chow algorithm were employed to couple pressure and velocity. Wall functions with the no-slip boundary condition were adopted. The Auto Timescale control option was selected with the conservative Length Scale and the Timescale Factor set to a default value of 1. Boundary conditions for the conducted numerical simulations are presented in Table 3.

The grid independence test was carried out with the use of the standard *k*-ε model. Three variants of discretization grids consisting of tetrahedral elements were tested. Their parameters are listed in Table 4. In each of the variants a boundary layer with a maximum thickness of 0.6 m was used in which the mesh size was equal to 0.01 m. In addition, mesh refinement was implemented on the surface of the inlet with a mesh edge length of 0.01 m. In the G1 variant, a default discretization grid with a mesh edge length of 0.3 m was used. In the G2 variant, the length of mesh edge was reduced to 0.1 m. In the G3 variant, additional mesh refinement was implemented in the supply jet axis with a refinement radius of 0.6 m and the length of the refined mesh edge being 0.01 m.

Table 3. Boundary conditions.

Boundary Condition	Value/Description
Analysis type	Steady state
Supply air speed	5.16 m/s
Inlet turbulence intensity	5%
Heat transfer	Isothermal
Air temperature	23 °C
Outlet relative pressure	1 Pa
Outlet pressure profile blend	0.05
Outlet pressure averaging	Average over whole outlet
Boundary condition	No slip wall
Wall roughness	Smooth wall
Reference domain pressure	101,325 Pa

Table 4. Tested discretization grid variants.

Discretization Grid Variant	Mesh Edge Length	Refinement Mesh Edge Length	Number of Elements
G1	0.3 m	-	4.10×10^4
G2	0.1 m	-	4.95×10^5
G3	0.1 m	0.01 m (refinement radius 0.6 m)	3.51×10^7

Numerical calculations carried out with the use of all discretization grid variants were validated with the use of experimental results [48] in the jet region (Figure 3). The best concurrence of the results was obtained with the use of G3 variant grid. The distribution of the mean axial velocity component $\overline{U}_x$ was the most similar to the measured one, both in terms of its maximum value and its profile. The results obtained with the use of the G2 grid variant were similar to the G3 variant, but the range of maximum values was higher and the air velocity profile was wider in the G2 variant. The G1 grid variant was not able to accurately reproduce the air velocity profile, which could have been affected by the large mesh size. Therefore, the G3 grid variant was adopted in the research on turbulence models.

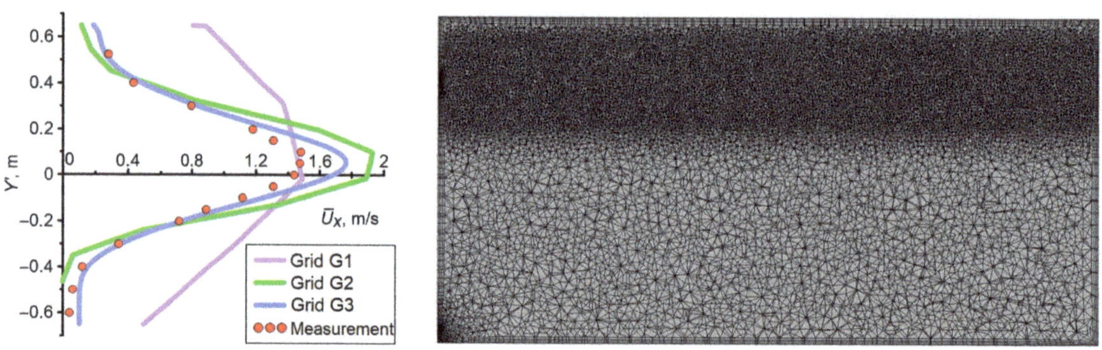

(a) (b)

Figure 3. Comparison of the measurement data with the CFD results obtained with the use of three discretization grid variants for the standard *k*-ε turbulence model in a jet region; distribution of the mean axial velocity component $\overline{U}_x$ for x/D_e = 20.8 (**a**) and the G3 discretization grid; cross-section in a plane passing through the center of the inlet (**b**).

In order to evaluate the numerical model's quality, a convergence assessment of the numerical solution was conducted. The residual is the most important determinant of numerical solution convergence as it directly reflects the accuracy of an equation's solution [50]. The root mean square residuals of pressure and velocity were selected to assess when convergence was reached. The value at which the pressure and velocity root mean square residuals stabilized for all the *k-ω* models was 1×10^{-7}, for the *k-ε* models it was 8×10^{-8} (i.e., the standard *k-ε* model) and 4×10^{-7} (i.e., the *RNG k-ε* model), and, for the one-equation *EVTM* model it was 1×10^{-7}. According to [50], root mean square residual levels of value 1×10^{-6} or lower show a very tight convergence and are sufficient for engineering applications.

The number of iterations after which the monitored parameters in the jet and the boundary layer region stabilized (velocity components, turbulent kinetic energy, and turbulent dissipation rate—except for models where turbulent kinetic energy is absent) was in the range of 1.2×10^3 for the *EVTM* model to 1×10^4 for the *SST* model. The relative wall distance value $y+$ was lower than 15 for all turbulence models analyzed.

2.3. Local, Gross, and Integral Parameters

The CFD results are most often validated by comparing profiles of local velocity parameters such as mean velocity $\overline{U}$, mean velocity components $\overline{U}_x$, $\overline{U}_y$, $\overline{U}_z$, and standard deviations of velocity fluctuations u_x^*, u_y^*, u_z^* in selected jet cross-sections. This method of validation can also be called point-to-point comparison. More reliable and representative validation can be carried out by comparing gross and integral jet parameters. Gross parameters, i.e., maximum mean velocity $\overline{U}_{xm}$, position of the maximum mean velocity y_m, and jet width R, and integral parameters, i.e., volume flux V and momentum flux M, characterize the airflow in the jet at a certain distance from the supply opening.

In the benchmark tests, the jet is supplied from the rectangular opening and affected by the ceiling; therefore, it cannot be treated as an axisymmetric jet. Due to the Coanda effect, the position of the point of maximum velocity y_m changes with the distance from the opening and has to be identified. The values of the mean axial air velocity component $\overline{U}_{xm}$, measured and calculated in several cross-sections of the turbulent jet region, were approximated using a quasi-Gaussian exponential curve:

$$\overline{U}_x = \overline{U}_{xm} \cdot exp\left[-(r/R_\alpha)^{7/4}\right] \tag{11}$$

The radial distance from the jet axis position equals:

$$r = \left[(z - z_m)^2 + (y - y_m)^2\right]^{1/2} \tag{12}$$

To describe the velocity distribution in an asymmetric air jet, the angular change in the velocity profile width should be considered. In the case of the CFD results, it was possible to analyze the radial changes in the velocity in the 180° range covered by the CFD data and 180° covered by the assumption of the flow symmetry in the z-plane. In this case, the jet profile width was calculated as a trigonometric series of six harmonic components:

$$R_\alpha = R(1 + a_1\, cos\alpha + a_2\, cos2\alpha + a_3\, cos3\alpha + a_4\, cos4\alpha + \cdots) \tag{13}$$

The angle α is found in this expression, the explicit form of which is given by:

$$\alpha = arctan2[(z - z_m), (y - y_m)]. \tag{14}$$

The set of parameters $\overline{U}_{xm}, R, y_m, z_m, a_1, \ldots a_6$, describing the distribution of the mean axial velocity component at different distances from the inlet opening x, were found by a least-squares method using the SOLVER procedure in EXCEL. Next, the air velocity in the quasi-free jet zone was approximated by the model of the jet from a point source of momentum. The set of equations for the PSM model with the profile exponent $n = 1.75$ is presented in Table 5. The gross parameters are jet spread coefficient α, position of the jet origin x_o/D_e, and coefficient of momentum loss K_M, which characterize the whole jet in the quasi-free region.

Table 5. Model of a free axisymmetric jet generated by a point source of momentum (PSM) model.

Definition	Equation	#
Jet spread	$R = a(x - x_o)$	(15)
$\overline{U}_x$ velocity profile	$\overline{U}_x/\overline{U}_{xm} = exp\left[-(r/R)^{7/4}\right]$	(16)
M/ρ in PSM (constant)	$M/\rho = 1.5210\,\overline{U}_{xm}^2 R^2$	(17)
Boundary momentum flux	$M_o/\rho = A_o\ U_o^2 = (\pi\ D_e^2/4)U_o^2$	(18)
Conservation of momentum flux	$(M/M_o)^{1/2} = K_M$	(19)
Decay of $\overline{U}_{xm}$	$\overline{U}_{xm}/U_o\ = K_M\ [0.7186\ /a]/[(x - x_o)/D_e]$	(20)

3. Results

3.1. Maps and Profiles of Mean Axial Velocity Component

Figure 4 shows contour maps of the mean axial air velocity component $\overline{U}_x$/s distribution in the supply air jet normalized by the inlet velocity U_o = 5.16 m/s. The maps were prepared using measurement data and the CFD calculation results with the use of *EVM* turbulence models. The area on the maps limited by 1% isoline is the background of the jet. The map for the zero-equation model significantly differs from the maps for the other *EVM* turbulence models and is characterized by a much shorter range for all of the isolines. Therefore, the results for the zero-equation model were excluded from further analyses.

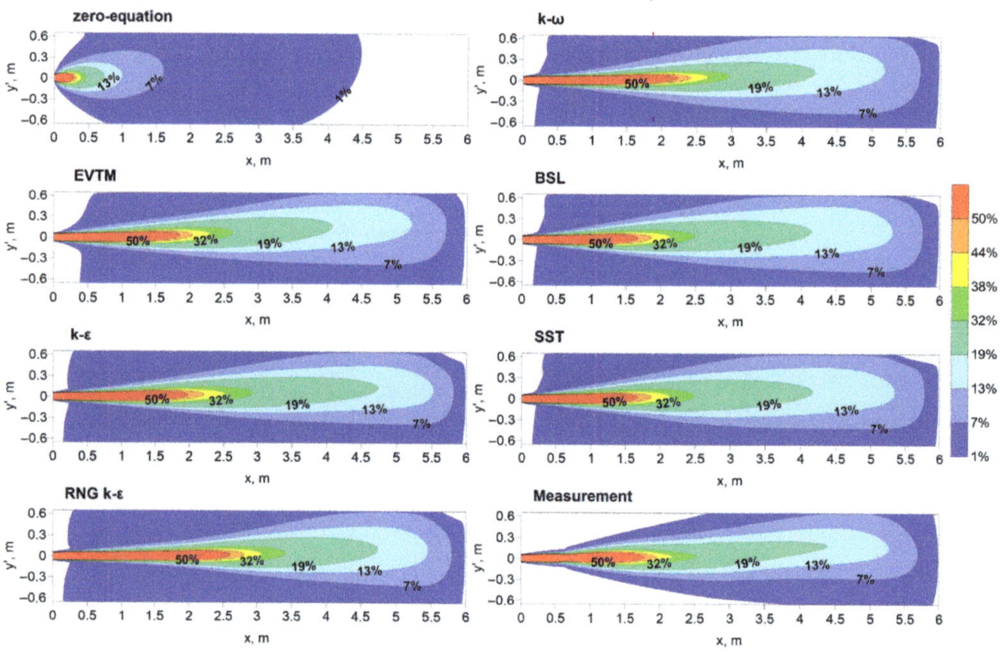

Figure 4. Contour maps of the normalized axial mean air velocity component $\overline{U}_x/U_o$ in the plane cross-section z = 0 for the *EVM* turbulence models and measurement data.

The vertical jet profiles of the axial mean air velocity component are deformed due to the deflection of the air jet towards the ceiling. Therefore, the profiles in the horizontal plane $y = 0$ at a distance from the inlet equal to x = 2.79 m (x/D_e = 20.8) were selected for comparison, see Figure 5. The results of measurements and CFD calculations were approximated by an exponential curve in a form corresponding to:

$$\overline{U}_x = \overline{U}_{xm}\,|_{x=2.79;\,y=0} \cdot e^{-(|z|/R|_{x=2.79;\,y=0})^{7/4}} \tag{21}$$

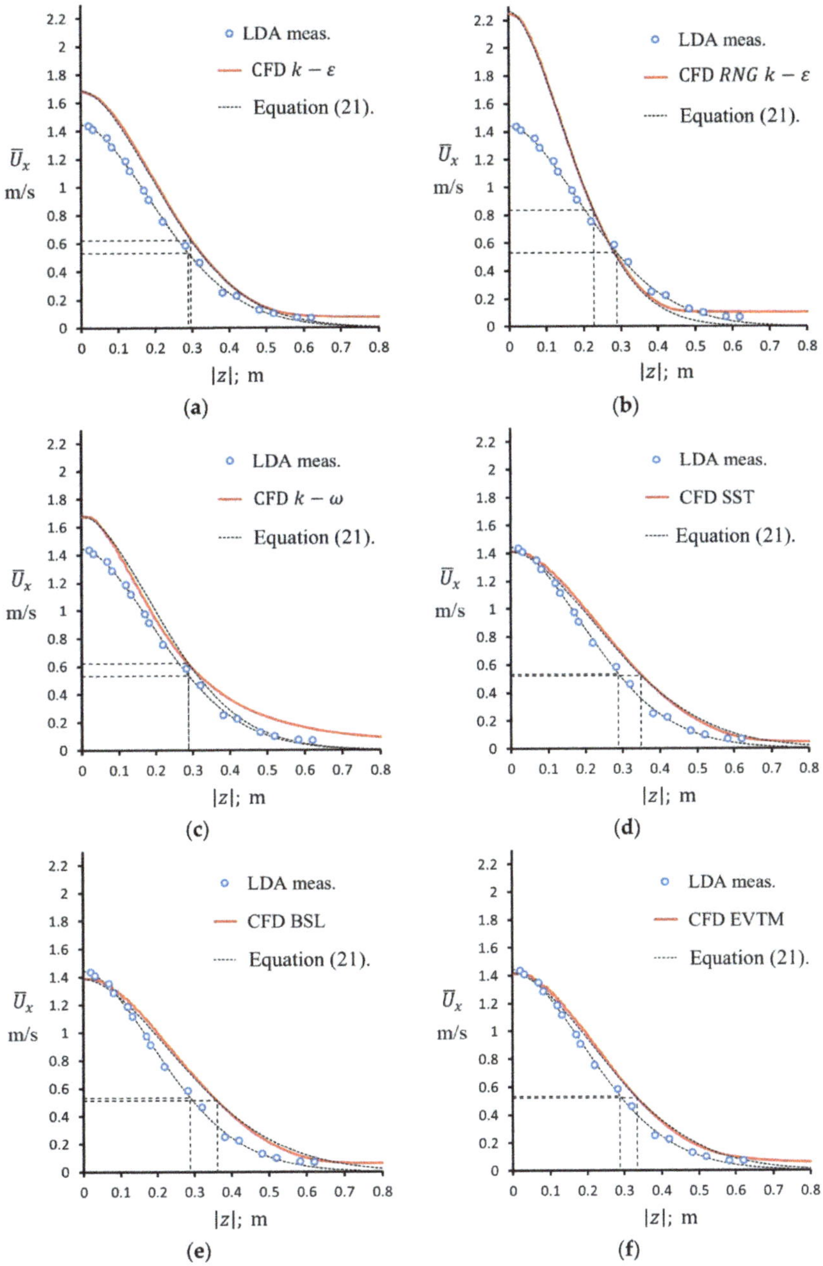

Figure 5. Axial mean air velocity component profiles in the horizontal plane with $y = 0$ and $x = 2.79$ m (i.e., $x/D_e = 20.8$). Turbulence models: (**a**) *k-ε*; (**b**) *RNG k-ε*; (**c**) *k-ω*; (**d**) *SST*; (**e**) *BSL*; (**f**) *EVTM*.

Two parameters were obtained as a result of the approximation for the compared profiles, i.e., the maximum value of the axial velocity component $\overline{U}_{xm}\,|_{|x=2.79;\ y=0}$ and the jet profile width $R|_{|x=2.79;\ y=0}$. Thus, it was possible to compare the analyzed profiles quantitatively. The approximation lines are marked in Figure 5 with dashed black lines.

3.2. Gross and Integral Parameters in the Quasi-Free Jet Zone

The distributions of the gross and integral parameters of the air jet for a certain distance from the inlet are shown in Figure 6. The parameters of the point source of momentum (PSM) model obtained by an approximation of the air velocity distribution in the quasi-free zone of the jet are presented in Table 6.

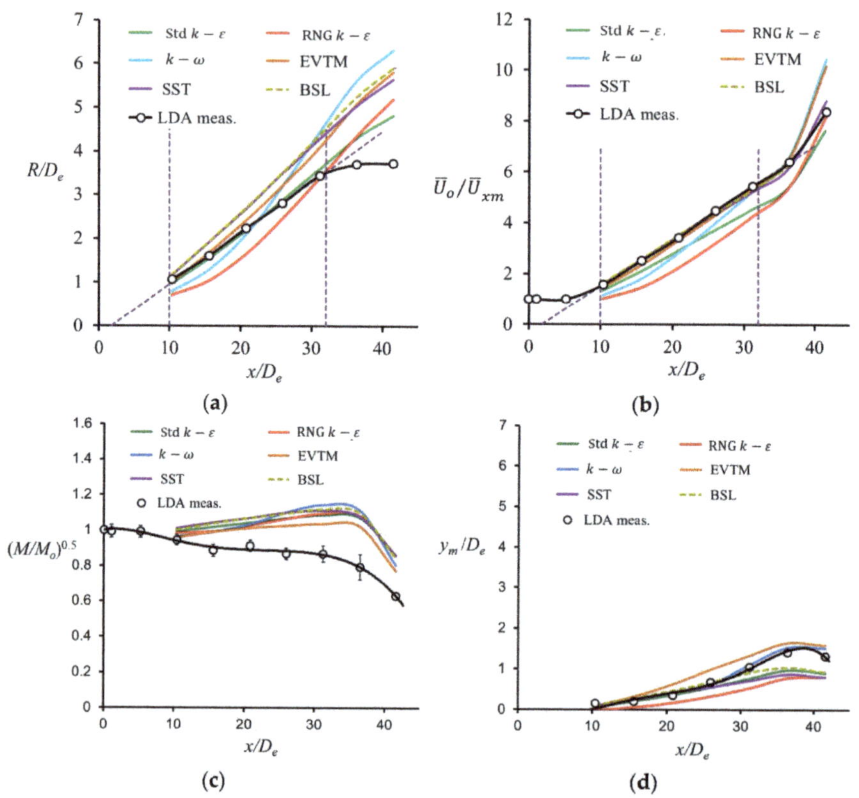

Figure 6. Changes in the gross and integral parameters of the jet for the distance from the inlet: (**a**) the jet width R/D_e; (**b**) the ratio of inlet velocity and maximum mean axial velocity $U_o/\overline{U}_{xm}$; (**c**) the square root of jet momentum flux $(M/M_o)^{1/2}$; and (**d**) the vertical position of the maximum mean velocity y_m/D_e.

Table 6. Parameters of the point source of momentum (PSM) model obtained by an approximation of the air velocity distribution in the quasi-free zone of the jet.

	x_o/D_e	a	K_M
LDA	1.8	0.117	88.4%
LES [32]	2.2	0.130	100.6%
k-ω	4.6	0.138	103.4%
Std k-ε	2.0	0.118	104.0%
RNG k-ε	4.0	0.105	102.9%
EVTM	3.0	0.137	100.6%
BSL	1.9	0.143	106.4%
SST	2.5	0.148	106.5%

A comparison of the *EVM* and *LES* results in the quasi-free jet region with the outcomes of the measurements using LDA is presented in Table 7. The quantities $\Delta(x_o/D_e)$, $\delta(a_u)$, and $\delta(K_M)$ represent the absolute differences in the position of the jet origin, relative differences in the jet spread coefficient, and relative differences in the momentum losses coefficient, i.e.,

$$\Delta(x_o/D_e) = (x_o/D_e)\,|^{\,\mathrm{EVM}} - (x_o/D_e)\,|^{\,\mathrm{LDA}}, \tag{22}$$

$$\delta(a_u) = (a\,|^{\,\mathrm{EVM}} - a\,|^{\,\mathrm{LDA}})/a\,|^{\,\mathrm{LDA}}, \tag{23}$$

$$\delta(K_M) = (K_M\,|^{\,\mathrm{EVM}} - K_M\,|^{\,\mathrm{LDA}})/K_M\,|^{\,\mathrm{LDA}}. \tag{24}$$

Table 7. Comparison of the *EVM* results in the quasi-free jet region with the outcomes of the measurement using the LDA.

Turbulence Model	Linear Jet Spread	Inverse Changes of Maximum Velocity	"Gaussian" Radial Profile of Velocity	$\Delta(x_o/D_e)$	$\delta(a)$	$\delta(K_M)$
LDA	+	+	+	0.0	0.0%	0.0%
LES [25]	+	+	+	0.4	11.1%	13.8%
k-ω	−	−	−	2.8	17.9%	17.0%
Std k-ε	+	+	+	0.2	0.9%	17.6%
RNG k-ε	−	−	−	2.2	−10.3%	16.4%
EVTM	+	+	+	1.2	17.1%	13.8%
BSL	+	+	+	0.1	22.2%	20.4%
SST	+	+	+	0.7	26.5%	20.5%

4. Discussion

The results of the LDA measurements confirmed that the tested jet at a distance x from 1.3 m to 4.2 m (x/D_e from 10 to 32) behaves like a quasi-free jet, which is proved by:

- The self-similarity of the mean velocity distribution, as given in Equation (11);
- The linear spread of the jet, as provided in Equation (15);
- The fact that they are inversely proportional to distance velocity decay, as given in Equation (20).

The maps for all the *EVM* turbulence models, see Figure 4, provided a qualitatively similar but not identical picture of the velocity field in the jet zone. Based on these maps, it is not possible to accurately determine the throw length of the jet because the isolines for less than 0.5 m/s terminate in the jet impingement zone, i.e., less than 1.8 m from the opposite wall. Comparing the contour maps $\overline{U}_x/U_o$ for the *BSL* and *SST* models, it is noteworthy that they are very similar. Examining the maps, it can also be seen that all the maps show a slight deflection of the jet towards the ceiling. Determining the other global parameters, such as x_o, R, and $\overline{U}_{xm}$, based on these maps, it is clear that they may include significant errors. Therefore, it can be concluded that the maps have little usefulness in the validation of the CFD results.

The profiles of the axial mean air velocity component in the middle of the quasi-free jet zone, when the horizontal plane $y = 0$ and the distance $x = 2.79$ m (i.e., $x/D_e = 21$), are presented in Figure 5. As shown in Figure 4, the maximum mean axial velocity $\overline{U}_{xm}\,|_{x=2.79;\, y=0}$ valued obtained for the *EVTM*, *BSL*, and *SST* models were very close to the measured one but, at this distance, the jet profile widths $R|_{x=2.79;\, y=0}$ were greater than the measured one by 10%, 24%, and 21%, respectively. The measured jet profile width $R|_{x=2.79;\, y=0}$ and the one calculated with the use of the *k-ω* and *Std k-ε* models differed very little, but the maximum mean axial velocity $\overline{U}_{xm}\,|_{x=2.79;\, y=0}$ was higher than the measured one by 20%. The results obtained for the *RNG k-ε* model varied the most from the measurements, both in terms of $\overline{U}_{xm}\,|_{x=2.79;\, y=0}$ and $R|_{x=2.79;\, y=0}$. The quantitative comparison of the velocity profiles presented in Figure 5 did not provide unambiguous and conclusive arguments that allowed for the assessment of the usefulness of the analyzed turbulence models. Therefore, an in-depth comparative analysis of the gross and integral parameters, as well as parameters for the free jet model generated by the point source of momentum (PSM), was needed. The results of this analysis are presented in Figure 6 and Tables 6 and 7.

In the quasi-free jet zone, the jet width increases linearly and the maximum velocity decreases inversely with the distance from the jet origin. Therefore, as seen in Figure 6a,b, the R/D_e and $U_o/\overline{U}_{xm}$ values changed linearly. Such a linear relationship can be observed for the *EVTM*, *Std k-ε*, *BSL*, and *SST* turbulence models but not for the *k-ω* and *RNG k-ε* models. The best agreement between the measurement and calculations was obtained for the *EVTM* and *Std k-ε* turbulence models. However, while the *EVTM* model accurately reproduced $\overline{U}_{xm}$, it overestimated R by 17%. The *Std k-ε* model performed in an opposite way, i.e., it overestimated $\overline{U}_{xm}$ by 17% and precisely predicted R. The changes in the momentum flux $(M/M_o)^{1/2}$ are shown in Figure 6c. All the turbulence models overestimated the measured value of this flux. The least overestimated results were obtained for the *EVTM* model. Figure 6d shows the changes in the vertical position of the maximum mean velocity y_m/D_e. All the turbulence models reproduced the deflection of the air jet towards the ceiling with satisfactory accuracy.

The highest discrepancies in the calculated and measured positions of the jet's origin x_o/D_e were obtained with the *k-ω* and *RNG k-ε* models, and the best compliance was obtained with the *Std k-ε* and *BSL* models. The coefficient a_u that characterized the jet spread was modeled with the best accuracy using the *Std k-ε* turbulence model.

In order to correctly model the air distribution in the occupied zone of the room, it is necessary to accurately reproduce the momentum flux of the supply air jet. The velocity in the occupied zone is directly proportional to the square root of the momentum flux $(M/\rho)^{1/2}$ [4]. Thus, the momentum loss coefficient K_M determined by the CFD calculations should be as close as possible to the measured one. Considering this criterion, optimal compliance was obtained for the *EVTM* turbulence model. In this case, the K_M coefficient and, consequently, the air velocity in the occupied zone were about 12% higher than the measured values.

The analysis presented in Table 7 shows that the *k-ω* and *RNG k-ε* turbulence models did not correctly reproduce the jet spread, the decay of the maximum velocity, and the velocity profile in the jet cross section and, therefore, cannot be recommended to simulate velocity distribution in a quasi-free jet region. The *BSL* and *SST* models calculated air distribution parameters with errors higher than 20%. The results obtained with the standard *Std k-ε* and *EVTM* models were most similar to the measurement data since their errors slightly exceeded 15%. Comparing the results of the CFD calculations using the *EVM* turbulence models with those of the *LES* calculations, a good agreement between the *EVTM* and *LES* results can be observed.

5. Conclusions

In the presented studies, numerical CFD modeling of an air jet supplying air from the sidewall to the room was carried out using seven *EVM RANS* turbulence models. Validation of CFD results was performed based on experimental data [48]. This case represents a typical office or residential room. A comparative analysis of gross and integral parameters, as well as parameters of the quasi-free jet model generated by a point source of momentum, was carried out. Based on the obtained results, the following conclusions can be drawn.

Jet spread, the decay of the maximum velocity, and the velocity profile in the jet cross-section were not correctly reproduced by the *k-ω* and *RNG k-ε* models. Therefore, they cannot be recommended for simulations of air velocity distribution in a quasi-free jet region.

The values of the global and integral parameters obtained with the standard *Std k-ε* and *EVTM* models were most similar to the experimental data. Their error values slightly exceeded 15%. These *EVM RANS* turbulence models can be effectively used in the HVAC industry to simulate air distribution in supplied air jets in the early conceptual stages of HVAC system design [36].

The greatest consistency in the momentum loss coefficient K_M was obtained for the *EVTM* turbulence model. Its value was overestimated by 12%, exactly the same value as in the case of the *LES* model.

The errors in the CFD simulations for the *BSL* and *SST* models were higher than 20%, which places these models in second place after the *Std k-ε* and *EVTM* models.

Further research on the use of the *EVM RANS* models could include a more thorough analysis of the air speed in the occupied zone to verify their applicability in the presence of people within a room.

Author Contributions: Conceptualization, M.H., P.C. and Z.P.; methodology, M.H. and Z.P.; software, P.C.; validation, M.H., P.C. and Z.P.; formal analysis, M.H. and Z.P.; data curation, P.C. and M.H.; writing—original draft preparation, M.H., P.C. and Z.P.; writing—review and editing, M.H., P.C. and Z.P.; supervision, Z.P. All authors have read and agreed to the published version of the manuscript.

Funding: This research received no external funding.

Data Availability Statement: Data are contained within the article.

Acknowledgments: The work was supported by the Polish Ministry of Education and Science within the research subsidy.

Conflicts of Interest: The authors declare no conflict of interest.

Nomenclature

a	jet spread coefficient (-)
B	boolean variable (-)
D	diameter (m)
H	upper limit value (-)
k	turbulence kinetic energy (m^2/s^2)
K	coefficient (-)
M	mean motion momentum flux in the axial direction ($kg \cdot m/s^2$)
n	number of samples (-)
r	radial distance from the jet axis (m)
R	radial width of the jet profile (m)
U	velocity (m/s)
W	speed (m/s)
x, y, z	Cartesian coordinates (m)
Greek symbols:	
ε	turbulence energy dissipation rate (m^2/s^3)
v	viscosity coefficient (m^2/s)
ρ	density (kg/m^3)
ω	turbulence vorticity (1/s)
Subscripts:	
e	equivalent
i	axis of coordinate system, $i = x, y, z$
m	molecular, maximum
M	momentum
o	inlet, origin
t	turbulent

References

1. *ANSI/ASHRAE Standard 55;* Thermal Environmental Conditions for Human Occupancy. ANSI: Washington, DC, USA, 2023.
2. *ISO 7730:2005 (En);* Ergonomics of the Thermal Environment—Analytical Determination and Interpretation of Thermal Comfort Using Calculation of the PMV and PPD Indices and Local Thermal Comfort Criteria. ISO: Geneva, Switzerland, 2005.
3. Hussein, T.; Löndahl, J.; Thuresson, S.; Alsved, M.; Al-Hunaiti, A.; Saksela, K.; Aqel, H.; Junninen, H.; Mahura, A.; Kulmala, M. Indoor Model Simulation for COVID-19 Transport and Exposure. *Int. J. Environ. Res. Public Health* **2021**, *18*, 2927. [CrossRef]

4. Cravero, C.; Marsano, D. Simulation of COVID-19 Indoor Emissions from Coughing and Breathing with Air Conditioning and Mask Protection Effects. *Indoor Built Environ.* **2022**, *31*, 1242–1261. [CrossRef]
5. Awbi, H.B. *Ventilation of Buildings*; Taylor & Francis: London, UK, 2003; ISBN 978-0-415-27056-4.
6. Jackman, P.J. *Air Movement in Rooms with Side-Wall Mounted Grilles: A Design Procedure*; Heating & Ventilation Research Association: London, UK, 1970.
7. *ANSI/ASHRAE Standard 113-2005*; Method of Testing for Room Air Diffusion. ANSI: Washington, DC, USA, 2005.
8. Yang, B.; Melikov, A.K.; Kabanshi, A.; Zhang, C.; Bauman, F.S.; Cao, G.; Awbi, H.; Wigö, H.; Niu, J.; Cheong, K.W.D.; et al. A Review of Advanced Air Distribution Methods—Theory, Practice, Limitations and Solutions. *Energy Build.* **2019**, *202*, 109359. [CrossRef]
9. Stern, F.; Wilson, R.V.; Coleman, H.W.; Paterson, E.G. *Verification and Validation of CFD Simulations*; Defense Technical Information Center: Fort Belvoir, VA, USA, 1999.
10. Chen, Q.; Srebric, J. A Procedure for Verification, Validation, and Reporting of Indoor Environment CFD Analyses. *HVAC&R Res.* **2002**, *8*, 201–216. [CrossRef]
11. Chen, Q. Editorial: Computer Simulations and Experimental Measurements of Air Distribution in Buildings: Past, Present, and Future. *HVAC&R Res.* **2007**, *13*, 849–851. [CrossRef]
12. Xiao, H.; Cinnella, P. Quantification of Model Uncertainty in RANS Simulations: A Review. *Prog. Aerosp. Sci.* **2019**, *108*, 1–31. [CrossRef]
13. Cravero, C.; De Domenico, D.; Marsano, D. Uncertainty Quantification Analysis of Exhaust Gas Plume in a Crosswind. *Energies* **2023**, *16*, 3549. [CrossRef]
14. Xia, L.; Zou, Z.-J.; Wang, Z.-H.; Zou, L.; Gao, H. Surrogate Model Based Uncertainty Quantification of CFD Simulations of the Viscous Flow around a Ship Advancing in Shallow Water. *Ocean Eng.* **2021**, *234*, 109206. [CrossRef]
15. Ansys CFX | Industry-Leading CFD Software. Available online: https://www.ansys.com/products/fluids/ansys-cfx (accessed on 24 April 2023).
16. Piomelli, U. Large Eddy Simulations in 2030 and beyond. *Phil. Trans. R. Soc. A* **2014**, *372*, 20130320. [CrossRef]
17. Zhai, Z.J.; Zhang, Z.; Zhang, W.; Chen, Q.Y. Evaluation of Various Turbulence Models in Predicting Airflow and Turbulence in Enclosed Environments by CFD: Part 1—Summary of Prevalent Turbulence Models. *HVAC&R Res.* **2007**, *13*, 853–870.
18. Zhang, Z.; Zhang, W.; Zhai, Z.J.; Chen, Q.Y. Evaluation of Various Turbulence Models in Predicting Airflow and Turbulence in Enclosed Environments by CFD: Part 2—Comparison with Experimental Data from Literature. *HVAC&R Res.* **2007**, *13*, 871–886.
19. Chen, Q.; Xu, W. A Zero-Equation Turbulence Model for Indoor Airflow Simulation. *Energy Build.* **1998**, *28*, 137–144. [CrossRef]
20. Wang, M.; Chen, Q. Assessment of Various Turbulence Models for Transitional Flows in an Enclosed Environment (RP-1271). *HVAC&R Res.* **2009**, *15*, 1099–1119.
21. Meslem, A.; Dia, A.; Beghein, C.; El Hassan, M.; Nastase, I.; Vialle, P.-J. A Comparison of Three Turbulence Models for the Prediction of Parallel Lobed Jets in Perforated Panel Optimization. *Build. Environ.* **2011**, *46*, 2203–2219. [CrossRef]
22. Hussain, S.; Oosthuizen, P.H.; Kalendar, A. Evaluation of Various Turbulence Models for the Prediction of the Airflow and Temperature Distributions in Atria. *Energy Build.* **2012**, *48*, 18–28. [CrossRef]
23. Stamou, A.; Katsiris, I. Verification of a CFD Model for Indoor Airflow and Heat Transfer. *Build. Environ.* **2006**, *41*, 1171–1181. [CrossRef]
24. Serra, N. Revisiting RANS Turbulence Modelling Used in Built-Environment CFD Simulations. *Build. Environ.* **2023**, *237*, 110333. [CrossRef]
25. Srebric, J.; Chen, Q. Simplified Numerical Models for Complex Air Supply Diffusers. *HVAC&R Res.* **2002**, *8*, 277–294. [CrossRef]
26. Cao, S.-J.; Meyers, J. Influence of Turbulent Boundary Conditions on RANS Simulations of Pollutant Dispersion in Mechanically Ventilated Enclosures with Transitional Slot Reynolds Number. *Build. Environ.* **2013**, *59*, 397–407. [CrossRef]
27. Tapsoba, M.; Moureh, J.; Flick, D. Airflow Patterns in a Slot-Ventilated Enclosure Partially Loaded with Empty Slotted Boxes. *Int. J. Heat Fluid Flow* **2007**, *28*, 963–977. [CrossRef]
28. Koskela, H.; Maula, H. A CFD Model of a Swirl Diffuser for Heating and Cooling Modes. *Int. J. Vent.* **2013**, *12*, 159–166. [CrossRef]
29. Susin, R.M.; Lindner, G.A.; Mariani, V.C.; Mendonça, K.C. Evaluating the Influence of the Width of Inlet Slot on the Prediction of Indoor Airflow: Comparison with Experimental Data. *Build. Environ.* **2009**, *44*, 971–986. [CrossRef]
30. Gurgul, S.; Fornalik-Wajs, E. On the Measure of the Heat Transfer Performance of RANS Turbulence Models in Single Round Jet Impingement. *Energies* **2023**, *16*, 7236. [CrossRef]
31. Chen, B.; Liu, S.; Liu, J.; Jiang, N.; Chen, Q. Application of Data-Driven RANS Model in Simulating Indoor Airflow. *Indoor Air* **2022**, *32*, e13123. [CrossRef] [PubMed]
32. Hurnik, M.; Ivanov, N.; Zasimova, M.; Popiolek, Z. Local and Gross Parameters of Air Distribution in a Room with a Sidewall Jet: CFD Validation Based on Benchmark Test. *Build. Environ.* **2022**, *207*, 108509. [CrossRef]
33. Kang, L.; van Hooff, T. Influence of Inlet Boundary Conditions on 3D Steady RANS Simulations of Non-Isothermal Mechanical Ventilation in a Generic Closure. *Int. J. Therm. Sci.* **2022**, *182*, 107792. [CrossRef]
34. Thysen, J.-H.; van Hooff, T.; Blocken, B.; van Heijst, G.J.F. CFD Simulations of Two Opposing Plane Wall Jets in a Generic Empty Airplane Cabin: Comparison of RANS and LES. *Build. Environ.* **2021**, *205*, 108174. [CrossRef]
35. Sánchez, M.N.; Giancola, E.; Blanco, E.; Soutullo, S.; Suárez, M.J. Experimental Validation of a Numerical Model of a Ventilated Façade with Horizontal and Vertical Open Joints. *Energies* **2020**, *13*, 146. [CrossRef]

36. Morozova, N.; Trias, F.X.; Capdevila, R.; Pérez-Segarra, C.D.; Oliva, A. On the Feasibility of Affordable High-Fidelity CFD Simulations for Indoor Environment Design and Control. *Build. Environ.* **2020**, *184*, 107144. [CrossRef]
37. Khayrullina, A.; van Hooff, T.; Blocken, B.; van Heijst, G. Validation of Steady RANS Modelling of Isothermal Plane Turbulent Impinging Jets at Moderate Reynolds Numbers. *Eur. J. Mech. B/Fluids* **2019**, *75*, 228–243. [CrossRef]
38. Lestinen, S.; Wesseling, M.; Kosonen, R.; Koskela, H.; Müller, D. Airflow Characteristics under Planar Opposed Ventilation Jets in a Controlled Indoor Environment. *E3S Web Conf.* **2019**, *111*, 01029. [CrossRef]
39. Kosutova, K.; van Hooff, T.; Blocken, B. CFD Simulation of Non-Isothermal Mixing Ventilation in a Generic Enclosure: Impact of Computational and Physical Parameters. *Int. J. Therm. Sci.* **2018**, *129*, 343–357. [CrossRef]
40. Kobayashi, T.; Sugita, K.; Umemiya, N.; Kishimoto, T.; Sandberg, M. Numerical Investigation and Accuracy Verification of Indoor Environment for an Impinging Jet Ventilated Room Using Computational Fluid Dynamics. *Build. Environ.* **2017**, *115*, 251–268. [CrossRef]
41. Moureh, J.; Yataghene, M. Large-Eddy Simulation of an Air Curtain Confining a Cavity and Subjected to an External Lateral Flow. *Comput. Fluids* **2017**, *152*, 134–156. [CrossRef]
42. van Hooff, T.; Blocken, B.; Tominaga, Y. On the Accuracy of CFD Simulations of Cross-Ventilation Flows for a Generic Isolated Building: Comparison of RANS, LES and Experiments. *Build. Environ.* **2017**, *114*, 148–165. [CrossRef]
43. Achari, A.M.; Das, M.K. Application of Various RANS Based Models towards Predicting Turbulent Slot Jet Impingement. *Int. J. Therm. Sci.* **2015**, *98*, 332–351. [CrossRef]
44. Hurnik, M.; Blaszczok, M.; Popiolek, Z. Air Distribution Measurement in a Room with a Sidewall Jet: A 3D Benchmark Test for CFD Validation. *Build. Environ.* **2015**, *93*, 319–330. [CrossRef]
45. Miltner, M.; Jordan, C.; Harasek, M. CFD Simulation of Straight and Slightly Swirling Turbulent Free Jets Using Different RANS-Turbulence Models. *Appl. Therm. Eng.* **2015**, *89*, 1117–1126. [CrossRef]
46. Heschl, C.; Inthavong, K.; Sanz, W.; Tu, J. Evaluation and Improvements of RANS Turbulence Models for Linear Diffuser Flows. *Comput. Fluids* **2013**, *71*, 272–282. [CrossRef]
47. Andersson, H.; Cehlin, M.; Moshfegh, B. A Numerical and Experimental Investigation of a Confluent Jets Ventilation Supply Device in a Conference Room. *Energies* **2022**, *15*, 1630. [CrossRef]
48. Hurnik, M.; Blaszczok, M.; Popiolek, Z. Air Speed and Velocity Measurements in a Room with a Sidewall Jet. *Data Brief* **2015**, *5*, 213–217. [CrossRef] [PubMed]
49. Blocken, B. LES over RANS in Building Simulation for Outdoor and Indoor Applications: A Foregone Conclusion? *Build. Simul.* **2018**, *11*, 821–870. [CrossRef]
50. *Ansys CFX-Solver Modeling Guide*; ANSYS, Inc.: Canonsburg, PA, USA, 2024.

energies

MDPI

Article

Impact of Shading Effect from Nearby Buildings on Energy Demand and Load Calculations for Historic City Centres in Central Europe

Agnieszka Sadłowska-Sałęga * and Krzysztof Wąs

Department of Rural Building, Faculty of Environmental Engineering and Geodesy, University of Agriculture in Krakow, 30-059 Krakow, Poland; krzysztof.was@urk.edu.pl

* Correspondence: agnieszka.sadlowska@urk.edu.pl

Abstract: EU legislation requires introducing a variety of measures to reduce energy consumption. Energy use decrease should be achieved through thermal retrofitting of the least energy-efficient buildings. In the case of cities like Kraków, this means the need to modernize (retrofitting as well as the incorporation of modern HVAC systems) historical buildings. Furthermore, urban morphology is an influencing factor with regards to the energy performance of buildings and therefore of cities (since the influence of shading from nearby buildings cannot be ignored). The paper presents the results of a study on the impact of shading on energy demand for heating and cooling in the historic centre of Krakow. A comparative analysis of the simulation calculation results was performed using the statistical climate as a boundary condition. In the case of a typical five-floor residential tenement house from the turn of the 20th century, an 8% increase in energy demand for heating and a 50% reduction in energy demand for cooling were estimated. As the analysis of the results shows, taking into account the shading from nearby buildings may be crucial for optimizing the volume (power of devices, diameters of ducts and pipes) of the HVAC installation.

Keywords: energy-efficient HVAC systems; historical buildings; retrofitting; shading effect; energy demand

1. Introduction

1.1. Literature Review

Urban morphology has been pointed out as one of the influencing factors with regards to the energy performance of cities. Studies show that in Central European cities, there are characteristic patterns illustrating strong investment in city centres and low diversity of land cover and use in historic city cores [1]. At the same time, it has been shown that in Poland, after EU accession, there have been more rapid changes in land use than in other Central European countries [2]. Additionally, in functional urban areas of Poland, there is an encroachment of urbanized areas into suburban areas, and thus an increase in the diversity of land cover and use in the access zones of cities [3]. As a result, so-called heat islands occur due to the uniformity of development and the disappearance of natural areas. The formation of urban heat islands, together with the ongoing climate change, has led to changes in energy demand for urbanized areas. For example, climate change in Europe is causing a significant increase in demand for cooling—even in regions (such as Poland) where cooling of buildings during the summer has so far been uncommon. This means an increased installation rate of air conditioning systems in buildings. European Union legislation requires implementing measures to reduce energy consumption for heating and cooling. The EU aims to reduce residential energy consumption by 16% by 2030 and 20–22% by 2035. National measures will have to ensure that at least 55% of the decrease in the average primary energy use is achieved through the renovation of the least energy-efficient buildings focusing on the thermal retrofitting of them [4]. In the case of cities like Kraków, this task is difficult to accomplish. This is because a large part of the city is listed for special protection as national heritage (a significant number of buildings in the city centre were built at the turn of the 20th century), whilst the medieval the old town was among the first sites chosen for UNESCO's original World Heritage List. The condition of the load-bearing structures of the buildings is usually satisfactory. However, secondary structures and heating systems are outdated and inefficient.

Restoration of historical buildings is one of the important trends of cultural heritage and has been studied and discussed for years [5–7]. Many new technologies have been developed to this day, but the retrofitting process has never been a simple undertaking owing to historical preservation codes, restrictions created by the existing building structure, and the threat that the latter poses to historical integrity [8,9]. In many cases, in residential buildings, thermal modernization comes down to replacing windows with more energy-efficient ones and insulation of the ceiling below an unheated attic. Incorporating modern HVAC systems into historical buildings is not a straightforward task as it is accompanied by a number of challenges [8]:

- old structures are made of different materials from those used in today's construction practices, which makes it difficult to incorporate a new system without causing harm to the rest of the structure;
- the thickness and the materials used for constructing partitions pose a challenge when trying to install ducts or piping fixtures;
- any external changes, such as the installation of vents, thermostats or other HVAC devices, have to be made in a way that will not alter the overall character of the building).

For this reason, the correct dimensioning of HVAC installation is crucial.

Heating and cooling load calculations are carried out to estimate the required capacity of the heating and cooling systems needed to maintain the required environment in the conditioned space. In most cases, calculations are made on the basis of steady-state methodology [10–12]. In these methods, the shading resulting from the use of shading devices (e.g., blinds) is considered in the cooling load calculations, but the surroundings of the building are not.

The positive effect of shading from nearby buildings on reducing solar gains was already known in ancient times in the Mediterranean basin [13]. In recent years, much attention has also been paid to investigating the influence (positive as well as negative) of shading from nearby buildings on the heating and cooling demand in temperate climates. Assuming that a reduction in the energy consumption of buildings is an effective means to build a low-carbon city, it has become essential for planners and designers to consider buildings as integral elements of the urban environment rather than stand-alone entities. This is related to the rational prediction of regional space cooling/heating loads and the design of distributed energy resource systems [14].

The analysis of the energy performance of buildings in relation to urban and street planning can become a theoretical basis for rational architectural layout and energy consumption reduction [15]. The influence of land development on the formation of the urban climate is expressed by various indicators. For example, the concept of the urban canyon is a model widely used in energy studies [16]. Its geometric layout may be described based on the following:

- its axis orientation;
- its cross-sectional dimensions: width of the street (W) and height of the building (H).

Aspect ratio (W/H) or inverted aspect ratio (H/W) are intuitive ways of characterizing the occlusion of an urban fabric. Canyon morphology has also been described quantifying the total amount of visible sky from a particular point, using other parameters such as the Sky View Factor (*SVF*) [17,18] or Sky Factor (*SF*) [19]. The larger the H/W, the smaller the SVF, meaning that the shading effect is stronger, resulting in a decrease in solar radiation reaching the ground and affecting thermal environment of street canyons.

There are also some index parameters that describe the relationship between buildings and the open area in a constant plot. The Floor Area Ratio (*FAR*) is one such parameter, which is the ratio of the built area to the lot area. A larger *FAR* value indicates a greater volume of the building. The Building Coverage Ratio (*BCR*) is the relationship between the ground floor area of enclosed buildings and the area of the land plot. Development scenarios where the *FAR* is same but the *BCR* is different will produce varying types of development, such as low rise or high rise, which can be incorporated into different layouts of buildings [14].

The H/W ratio is associated with thermal comfort in urban street canyons [20] (in some cases, with a different H/W the difference in dry-bulb temperature can reach 10 °C [21]), which affects the energy demand of buildings. Huang et al. [22], Oke [23] and Arnfield [24] proved in their works that H/W is a significant parameter influencing how much solar radiation reaches buildings. For example, for Hong Kong, the increase in energy demand due to shading from nearby buildings was determined to be 2% [25]. Other studies show that reductions in space cooling reach up to 30% [25,26]. Research using a hypothetical nine-building block that contained a three-storey commercial reference building [27] in eight cities in the U.S. indicated that the building's energy consumption can increase by 60.4% compared with a building without shading.

Over the past decades, detailed individual Building Energy Models (BEMs) have become established modes of analysis for building designers and energy policy makers. More recently, these models have evolved into so-called Urban Building Energy Models (UBEMs) [28,29] taking into account the energy performance of neighbourhoods, i.e., several dozens to thousands of buildings. This term is attributed explicitly to bottom–up (intended to focus on individual buildings [30]) engineering models. In the case of large-scale models, even a slight increase in resolution for one or more aspects can lead to a noticeable growth in the computational burden due to the issue of dimensionality [30]. Considering that the modernization of historic buildings is essentially a case study, the solution may be to use BEMs that take into account the building's immediate surroundings. Modern calculation tools such as EnergyPlus or WUFI®plus allow for the accurate assessment of heat gains and losses in buildings. They also have the ability to perform calculations that take into account shading from nearby buildings [26,31].

There are different ways of taking the shading into account. The most common is the adaption of radiation gains by reduced absorption and emission coefficients. Another way is calculating the radiation load by determining the position and size of the shading object in the field of view of the receiving surface [32].

1.2. Aim of the Study

Climate is a major factor that affects shading, apart from the index parameters and building layouts [14,33]. Most of the research in this field concerns warm climates or areas with modern buildings (e.g., [34]). Therefore, the aim of the study was to investigate the influence of shading from nearby buildings on the energy demand for heating and cooling as well as heating and cooling loads (for the entire building and individual floors of the building treated as separate calculation zones), in a temperate climate of Central Europe in the case of historical city centres, using Kraków as an example.

2. Materials and Methods

Calculations were performed using WUFI®plus software (ver. 3.5), enabling comprehensive thermal and moisture analysis of entire buildings, taking into account heat exchange with the ground [35]. As studies comparing contemporary BEMs (such as EnergyPlus, WUFI®Plus or ESP-r) show, there are no significant differences in their general accuracy [36,37]. However, the studies described in [38] have shown that in terms of goodness of fit, the WUFI®Plus model attains a significantly higher temperature fit as well as a higher water vapour pressure fit when compared to the EnergyPlus model. WUFI®Plus is the most complete heat and moisture simulation tool in the WUFI software family. It has been tested many times for use in calculations of historical buildings (including many times by the authors of this study [39–44]). It is a recognized tool used in so-called preventive conservation and collection care [39,44,45]. One of its advantages, in the case of calculations for historical buildings, is the extensive database of historical building materials.

The analyses carried out according to the block chart (Figure 1) include comparison of the results of energy demand calculations in transient conditions with and without taking into account the shading from nearby buildings, supplemented by a shortened comparison of the results of the heating and cooling loads in transient conditions with steady-state calculation (without shading from nearby buildings).

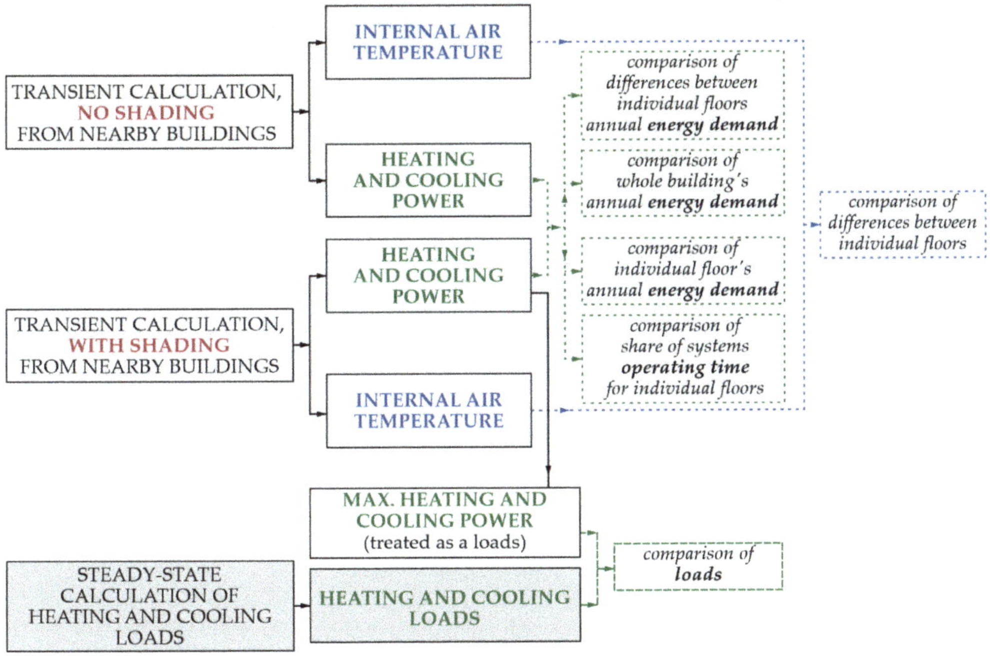

Figure 1. The block chart of conducted analyses.

For every simulation time step, the shading of the building components and windows is calculated depending on solar position and the building's visualized 3D geometry [46].

2.1. Case Study

The paper is based on the analysis of a generic multi-family residential tenement house representative of the old, historical centre of Kraków. Tenement houses in this part of the city are most often connected to each other in a closed quadrangle block, which has a large courtyard inside (Figure 2). They have a masonry structure and are most often made of full ceramic bricks. The thickness of the structure usually exceeds 40 cm (in some cases can even reach up to 100 cm). The ceilings are made using various technologies, including wooden structures, structures with steel beams or monolithic structures. The buildings usually have 3 to 5 floors above ground level.

Windows in these types of buildings are box structures with two layers of glazing, which, over the last two decades, have been gradually replaced with double- or triple-glazed composite structures as a result of thermal modernization activities. Also, ceilings below unheated attics or above unheated basements as well as the roof slopes have been additionally insulated to reduce energy losses. On the other hand, external walls very often cannot be insulated from the outside, due to conservation restrictions, and only insulation from the inside is allowed, which is rarely implemented by owners.

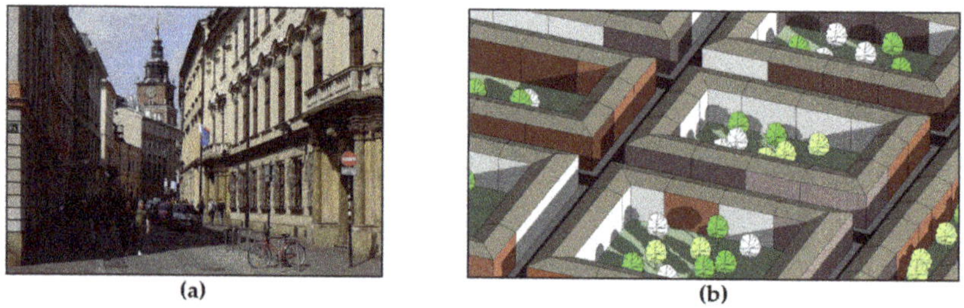

Figure 2. View of the streets of Kraków (reproduced from [47]) (a) and typical layout of Kraków's buildings in its historic centre (b).

The distance between individual blocks of tenement houses in the analysed area ranges from 16 to 20 m, excluding larger road arteries. The discussed arrangement is presented in the layout in Figure 2.

2.2. Three-Dimensional Model of the Building

For the analyses of heat gains and losses, a generic building was created along with a fragment of a typical urban layout:

1. A two-dimensional generic urban layout characteristic of Kraków's development at the turn of the 20th century was prepared based on publicly available maps (Figure 3a).
2. On this basis, a three-dimensional model of the analysed building and the surrounding buildings was modelled using the SketchUp tool (Figure 3b).
3. The last stage was the implementation of the developed model in WUFI®Plus, a tool for dynamic thermal and moisture analysis (Figure 3c).

The final building model is presented in Figure 4. It should be noted that the building model was created in two variants. Variant 1 does not take into account the nearby buildings (Figure 4a), and therefore does not take into account the shading generated by these buildings. Variant 2 takes into consideration the nearby buildings (Figure 4b). The modelled building has 5 floors above ground with a room height of 3.7 m. Each floor has an area of 672 m^2, which translates into a total building area of 3360 m^2. The tenement house has an L-shape with dimensions of 27 × 36 m and its height is 23 m (Figure 5). The building is located in the south-western corner of the tenement house block, among other buildings of the same height.

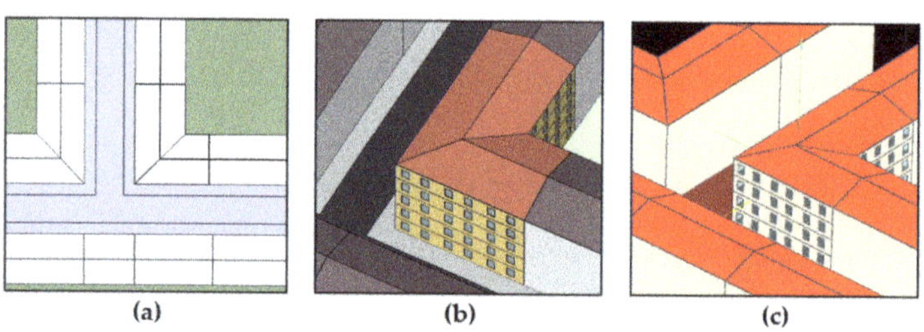

Figure 3. The process of creating a building model: a two-dimensional generic urban layout (a), 3D model in SketchUp (b) and 3D model in WUFI®plus (c).

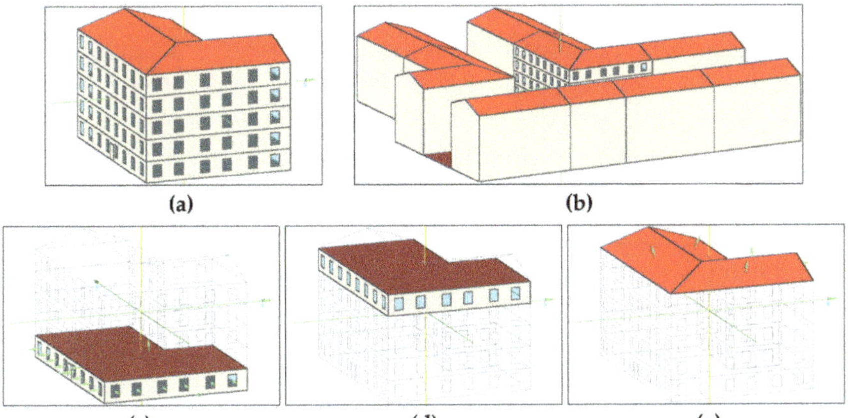
(a) (b)
(c) (d) (e)

Figure 4. 3D building model in WUFI®plus: without nearby buildings (**a**) and with them (**b**), as well as division of the building into zones: first floor (**c**), fifth floor (**d**) and unused attic (**e**).

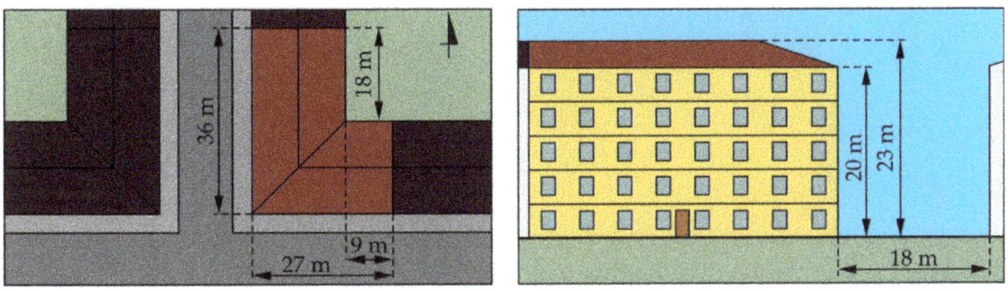

(a) (b)

Figure 5. Building dimensions: length and width (**a**) and height (**b**).

The building was divided into a total of seven simulation zones (Figure 4c–e). Each of the five floors was considered a separate zone: the five inhabited floors, unused attic. Adjacent buildings were combined into one zone. Each floor was divided into 5 apartments and a staircase. Internal partitions were included in the model as so-called non-visualized components. As a result, they affected the building's accumulation capacity.

All partitions, except ceilings and floors, were made with the use of masonry technology. The thickness of the brick wall was assumed to be 48 cm for load-bearing walls and 16 cm for partition walls. Additionally, a 1 cm layer of plaster was assumed on both sides of each of the walls. The ceilings were assumed to be of reinforced concrete construction with a thickness of 30 cm. In the case of the unheated attic, additional insulation was installed with a layer of 10 cm thick polystyrene. The *U*-factor for the external walls is 1.005 $W \cdot m^{-2} \cdot K^{-1}$. The double-glazed windows have dimensions of 1.6 × 2.2 m. Facing the street, there are 8 windows on the western elevation and 6 on the southern one. Facing the courtyard, there are 2 windows on the northern elevation and 4 on the eastern one. The *U*-factor for the windows is 1.6 $W \cdot m^{-2} \cdot K^{-1}$ and the solar heat gain coefficient (*SHGC*) is 0.7. The basic parameters of the partitions are presented in Tables 1–5.

Only shading from nearby buildings is included in the calculations. The model does not include any shading devices or elements. Due to the historic nature of the buildings in the center of Kraków and the conservation care, it is not possible to install shading elements on the outside of the building.

Table 1. Partition design parameters of the load-bearing walls.

Layers	Thickness [m]	Thermal Conductivity [W·m^{-1}·K^{-1}]	*U*-Value [W·m^{-2}·K^{-1}]
Plaster	0.010	0.80	
Solid brick	0.480	0.60	1005
Plaster	0.010	0.80	

Table 2. Partition design parameters of the partition walls.

Layers	Thickness [m]	Thermal Conductivity [W·m^{-1}·K^{-1}]	*U*-Value [W·m^{-2}·K^{-1}]
Plaster	0.010	0.80	
Solid brick	0.160	0.60	1.709
Plaster	0.010	0.80	

Table 3. Partition design parameters of the ceilings.

Layers	Thickness [m]	Thermal Conductivity [W·m^{-1}·K^{-1}]	*U*-Value [W·m^{-2}·K^{-1}]
Reinforced concrete	0.300	1.60	2.581

Table 4. Partition design parameters of the ceilings to unheated attic.

Layers	Thickness [m]	Thermal Conductivity [W·m^{-1}·K^{-1}]	*U*-Value [W·m^{-2}·K^{-1}]
Polystyrene	0.100	0.04	0.346
Reinforced concrete	0.300	1.60	

Table 5. Partition design parameters of the ground slab.

Layers	Thickness [m]	Thermal Conductivity [W·m^{-1}·K^{-1}]	*U*-Value [W·m^{-2}·K^{-1}]
Concrete screed	0.080	1.60	
Extruded polystyrene	0.120	0.03	0.230
Bitumen	0.005	0.17	
Concrete	0.150	1.60	

2.3. Boundary Conditions

Selecting the right boundary conditions for calculating energy demand is a key issue. In the era of rapid climate change that is currently taking place, it is difficult to use so-called TMYs (Typical Meteorological Years). In order to minimize the possibility of making an incorrect selection of boundary conditions, three statistical climates were adopted for Kraków based on data from different periods [48]:

- TMY_1 based on data from 2004 to 2018;
- TMY_2 based on data from 2007 to 2021;
- TMY_3 based on data from 2009 to 2023.

These climates differ slightly from each other (Figure 6a). Based on the analysis of the main outdoor air parameters influencing heat exchange through the building envelope, it can be concluded that:

1. The average outside air temperature in Kraków has been systematically increasing over the last few years. The increase in this average is mainly caused by the increase in air temperatures in the summer. The difference in maximum temperature between TMY_3 and TMY_1 is 3 °C (Table 6). There was a more than three-fold increase in the share of temperatures above 30 (from 20 h for TMY_1 to 65 h for TMY_3).

2. Maximum solar radiation in all TMYs is at a similar level (Figure 6b). However, the number of hours with radiation has increased by about 5%.

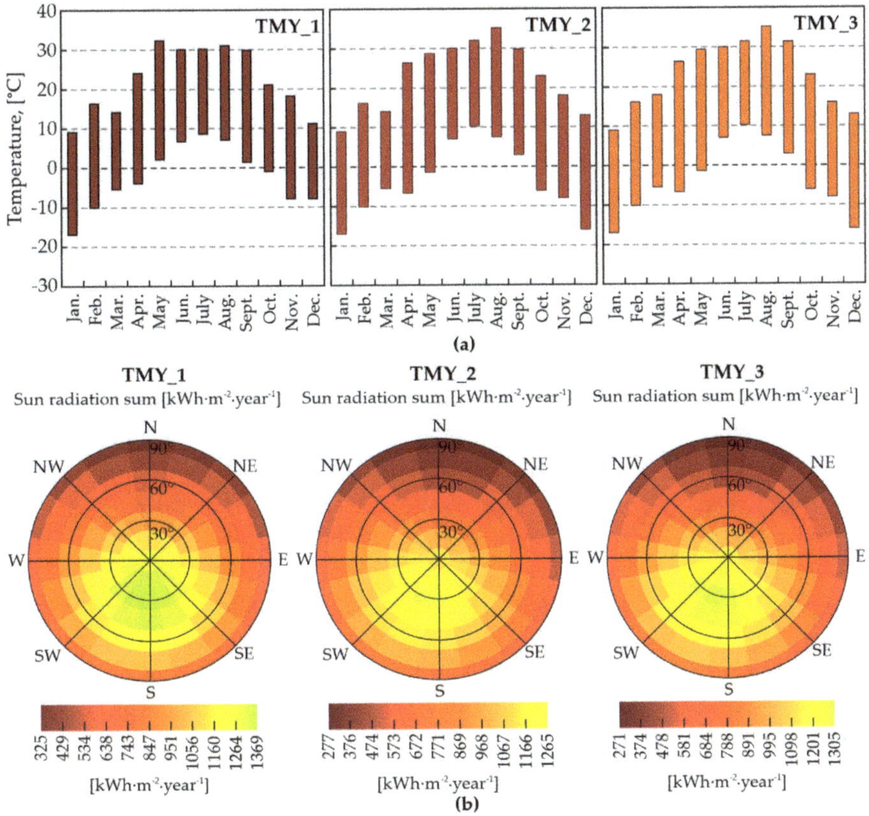

Figure 6. Outside air temperature extremes in individual months (**a**) as well as sun radiation sum depending on the angle of the surface and its orientation (**b**) for chosen TMYs.

Table 6. Basic statistics for temperature and solar radiation for TMYs.

Parameter	TMY_1	TMY_2	TMY_3
Maximum temperature [°C]	32.3	35.2	35.2
Minimum temperature [°C]	−17.0	−16.7	−17.0
Median for temperature [°C]	9.4	9.5	9.7
Maximum solar radiation [kW·m^{-2}·K^{-1}]	895	858	869
Hours of sun	4402	4614	4609

2.4. Internal Heat Gains

Internal heat gains were included in the calculations. Daily profiles from the database were adopted (for a 4-person family—see Figure 7). The maximum heat gains for individual simulated zones reach 9390 W on weekdays and 9735 W on weekends.

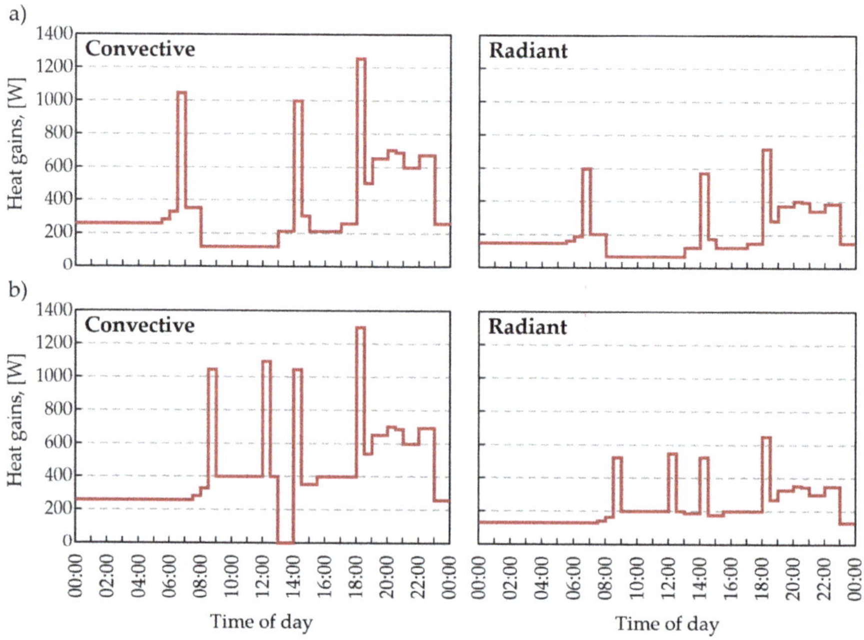

Figure 7. Heat gains of a single apartment for weekdays (**a**) and weekends (**b**).

2.5. HVAC and Ventilation Systems

To determine the energy demand for heating and cooling, the calculations assumed the operation of ideal (with infinite power) heating and cooling systems. For the thermal comfort of the users, the minimum permissible temperature was set at 20 °C and the maximum at 26 °C. The selected temperature range corresponds to the 2nd comfort class according to the standard EN 16798-1:2019 [49] (for residential buildings—sedentary activity about 1.2 met).

In terms of ventilation, it was assumed that the air exchange rate (*ACH*) resulting from infiltration was 0.1 and from gravity ventilation was 0.5.

2.6. Calculation Variants

In order to obtain complete data, calculations were performed for 6 variants for each of the statistical years (Table 7). In the methodology of calculating the heating load for design purposes, internal heat sources are not taken into account. Therefore, in the basic variants, only the empty building envelope is taken into account. Thus, ultimately 18 versions of calculations were performed.

Table 7. Variants of calculation.

Variant	Internal Heat Gains	HVAC Systems		Shading from Nearby Buildings
		Heating	Cooling	
V_1A	NO	NO	NO	NO
V_2A	NO	YES	NO	NO
V_3A	YES	YES	YES	NO
V_1B	NO	NO	NO	YES
V_2B	NO	YES	NO	YES
V_3B	YES	YES	YES	YES

3. Results

For the purpose of the data analysis, the building was divided into individual calculation zones corresponding to individual floors of the building (from Z_1 corresponding to the 1st floor to Z_5 corresponding to the 5th floor).

3.1. Internal Temperature

As a result of the calculations, the internal air temperature annual variations were obtained. Considering the only empty building envelope, in winter, regardless of TMY, the temperature inside the building drops below 0 °C. In summer, however, it does not exceed 25 °C. Figure 8 shows the example of the calculation results for TMY_3 (for Z_1 and Z_5, i.e., two extreme curves).

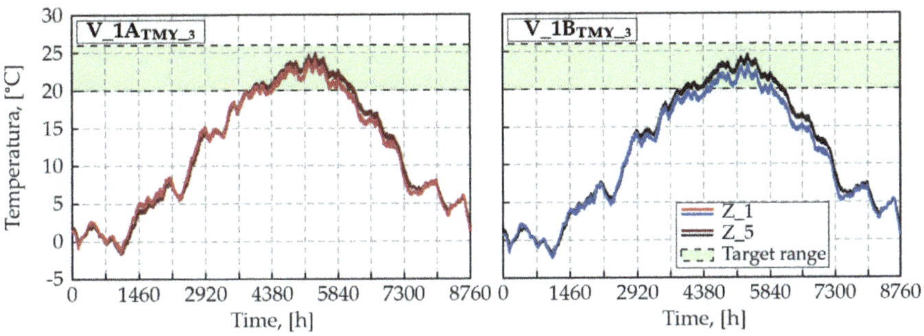

Figure 8. Internal air temperature of the first and fifth zones based on calculations that do not take into account (**left**) and take into account (**right**) shading from nearby buildings (TMY_3).

Depending on the building floor, for V_1A, the minimum temperatures for TMY_1 ranged from −2.1 °C to −1.9 °C, for TMY_2 from −2.5 °C to −2.4 °C, and for TMY_3 from −1.8 °C to −1.5 °C (Figure 9). After taking into account the shading from the nearby buildings (V_1B), these temperatures decreased even further: for TMY_1 it ranged from −2.7 °C to −2.4 °C, for TMY_2 from −3.4 °C to −2.9 °C, and for TMY_3 from −2.4 °C to −2.1 °C. For V_1A, the maximum temperatures for TMY_1 ranged from 23.9 °C to 24.7 °C, for TMY_2 and TMY_3 from 24.2 °C to 25.0 °C. For V_1B, similarly to V_1A, after taking into account shading, the temperatures decreased: for TMY_1 it ranged from 23.0 °C to 24.4 °C, and for TMY_2 and TMY_3 from 23.2 °C to 24.6 °C.

As can be seen from Figure 8 as well as Figure 9, the results obtained for the individual zones differ slightly from each other. However, they are highly correlated. The correlation coefficient between individual zones, for all the considered variants, is 0.999.

In the case of the analysed type of building in the climate conditions of Kraków, cooling is typically not used in the summer. In the case of passive microclimate shaping (without taking into account any internal heat gains), the temperature in this period is within the target range (Figure 8). However, after taking into account internal heat gains (for an additional calculation variant), as a result of the simulation, regardless of the selected TMY, the temperatures on all floors of the building exceeded the assumed upper limit of this range, i.e., 26 °C.

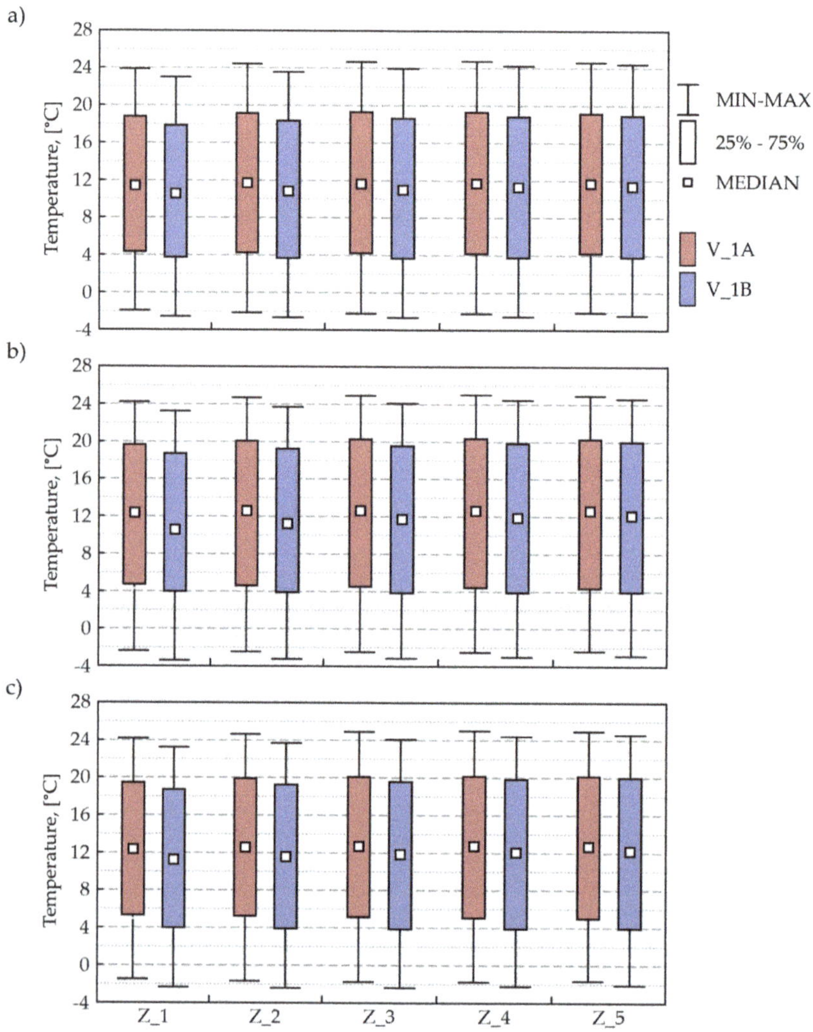

Figure 9. Basic descriptive statistics (location parameters) for internal air temperature (V_1A and V_1B) for chosen TMYs.

For variants where both heating and cooling systems are assumed, room temperature is maintained within the specified range. Due to the assumption of ideal systems, it is exactly 20 °C during the heating period and exactly 26 °C during the cooling period (Figure 10). After taking into account shading, the differences in indoor air temperature between individual zones are more visible (this applies to transitional periods when HVAC installations are not operating)—the temperatures on the lower floors are lower, with the average difference between Z_5 and Z_1 being 0.15 °C higher when taking into account shading (the maximum difference is 0.4 °C higher).

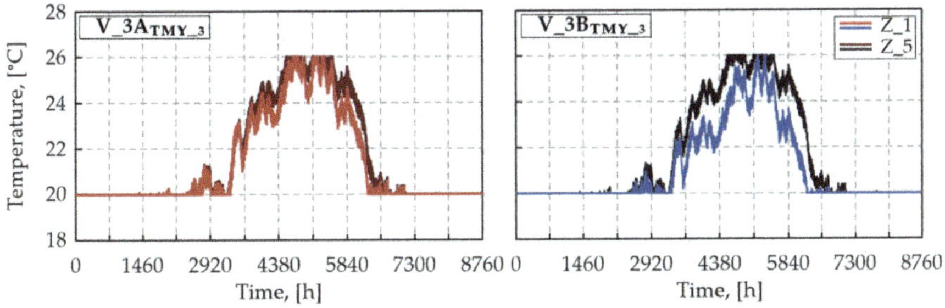

Figure 10. Internal temperature for Z_1 and Z_5 based on the calculations that take into account HVAC systems (heating and cooling), V_3A and V_3B for TMY_3.

3.2. Solar Heat Gains

As a result of the simulations, the solar heat gains were also determined. In the case of calculations without shading, for each floor (calculation zone) the maximum hourly solar heat gains were 13 kW for TMY_1 and 14 kW for TMY_2 as well as TMY_3. As can be seen in Figure 11, the variability patterns for individual zones are almost identical (within individual TMYs)—with a standard deviation of 3.65 kW. In the case of variants in which shading was taken into account, the solar heat gains differ depending on the zone. For this case, the maximum hourly heat gains and standard deviation are listed in Table 8.

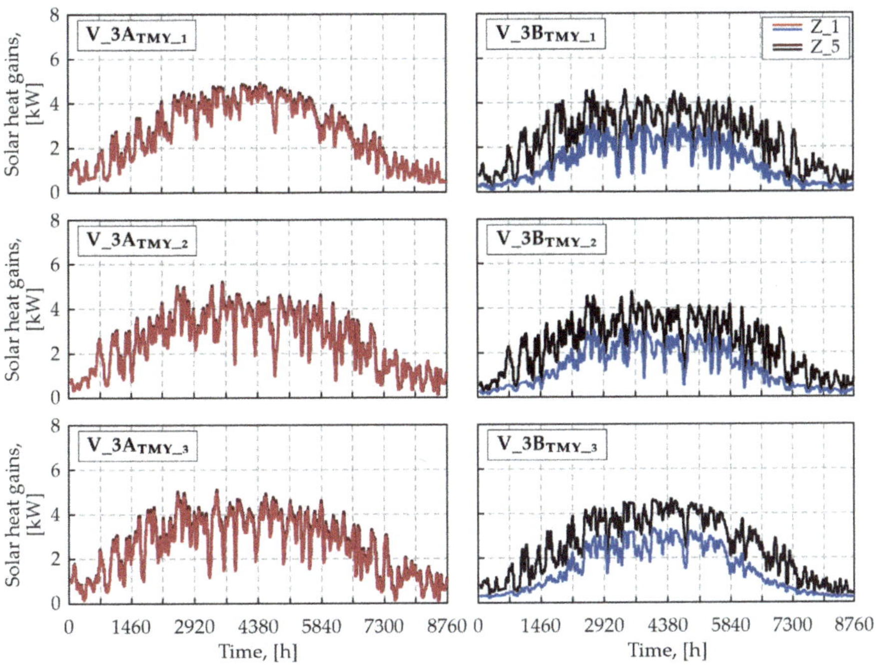

Figure 11. Moving average (48 h) of solar heat gains for Z_1 and Z_5 (the concept of moving average was used for the sake of clear graphical interpretation of the results), V_3A and V_3B for chosen TMYs.

Table 8. Maximum solar heat gains and standard deviation for V_3B.

External Climate	Maximum Solar Heat Gains, [kW]					Standard Deviation, [kW]				
	Z_1	Z_2	Z_3	Z_4	Z_5	Z_1	Z_2	Z_3	Z_4	Z_5
TMY_1	10.7	12.8	13.1	13.6	13.8	2.1	2.5	2.7	3.2	3.4
TMY_2	10.7	12.8	13.1	13.6	13.8	2.0	2.4	2.6	3.1	3.3
TMY_3	10.7	11.4	11.6	11.9	12.2	2.3	2.7	2.9	3.3	3.5

The medians of hourly solar heat gain values obtained for variant A were 5.1 kW for TMY_1, and 4.8 kW for TMY_2 and TMY_3. In the case of variant B, they differed depending on the zone:

- for TMY_1: 2.2 kW (for Z_1), 2.5 kW (Z_2), 2.8 kW (Z_3), 4.0 kW (Z_3) and 4.4 kW (Z_5);
- for TMY_2 and TMY_3: 1.9 kW (for Z_1), 2.1 kW (Z_2), 2.5 kW (Z_3), 3.6 kW (Z_3) and 4.1 kW (Z_5).

In case of the total (annual) value of the solar heat gains for the variants without considering shading, they amounted to 114,499 kW for TMY_1, 109,973 kW for TMY_2 and 112,496 kW for TMY_3. After considering the shading, they dropped to a level of 85,556 kW for TMY_1, 78,292 kW for TMY_2 and finally 80,107 for TMY_3.

The solar heat gains were also characterized by different levels and frequency of changes in individual zones. For the variants in which shading was not taken into account, the coefficient of variation is at the same level, i.e., 1.4. For the variants in which shading was taken into account, it is 1.57–1.60 for Z_1, 1.57–1.58 for Z_2, 1.54–1.55 for Z_3, 1.56–1.47 for Z_4 and 1.42–1.44 for Z_5 (Figure 12). This means that the lower the floor, the greater the variability in the solar heat gains.

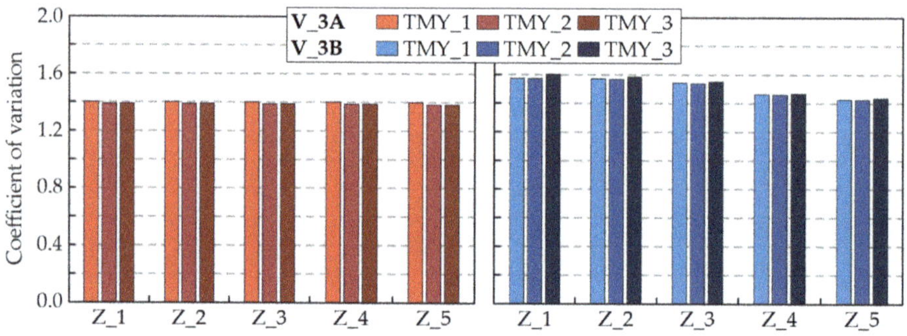

Figure 12. Coefficient of variation of solar heat gains for V_3A and V_3B for chosen TMYs.

3.3. Energy Demand

The energy demand for heating was determined based on variants V_2A and V_2B, while the energy demand for cooling was determined based on variants V_3A and V_3B. Heating and cooling power (per hour) during the calendar year for TMY_3 is illustrated in Figure 13.

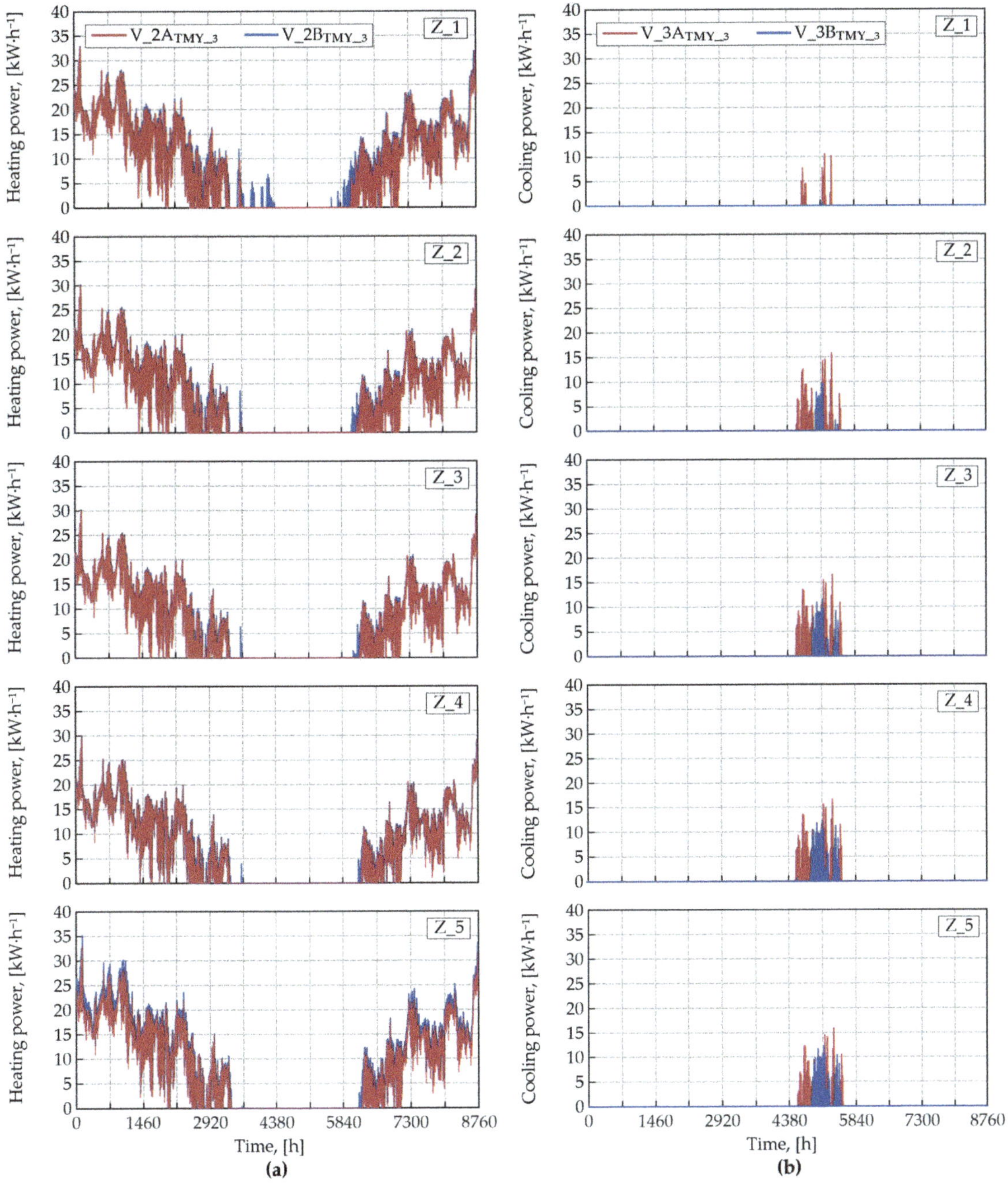

Figure 13. Heating power (V_2A and V_2B) (**a**) and cooling power (V_3A and V_3B) (**b**) for TMY_3.

3.3.1. Heating

In the case of the calculations without shading, no differences were noted between the maximum heating power for zones Z_2, Z_3 and Z_4. It was 30.3 kW$\cdot$h^{-1} for TMY_1, 31.0 kW$\cdot$h^{-1} for TMY_2 and 30.0 kW$\cdot$h^{-1} for TMY_3 (Table 9). In the case of the 1st floor

(Z_1), the increase in heating power by an average of 2.63 kW·h^{-1} (average calculated from all TMYs) was influenced by the heat exchange through the floor to the ground. For Z_5, the increase of an average of 2.9 kW·h^{-1} was influenced by the heat exchange through the ceiling to the unheated attic. In the case of the calculations with shading, relatively small changes (of approx. 0.2 kW·h^{-1}) were observed between zones Z_2 and Z_4. Similarly, an increase in heating power was observed for the extreme zones. It amounted to an average of 2.63 kW·h^{-1} and 5.19 kW·h^{-1} for the first and top floors, respectively. It should be emphasized that the differences between individual TMYs are greater compared to the variants that do not take into account shading.

Table 9. Maximum heating power.

External Climate	Without Shading, [kW·h^{-1}]					With Shading, [kW·h^{-1}]				
	Z_1	Z_2	Z_3	Z_4	Z_5	Z_1	Z_2	Z_3	Z_4	Z_5
TMY_1	32.9	30.3	30.3	30.3	33.1	32.9	30.3	30.2	30.1	34.9
TMY_2	33.6	31.0	31.0	31.0	34.1	34.1	31.5	31.4	31.1	36.7
TMY_3	32.7	30.0	30.0	30.0	32.8	32.9	30.2	30.2	30.0	35.2

Obviously, the heating system does not operate at maximum power all the time. The values of the median of heating power for zones Z_2 to Z_4 are similar, regardless of TMY and whether shading has been taken into account or not. They remain within the range 12.4 to 12.8 kW·h^{-1} (Figure 14a). In the case of Z_1, the median is in the range of 14.4 to 14.7 kW·h^{-1} (except for variant V_2B for TMY_1 where it is 13.9 kW·h^{-1}). For zone Z_5, there is a clear difference between the variant V_2A (depending on TMY 14.5–14.6 kW·h^{-1}) and the one in which shading was taken into account depending on TMY 15.6–15.7 kW·h^{-1}).

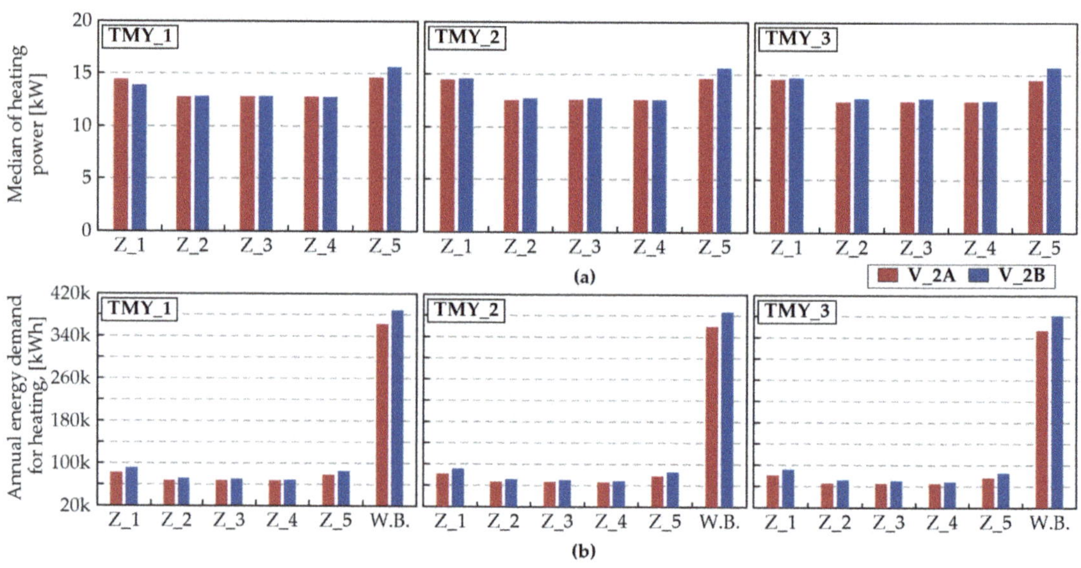

Figure 14. Median of heating power (**a**) and annual energy demand for heating (**b**), (V_2A and V_2B) for chosen TMYs for zones (Z_1–Z_5) and whole building (W.B.).

Taking into account the annual energy demand for heating, the differences between variants V_2A and V_2B are visible. Depending on TMY, for V_2A they remain in the following ranges (Figure 14b):

- 80,418–82,384 kWh for Z_1;
- 65,576–67,474 kWh for Z_2–Z_4;
- 76,384–78,138 kWh for Z_5.

These give the annual heating energy demand for the entire building (W.B in Figure 14b) at the level of 353,933–362,456 kW. For V_2B, where the differences between zones Z_2–Z_4 are visible, the results were as follows:

- 88,610–91,427 kWh for Z_1;
- 71,451–72,388 kWh for Z_2;
- 69,957–70,962 kWh for Z_3;
- 67,575–68,664 kWh for Z_4;
- 84,066–85,456 kWh for Z_5.

These figures give the annual heating energy demand for the entire building at the level of 381,659–388,248 kWh. The lowest demand was recorded for TMY_3 and the highest for TMY_1.

3.3.2. Cooling

In the case of cooling power, regardless of whether the shading from nearby buildings is taken into account or not, there are differences depending on the floors. They are the result of increased heat exchange between the outer floors and the surroundings (ground and unheated attic), which influences the heat exchange between the middle floors. The maximum cooling power is presented in Table 10. In the absence of shading, it assumes values in the range 7.0–12.8 kW·h^{-1} for TMY_1, 10.8–17.0 kW·h^{-1} for TMY_2 and 10.5–16.6 kW·h^{-1} for TMY_3. After taking into account shading, the cooling power decreased significantly for the lower floors and only slightly for the top floor. The ranges in this case are 0.7–12.3 kW·h^{-1} for TMY_1, 4.7–15.7.0 kW·h^{-1} for TMY_2 and 2.8–15.7 kW·h^{-1} for TMY_3.

Table 10. Maximum cooling power.

External Climate	Without Shading, [kW·h^{-1}]					With Shading, [kW·h^{-1}]				
	Z_1	Z_2	Z_3	Z_4	Z_5	Z_1	Z_2	Z_3	Z_4	Z_5
TMY_1	7.0	12.3	12.8	12.8	12.3	0.7	9.7	11.5	12.2	12.3
TMY_2	10.8	15.8	17.0	16.6	15.8	4.7	12.1	14.8	15.7	15.4
TMY_3	10.5	15.7	16.5	16.6	15.9	2.8	12.0	14.8	15.7	15.3

In the case of the calculations for cooling, the differences related to the TMY, used as the boundary conditions, are clearly visible. However, in each case there is a clearly visible reduction in cooling power for the variants taking into account shading (Figure 15a). The greatest differences between variants A and B were noted for the lower floors:

1. for TMY_1:
    - for variant without shading (variant V_3A): 1.9 kW·h^{-1} for Z_1, 7.2 kW·h^{-1} for Z_2, and 7.6 kW·h^{-1} for Z_3 as well as Z_4, and 7.1 kW·h^{-1} for Z_5;
    - for variant with shading (variant V_3B): 0.7 kW·h^{-1} for Z_1, 3.7 kW·h^{-1} for Z_2, 5.8 kW·h^{-1} for Z_3, 7.0 kW·h^{-1} for Z_4 and 6.8 kW·h^{-1} for Z_5,
2. for TMY_2:
    - for variant V_3A: 3.2 kW·h^{-1} for Z_1, 4.7 kW·h^{-1} for Z_2, 4.3 kW·h^{-1} for Z_3, 4.2 kW·h^{-1} for Z_4 and 4.5 kW·h^{-1} for Z_5;
    - for variant V_3B: 2.1 kW·h^{-1} for Z_1, 3.5 kW·h^{-1} for Z_2, 3.8 kW·h^{-1} for Z_3, 4.2 kW·h^{-1} for Z_4 and 4.4 kW·h^{-1} for Z_5,
3. for TMY_3:
    - for variant V_3A: 2.6 kW·h^{-1} for Z_1, 3.6 kW·h^{-1} for Z_2, 4.1 kW·h^{-1} for Z_3, 4.2 kW·h^{-1} for Z_4 and 3.9 kW·h^{-1} for Z_5;

- for variant V_3B: 0.7 kW·h^{-1} for Z_1, 2.6 kW·h^{-1} for Z_2, 3.4 kW·h^{-1} for Z_3, 3.6 kW·h^{-1} for Z_4 and 3.4 kW·h^{-1} for Z_5.

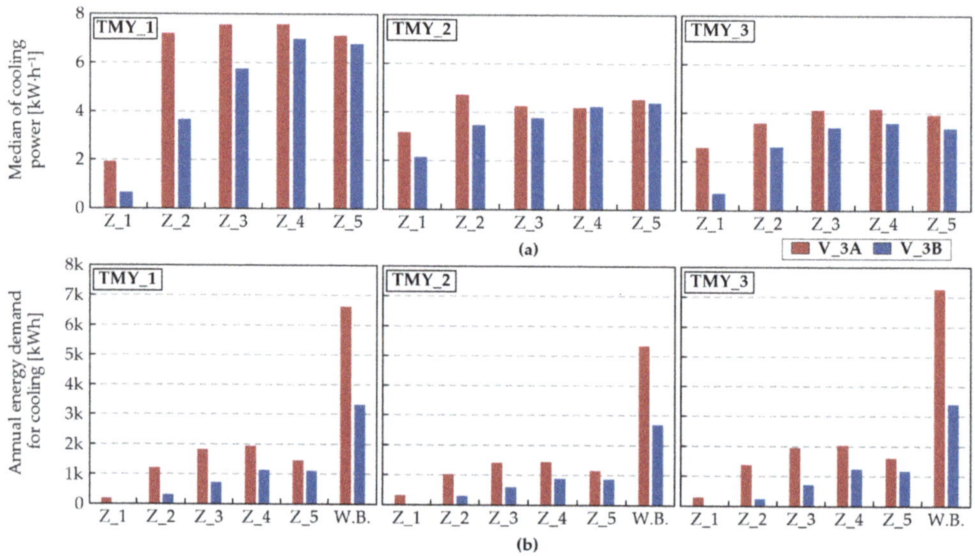

Figure 15. Median of cooling power (**a**) and annual energy demand for cooling (**b**), (V_3A and V_3B) for chosen TMYs for zones (Z_1–Z_5) and whole building (W.B.).

Taking into account the annual energy demand for cooling, the differences between variants V_3A and V_3B are clearly visible. Depending on the TMY, for V_3A they amount to the following (Figure 15b):

- 193–322 kWh for Z_1;
- 1023–1382 kWh for Z_2;
- 1408–1952 kWh for Z_3;
- 1442–2042 kWh for Z_4;
- 1151–1601 kWh for Z_5.

These figures give the annual cooling energy demand for the entire building (W.B at Figure 15b) at the level of 5347–7269 kWh. For V_3B, where the differences between zones Z_2–Z_4 are visible, the following figures were obtained:

- 1–25 kWh for Z_1;
- 248–323 kWh for Z_2;
- 603–737 kWh for Z_3;
- 892–1237 kWh for Z_4;
- 783–1178 kWh for Z_5.

These figures give the annual cooling energy demand for the entire building at the level of 2702–3398 kWh. The lowest demand was recorded for TMY_2 and the highest for TMY_3.

4. Discussion of the Results

The analysed building is located in a temperate climate, where heating is a key issue. However, calculations have confirmed that in the era of climate warming, in order to maintain thermal comfort conditions, cooling is also necessary. Based on the projected climate changes in Central Europe, regardless of the climate change scenario, increasingly longer heat waves with increasingly higher temperatures are predicted. As a consequence, higher temperatures are expected to cause a shift in the geographical distribution of climate zones [50]. According to data collected in various parts of the world, climate change has caused an increase in the frequency, duration and intensity of heat waves [51,52]. Therefore, a multi-criteria approach to managing and optimizing energy consumption in buildings is important. As the analysis of the results shows, taking into account the shading from nearby buildings may be crucial for optimizing the volume (power of devices, diameters of ducts and pipes) of the HVAC installation.

In the analysed case, the value of the inverted aspect ratio H/W for the modelled building with its surroundings is 1.11 for both the north–south and east–west axes. This value, as given by [53], is similar to the value for regular urban canyons, where the H/W values are 0.5 for avenue-type canyons, 1.0 for regular canyons with a similar building height and street width, and values above 2.0 for deep canyons. In [54], it was pointed out that H/W is more important than orientation. For this reason, changes in building orientation were not tested.

In the discussed case, the annual energy demand for heating, after considering the shading, increased by 7% for TMY_1 (25,793 kWh) and by 8% (27,505 kWh) for TMY_2 and TMY_3 (27,726 kWh). Considering the division of the building into zones, the calculations show that the increase in energy demand for heating also depends on the floor. It is as follows:

- for TMY_1: 11% for Z_1, 7% for Z_2, 5% for Z_3, 2% for Z_4 and 9% for Z_5;
- for TMY_2: 10% for Z_1, 8% for Z_2, 6% for Z_3, 3% for Z_4 and 10% for Z_5;
- for TMY_3: 10% for Z_1, 8% for Z_2, 7% for Z_3, 3% for Z_4 and 10% for Z_5.

For all TMYs, the results for individual zones are similar. The greatest increase in demand for heating is observed on the outer floors, while the smallest is observed on the middle floor, i.e., where the climate is stabilized by the adjacent heated zones.

In the case of energy demand for cooling, as was to be expected, a decrease was noted after considering shading: 50% for TMY_1 (3302 kWh), 49% for TMY_2 (2645 kWh) and finally 53% for TMY_3 (3871 kWh). Similarly to the heating, in this case there were differences for the zones as follows:

- for TMY_1: 99% for Z_1, 73% for Z_2, 59% for Z_3, 41% for Z_4 and 24% for Z_5;
- for TMY_2: 92% for Z_1, 70% for Z_2, 57% for Z_3, 38% for Z_4 and 24% for Z_5;
- for TMY_3: 95% for Z_1, 82% for Z_2, 63% for Z_3, 39% for Z_4 and 26% for Z_5.

For all TMYs, the results for individual zones are similar. The greatest decrease in demand for cooling is observed on the bottom floors, while the smallest is observed on the top floor.

The 8% increase in energy demand for heating and 50% decrease in energy demand for cooling obtained from the calculations are confirmed in the literature. These results cannot be extrapolated directly to those obtained for hot or cold climates, which are the focus of most publications (e.g., shading studies showed that energy demand decreases by 42% in summer in Miami and increases by 22% in winter in Minneapolis [31]), but according to [27], for a building analysed in a moderate climate with H/W equal to 2.0, the increase in the energy demand for heating, after considering shading, reaches 26.3% and the decrease in the energy demand for cooling reaches 53.6%. Therefore, the depth of the urban canyon will have a significant impact on the differences in the building's demand for energy needed for heating and cooling. This relationship in Poland was also confirmed in [55] (pp. 107–116).

As shown by the detailed analysis of the maximum energy demand, considering shading has less impact on the size of the designed installation in the case of heating than cooling (because these systems are designed for maximum heat losses and gains). For heating, the average increase in the maximum energy demand for all the tested outdoor climates was 0.6% for all zones except the top floor (Z_5), for which the increase was 6.8%. For cooling, greater variability was observed. Related to both the change in zone and external climate, maximum energy demand decreased by the following:

- for TMY_1: 87% for Z_1, 21% for Z_2, 10% for Z_3, 5% for Z_4 and for Z_5 remained unchanged;
- for TMY_2: 57% for Z_1, 23% for Z_2, 13% for Z_3, 5% for Z_4 and 3% for Z_5;
- for TMY_3: 73% for Z_1, 24% for Z_2, 10% for Z_3, 5% for Z_4 and 3% for Z_5.

The differences in the results for individual calculation zones confirm the necessity of such divisions, even for relatively low buildings. As shown in the results of previous research [30], division into individual storeys is sufficient. It should be emphasized that in the case of calculations used in the dimensioning of air-conditioning installations, the division of individual storeys into smaller calculation zones should be considered (due to the high dependence of heat gains on solar radiation).

It should be noted that the annual energy demand for both processes is influenced by the operating time of the devices, in addition to the hourly energy demand. As can be seen in the example of TMY_3 (Figure 16), considering shading resulted in an increase in the operating time of heating devices and a reduction in the operating time of cooling devices. In case of heating the percentage of equipment operating time during the year increased as follows:

- for TMY_1: from 59.1% to 62.8% for Z_1, from 54.0% to 57.2% for Z_2, from 53.5% to 57.2% for Z_3, from 53.4% to 54.4% for Z_4 and from 55.6% to 56.2% for Z_5;
- for TMY_1: from 57.5% to 62.8% for Z_1, from 52.8% to 56.9% for Z_2, from 50.0% to 55.5% for Z_3, from 52.1% to 54.2% for Z_4 and from 53.7% to 56.2% for Z_5;
- for TMY_1: from 58.1% to 63.7% for Z_1, from 52.7% to 57.8% for Z_2, from 51.9% to 56.7% for Z_3, from 52.9% to 55.4% for Z_4 and from 53.7% to 57.9% for Z_5.
- In the case of cooling, the changes in percentage are as follows:
- for TMY_1: from 0.9% to 0.1% for Z_1, from 3.3% to 0.8% for Z_2, from 4.9% to 2.1% for Z_3, from 5.2% to 3.3% for Z_4 and from 4.2% to 3.2% for Z_5;
- for TMY_1: from 0.9% to 0.1% for Z_1, from 2.2% to 0.8% for Z_2, from 3.2% to 1.5% for Z_3, from 3.3% to 2.1% for Z_4 and from 2.5% to 2.1% for Z_5;
- for TMY_1: from 1.0% to 0% for Z_1, from 3.6% to 1.1% for Z_2, from 4.7% to 2.3% for Z_3, from 4.8% to 3.1% for Z_4 and from 4.0% to 3.1% for Z_5.

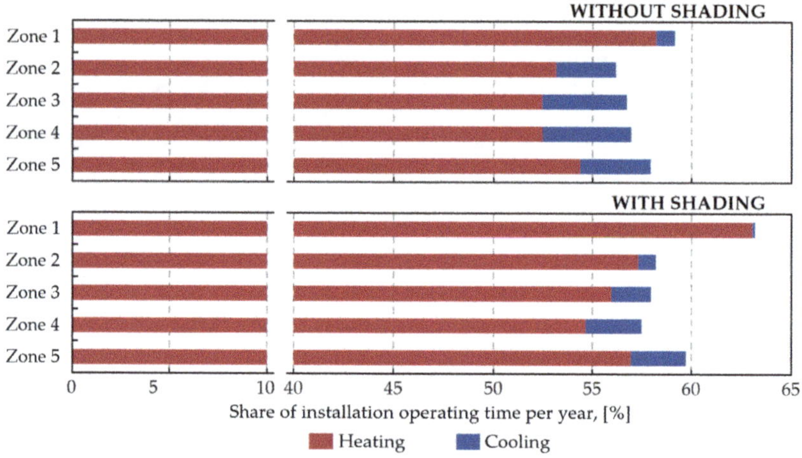

Figure 16. Percentage of heating and cooling devices operating time per year for TMY_3.

The maximum hourly energy demand values can be compared to the heating and cooling power. Design heat load is defined as the heat flow (power) required to achieve the specified internal design temperature under external design conditions. For the whole building, it is calculated from all transmission heat losses to the exterior and the ventilation heat loss of the building and with or without additional heating power. As a result of the calculations carried out (according to [11,12]), without taking into account the heating power, for the middle floor (Z_3) of the analysed building, a value of 38.3 kW was obtained. This means that in the case of steady-state calculations, 22–28% higher heating load (depending on TMY) values were obtained. It should be emphasized, however, that the calculations were performed at the currently applicable calculation temperature for winter, which is −20 °C for Kraków. In the case of calculations made assuming a designed outside temperature of −17 °C (minimum temperature for the assumed statistical climates) the heating load decreased to 35.4 kW. In this case, the heating load obtained was higher than for transient calculations by 13–18%.

There are many methods for sizing HVAC systems. A review of the calculation methods used for this purpose can be found in [56,57]. In Poland, the calculation methodology proposed in [58–60] is commonly used. Based on calculations made according to this method, assuming internal heat gains at the level of 9.8 kW (which corresponds to the maximum heat gains assumed in transient calculations), the cooling load for Z_3 is 51 kW. Therefore, it can be stated that the transient calculations gave a result that was 64–73% lower for the variant without shading and 71–77% with shading. Such large differences result from the fact that constant (lasting throughout the day) internal heat gains were assumed in the calculations and also from the estimated determination of the building's thermal storage capacity.

Due to the fact that the analysed building is generic, it was impossible to validate and calibrate the model. The calculations also assume that HVAC systems operate without any losses. Therefore, the energy demand results cannot be considered final. The accuracy of the predictions is also influenced by the adoption of a simplified scheme of internal heat and moisture gains. As the analyses show, their impact on the demand for energy for cooling is significant. In variants where internal heat gains were not taken into account (e.g., V_1A and V_1B), the calculated internal temperature in the summer period did not exceed the permissible range (Figure 8). Therefore, according to these variants, for the assumed external climates, there was no need for cooling. Nevertheless, adopting the above-described simplifications allows for the formulation of generalized conclusions.

It should also be emphasized that the use of statistical climates affects the calculated energy demand (for heating and cooling) and loads. Observations of the actual climate in Kraków show that in recent years it has been significantly warmer than statistical climates, with long heat waves (in studies concerning Poland, a heat wave is often defined as a series of at least 3 days with maximum air temperature above 30 °C on each day) and tropical nights (by definition, a tropical night is when the air temperature does not drop below 20 °C throughout the night). For example, according to TMY_3, no tropical night is expected in July. In reality, a series of 6 nights in a row with temperatures not falling below 18 °C (in Polish conditions—a very warm night) was recorded in July 2024 (Figure 17). Three of them were tropical.

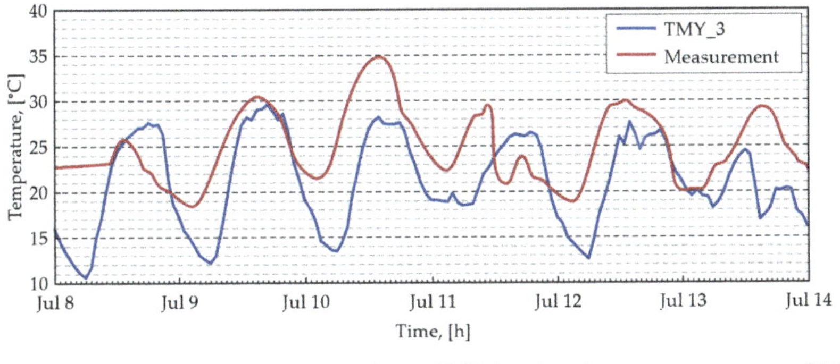

Figure 17. External air temperature according to TMY_3 against the temperature measured in Kraków on 8–13 July 2024.

5. Conclusions

The above analysis shows that considering shading from nearby buildings in the calculations of energy and power demand can significantly affect these calculations even in moderate climates. It also shows that the division of buildings into individual floors is equally important. The accuracy of the calculations not only impacts the system size but also influences the building's performance over the long run since oversized or undersized HVAC systems can exhibit less than optimal operation [56].

In temperate climates, the influence of shading in urban areas cannot be neglected in the calculation of energy demand for heating and cooling as well as in load calculation. In the case of a typical five-floor residential tenement house from the turn of the 20th century in Kraków, an 8% increase in energy demand for heating and a 50% reduction in energy demand for cooling were estimated.

The use of steady-state calculations (without considering shading from nearby buildings) may lead to oversizing the air conditioning systems on the lowest floors of the building. However, determining the exact oversizing requires further research.

In calculations, it is advisable to introduce separate calculation zones for each floor.

Author Contributions: Conceptualization: A.S.-S.; methodology: A.S.-S. and K.W.; formal analysis: A.S.-S. and K.W.; data curation: A.S.-S. and K.W.; writing—original draft preparation: A.S.-S.; writing—review and editing: A.S.-S. and K.W.; visualization: A.S.-S. and K.W. All authors have read and agreed to the published version of the manuscript.

Funding: This research received no external funding.

Data Availability Statement: The data presented in this study are available on request from the corresponding author.

Conflicts of Interest: The authors declare no conflicts of interest.

References

1. Kudas, D.; Wnęk, A.; Hudecová, L'.; Fencik, R. Spatial Diversity Changes in Land Use and Land Cover Mix in Central European Capitals and Their Commuting Zones from 2006 to 2018. *Sustainability* **2024**, *16*, 2224. [CrossRef]
2. Wnęk, A.; Kudas, D.; Stych, P. National Level Land-Use Changes in Functional Urban Areas in Poland, Slovakia and Czechia. *Land* **2021**, *10*, 39. [CrossRef]
3. Kudas, D.; Wnęk, A.; Tátošová, L. Land Use Mix in Functional Urban Areas of Selected Central European Countries from 2006 to 2012. *Int. J. Environ. Res. Public Health* **2022**, *19*, 15233. [CrossRef] [PubMed]
4. *EU/2024/1275*; Energy Performance of Buildings Directive. European Parliament and of the Council: London, UK, 2024.
5. Martínez-Molina, A.; Tort-Ausina, I.; Cho, S.; Vivancos, J.-L. Energy efficiency and thermal comfort in historic buildings: A review. *Renew. Sustain. Energy Rev.* **2016**, *61*, 70–85. [CrossRef]
6. Prömmel, A. Increased energy efficiency in existing buildings. *Habitat Int.* **1978**, *3*, 569–575. [CrossRef]
7. Johnsen, K. Energy conservation in the built environment. *Build. Environ.* **1980**, *15*, 15.
8. Tejani, A. Integrating Energy-Efficient HVAC Systems into Historical Buildings: Challenges and Solutions for Balancing Preservation and Modernization. *ESP JETA* **2021**, *1*, 83–97.
9. Szecskő, H.; Horváth, T. Challenges and opportunities in the energetic modernization of historic residential buildings in Budapest downtown. In Proceedings of the International Conference on New Energy and Future Energy Systems (NEFES 2024), Győr, Hungary, 29 July–1 August 2024.
10. *EN ISO 52016-1*; Energy Performance of Buildings. Calculation of the Energy Needs for Heating and Cooling, Internal Temperatures and Heating and Cooling Load in a Building or Building Zone—Part 1: Calculation Procedures. European Committee for Standardization: Brussels, Belgium, 2017.
11. *Environmental Design: CIBSE Guide A*; Chartered Institution of Building Services Engineers: London, UK, 2021.
12. *EN 12831:2003*; Heating Systems in Buildings—Method for Calculation of the Design Heat Load. European Committee for Standardization: Brussels, Belgium, 2003.
13. Vartholomaios, A. A parametric sensitivity analysis of the influence of urban form on domestic energy consumption for heating and cooling in a Mediterranean city. *Sustain. Cities Soc.* **2017**, *28*, 135–145. [CrossRef]
14. Liu, H.; Pan, Y.; Yang, Y.; Huang, Z. Evaluating the impact of shading from surrounding buildings on heating/ cooling energy demands of different community forms. *Build. Environ.* **2021**, *206*, 108322. [CrossRef]
15. Ichinose, T.; Lei, L.; Lin, Y. Impacts of shading effect from nearby buildings on heating and cooling energy consumption in hot summer and cold winter zone of China. *Energy Build.* **2017**, *136*, 199–210. [CrossRef]
16. Santamouris, M.; Asimakopoulos, D. *Energy and Climate in the Urban Built Environment*, 1st ed.; James & James: London, UK, 2001.

17. Johnson, G.; Watson, I. The determination of View-Factors in Urban Canyons. *J. Appl. Meteorol. Climatol.* **1984**, *23*, 329–335. [CrossRef]
18. Matzarakis, A.; Matuschek, O. Estimation of Sky View Factor in Urban Environments. Available online: https://www.researchgate.net/publication/267944926_Estimation_of_Sky_View_Factor_in_urban_environments (accessed on 11 November 2024).
19. Beckers, B. Geometrical Interpretation of Sky Light in Architecture Projects. Available online: http://www.heliodon.net/downloads/Beckers_2007_Helio_001_en.pdf (accessed on 11 November 2024).
20. Ali-Toudert, F.; Mayer, H. Effects of asymmetry, galleries, overhanging facedes and vegetation on thermal comfort in urban street canyons. *Sol. Energy* **2007**, *81*, 742–754. [CrossRef]
21. Johansson, E. Influence of the urban geometry on outdoor thermal comfort in a hot dry climate: A study in Fez, Morocco. *Build. Environ.* **2006**, *41*, 1326–1338. [CrossRef]
22. Huang, Y.; Akbari, H.; Taha, H.; Rosenfeld, A. The potential of vegetation in reducing summer cooling loads in residential buildings. *J. Appl. Meteorol.* **1987**, *26*, 1103–1116. [CrossRef]
23. Oke, T. Street design and urban canopy layer climate. *Energy Build.* **1988**, *11*, 103–113. [CrossRef]
24. Arnfield, J. Street design and urban canyon solar access. *Energy Build.* **1990**, *14*, 117–131. [CrossRef]
25. Lam, J. Shading effects due to nearby buildings and energy implications. *Energy Convers. Manag.* **2000**, *41*, 647–659. [CrossRef]
26. Danny, H.; Li, W.; Wong, S. Daylighting and energy implications due to shading effects from nearby buildings. *Appl. Energy* **2007**, *84*, 1199–1209.
27. Han, Y.; Taylor, J.; Pisello, A. Exploring mutual shading and mutual reflection inter-building effects on building energy performance. *Appl. Energy* **2017**, *185*, 1556–1564. [CrossRef]
28. Ali, U.; Shamsi, M.H.; Hoare, C.; Mangina, E.; O'Donnell, J. Review of urban building energy modeling (UBEM) approaches, methods and tools using qualitative and quantitative analysis. *Energy Build.* **2021**, *246*, 111073. [CrossRef]
29. Frayssinet, L.; Merlier, L.; Kuznik, F.; Hubert, J.L.; Milliez, M.; Roux, J.J. Modeling the heating and cooling energy demand of urban buildings at city. *Renew. Sustain. Energy Rev.* **2018**, *81*, 2318–2327. [CrossRef]
30. Faure, X.; Johansson, T.; Pasichnyi, O. The Impact of Detail, Shadowing and Thermal Zoning Levels on Urban Building Energy Modelling (UBEM) on a District Scale. *Energies* **2022**, *15*, 1525. [CrossRef]
31. Pisello, A.; Taylor, J.; Xu, X. Inter-building effect: Simulating the impact of a network of buildings on the accuracy of building energy performance predictions. *Build. Environ.* **2012**, *58*, 37–45. [CrossRef]
32. Bludau, C. Shading of flat roofs. In Proceedings of the 2nd International Conference on Moisture in Buildings (ICMB23), Online, 3–4 July 2023.
33. Wen, J.; Xie, Y.; Yang, S.; Yu, J.; Lin, B. Study of surrounding buildings' shading effect on solar radiation through. *Sustain. Cities Soc.* **2022**, *86*, 104143. [CrossRef]
34. Natanian, J.; Wortmann, T. Simplified evaluation metrics for generative energy-driven urban design. *Energy Build.* **2021**, *240*, 110916. [CrossRef]
35. Nawalany, G.; Sokołowski, P.; Lendelova, J.; Zitnak, M.; Jakubowski, T.; Atilgan, A. Numerical analysis of the heat exchange model with the ground on the example of a complex of industrial halls. *Energy Build.* **2023**, *300*, 113689. [CrossRef]
36. Wijesuriya, S.; Tabares-Velasco, P.C.; Biswas, K.; Heim, D. Empirical validation and comparison of PCM modeling algorithms commonly used in building energy and hygrothermal software. *Build. Environ.* **2020**, *73*, 106750. [CrossRef]
37. Staszczuk, A.; Kuczyński, T.; Wojciech, M.; Ziembicki, P. Comparative Calculation of Heat Exchange with the Ground in Residential Building Including Periodes of Heat Waves. *CEER* **2016**, *21*, 109–119. [CrossRef]
38. Coelho, G.; Silva, H.; Henriques, F. Development of a three-dimensional hygrothermal model of a historic building in WUFI®Plus vs EnergyPlus. In *MATEC Web of Conferences 282, Proceedings of the International Conference of 4th Central European Symposium on Building Physics (CESBP 2019), Prague, Czech Republic, 2–5 September 2019*; EDP Sciences: Les Ulis, France, 2019; p. 02079.
39. Lissner, J.; Kaiser, U.; Kilian, R. *Built Cultural Heritage in Times of Climate Change*; Fraunhofer-Center for Central and Eastern Europe MOEZ: Leipzig, Germany, 2014.
40. Sadłowska-Sałęga, A.; Radoń, J.; Sobczyk, J.; Wąs, K. Influence of microclimate control scenarios on energy consumption in the Gallery of the 19th-Century Polish Art in the Sukiennice (the former Cloth Hall) of The National Museum in Krakow. *IOP Conf. Ser. Mater. Sci. Eng.* **2018**, *415*, 012026. [CrossRef]
41. Radoń, J.; Sadłowska-Sałęga, P.; Wąs, K.; Gryc, A.; Kupczak, A. Energy use optimization in the building of National Library. *IOP Conf. Ser. Mater. Sci. Eng.* **2018**, *415*, 012029. [CrossRef]
42. Sadłowska-Sałęga, A.; Radoń, J. Feasibility and limitation of calculative determination of hygrothermal conditions in historical buildings: Case study of st. Martin church in Wiśniowa. *Build. Environ.* **2020**, *186*, 107361. [CrossRef]
43. Nawalany, G.; Sokołowsk, P.; Michalik, M. Analysis of the Operation of an Unheated Wooden Church to the Shaping of Thermal and Humidity Conditions Using the Numerical Method. *Energies* **2021**, *14*, 5200. [CrossRef]
44. Sobczyk, J. Weryfikacja Metod Analizy Mikroklimatu Historycznych Budynków Muzealnych (Ang. Verification of Methods of Microclimate Analysis of Historical Museum Buildings). Ph.D. Dissertation, University of Agriculture in Kraków, Kraków, Poland, 2024.

45. Sadłowska-Sałęga, A.; Sobczyk, J.; del Hoyo-Meléndez, J.M.; Wąs, K.; Radoń, J. Preservation strategy and optimization of the microclimate management system for the Chapel of the Holy Trinity in Lublin. In Proceedings of the International Conference of Energy Efficiency in Historic Buildings (EEHB2018), Visby, Sweden, 26–27 September 2018.
46. Winkler, M.; Antretter, F.; Radoń, J. Critical discussion of a shading calculation method for low energy building and passive house design. *Energy Procedia* **2017**, *137*, 33–38. [CrossRef]
47. Wikimedia Commons. UlicaBracka-WidokNaPółnoc-POL, Kraków.jpg by Mach240390. Available online: https://commons.wikimedia.org/wiki/File:UlicaBracka-WidokNaP%C3%B3%C5%82noc-POL,_Krak%C3%B3w.jpg (accessed on 16 November 2024).
48. Repository of Building Simulation Climate Data. Available online: https://climate.onebuilding.org/ (accessed on 11 November 2024).
49. *EN 16798-1:2019*; Energy Performance of Buildings—Ventilation for Buildings, Part 1: Indoor Environmental Input Parameters for Design and Assessment of Energy Performance of Buildings Addressing Indoor Air Quality, Thermal Environment, Lighting and Acoustic. European Committee for Standardization: Brussels, Belgium, 2019.
50. Consequences of Climate Change. Available online: https://climate.ec.europa.eu/climate-change/consequences-climate-change_en (accessed on 11 November 2024).
51. Climate Change Indicators: Heat Waves. Available online: https://www.epa.gov/climate-indicators/climate-change-indicators-heat-waves (accessed on 11 November 2024).
52. *Extreme Weather Events in Europe: Preparing for Climate Change Adaptation*; Norwegian Meteorological Institute: Oslo, Norway, 2013.
53. Vardoulakis, S.; Fisher, B.; Pericleous, K.; Gonzalez-Flesca, N. Modelling air quality in street canyons: A review. *Atmos. Environ.* **2003**, *37*, 155–182. [CrossRef]
54. Mangan, S.; Oral, G.; Kocagil, I. The impact of urban form on building energy and cost efficiency in temperate-humid zones. *J. Build. Eng.* **2021**, *33*, 101626. [CrossRef]
55. Zielonko-Jung, K. Kształtowanie budynków energooszczędnych w przestrzeniach miejskich. In *Budownictwo Energooszczędne w Polsce—Stan i Perspektywy (Energy-Efficient Construction in Poland—Status and Prospects)*; Wesołowska, M., Podhorecki, A., Eds.; Wydawnictwa Uczelniane Uniwersytetu Technologiczno-Przyrodniczego w Bydgoszczy: Bydgoszcz, Poland, 2015; pp. 107–116.
56. Mao, C.; Baltazar, J.; Haberl, J. Comparison of ASHRAE peak cooling load calculation methods. *Sci. Technol. Built Environ.* **2018**, *25*, 189–208. [CrossRef]
57. Mao, C.; Baltazar, J.C.; Haberl, J.S. Literature review of building peak cooling load methods in the United States. *Sci. Technol. Built Environ.* **2017**, *24*, 228–237. [CrossRef]
58. Malicki, M. *Wentylacja i Klimatyzacja*; PWN: Warsaw, Poland, 1980.
59. Pełech, A. Wentylacja i klimatyzacja. In *Podstawy*; Oficyna Wydawnicza Politechniki Wrocławskiej: Wrocław, Poland, 2013.
60. Recknagel, H.; Sprenger, E. *Recknagel Taschenbuch für Heizung- und Klimatechnik 92/93*; R. Oldenbourg Verlag GmbH: Munich, Germany, 1991.

energies

MDPI

Article

Prioritizing Energy Performance Improvement Factors for Senior Centers Based on Building Energy Simulation and Economic Feasibility

Arisae Nam [1] and Young Il Kim [2,*]

1 Department of Architectural Engineering, Graduate School, Seoul National University of Science and Technology, Seoul 01811, Republic of Korea; nalgae1505@naver.com
2 School of Architecture, Seoul National University of Science and Technology, Seoul 01811, Republic of Korea
* Correspondence: yikim@seoultech.ac.kr

Abstract: This study examined energy performance improvement factors by analyzing both energy performance and the economic impacts to reduce energy costs for senior centers. A fact-finding survey was conducted on 20 senior centers in a metropolitan area, identifying key energy improvement factors. Energy simulations of the buildings were performed using ECO2, an officially certified energy assessment program in Korea, comparing the energy requirements before and after the improvements. The energy demand, energy consumption, and floor area were analyzed, with the J, K, and S standard models selected based on the median values of these factors. To assess the impact of the improvements, blower door tests were conducted on two senior centers before and after window upgrades. Based on the ECO2 simulations and the blower door test results, improvement priorities were identified in the following order: windows, exterior walls, boilers, roofs, and doors. Finally, an economic feasibility analysis applied the construction and heating costs to the standard models. Over a 40-year period, only boiler improvements generated a net profit. Without government support, this study recommends prioritizing boiler upgrades when selecting energy performance improvements.

Keywords: senior center; building energy simulation; primary energy; life cycle cost; blower door test

1. Introduction

Amid the global phenomena of climate change and abnormal warming, energy demand for buildings has been increasing. Following the rise in international oil prices due to the Russia–Ukraine war, issues regarding energy vulnerability among low-income groups have emerged in South Korea [1,2].

In addition, international oil prices have increased due to the Russia–Ukraine war, and in Korea, the issues of energy-vulnerable groups emerged due to the burden of heating energy costs [1,2]. The number of patients affected by hypothermia and frostbite, particularly elderly patients in their 60s and older, increased compared to 2021 due to the cost burden of heating energy consumption and abnormal temperature phenomena [3,4].

Senior centers, which are spaces for the elderly (over the age of 60), are often established in renovated buildings. Old senior centers have high air infiltration due to leakage areas in the building, which increases the heating load and increases energy consumption.

In order to achieve energy welfare for vulnerable groups through the improvement of energy performance in senior centers, the Seoul City implemented energy performance improvement projects at 15 locations in 2020, 7 locations in 2021, and 12 locations in 2022. This initiative also contributed to the reduction of greenhouse gas emissions from buildings by encouraging private sector participation. By prioritizing energy performance improvements when specifying the scope of renovations, the effectiveness of the projects was enhanced, further contributing to the reduction of greenhouse gas emissions from buildings.

This study confirmed the need to improve the energy performance of senior centers, which are mainly used by elderly people over 60 years of age. Among the energy performance improvement projects to be activated to achieve the domestic greenhouse gas reduction target, this study intends to present the effects of each energy performance improvement factor of the senior centers. In addition, an economic feasibility analysis was conducted for each energy performance improvement factor to reduce the energy consumption burden of these centers and their users.

2. Objective

As mentioned in the introduction, it is necessary to improve the energy performance of senior centers, which are mainly used by elderly people in their 60s or older. Prior to an analysis, this study selected the target area by checking the regional status of old senior centers. Presently, the "2nd Green Building Basic Plan" categorizes buildings that have aged more than 20 years since the building completion as old buildings [5]. As shown in Figure 1, the statuses of facilities for the elderly and children, built more than 20 years ago, were analyzed for each region in Korea. [6].

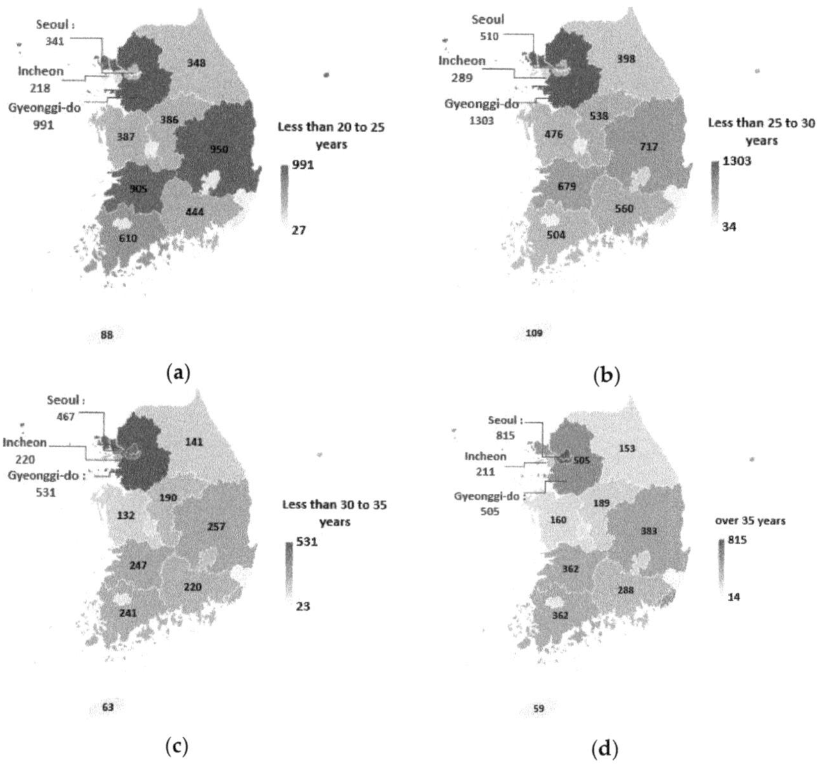

Figure 1. Number of years since completion of the buildings for the elderly and children in Korea: (**a**) number of buildings aged 20–25 years since completion; (**b**) number of buildings aged 25–30 years since completion; (**c**) number of buildings aged 30–35 years since completion; and (**d**) number of buildings aged > 35 years since completion.

An analysis of the current status of the facilities for the elderly as of 2021 revealed that old buildings aged 20 to 35 years after completion were most concentrated in the Gyeonggi-do province. In addition, it was confirmed that there are many old buildings in Seoul that are over 35 years old.

In Korea, Seoul, Gyeonggi-do, and Incheon are categorized as metropolitan areas. Therefore, this study investigated the effects of enhancing the energy performance of senior centers, specifically targeting the metropolitan areas where buildings for the elderly and children, aged over 20 years, are concentrated. In addition, we conducted an economic feasibility analysis of the improvement factors, and the effect of reducing heating energy charges on the improvement factors was presented. Hence, the objective of this study is to delineate the hierarchical prioritization of the improvement factors aimed at enhancing the energy efficiency of senior centers. This will be accomplished through a comprehensive analysis of the energy performance impact and economic viability associated with each factor.

3. Research Process

To analyze senior center energy performance improvement effects and conduct an economic analysis, we surveyed the energy performance improvement status of the senior centers in the metropolitan area, conducted an ECO2 simulation to diagnose their energy performance, selected standard models, performed an economic analysis, and conducted blower door tests before and after window improvement. In addition, an economic feasibility analysis was performed to analyze the energy cost reduction effect of implementing various building energy performance improvement factors. The research process is depicted in Figure 2.

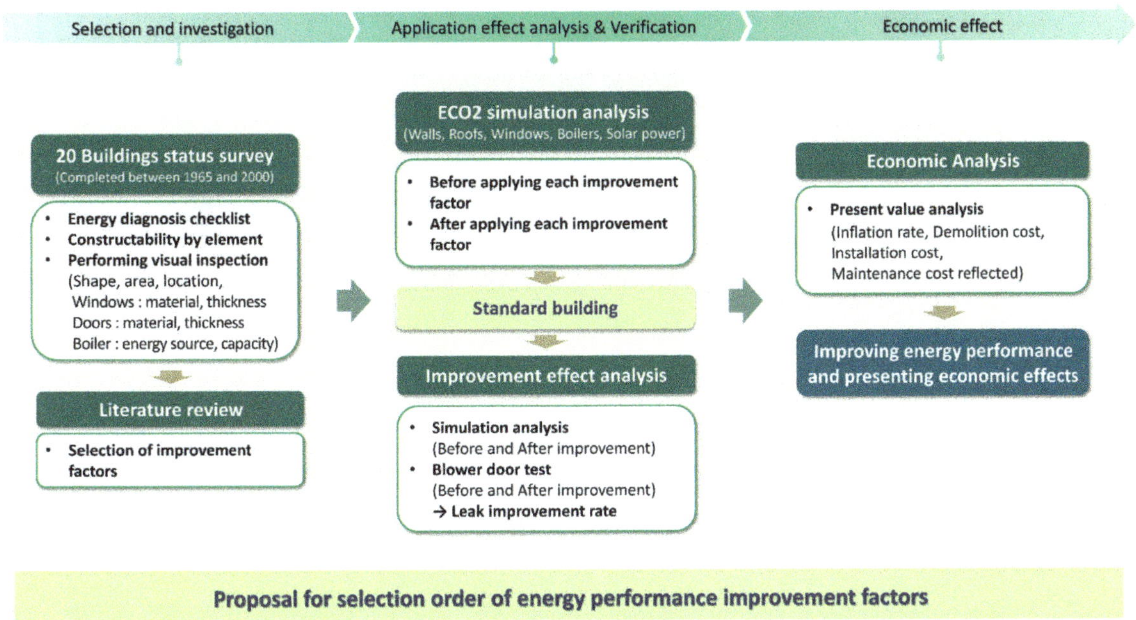

Figure 2. Research process for analyzing the effects and economics of improving the energy performance of senior centers.

Initially, a comprehensive field survey encompassing 20 senior centers, subject to governmental support aimed at enhancing energy performance within the metropolitan region, was undertaken. We collected the existing design documents to analyze energy consumption and other factors, followed by an on-site survey. The on-site survey included user questionnaires, an assessment of the condition and aging status of each building component, an identification of the areas and scopes for improvement, an asbestos inspection, and a visual structural safety assessment. In addition, the energy performance improvement factors were selected via a literature review. Secondly, the energy efficiency of these buildings was evaluated through ECO2 simulations, wherein each selected factor contributing to energy performance enhancement was analyzed. Thirdly, the standard model was chosen by assessing key parameters including energy demand, consumption, primary energy consumption, and the building area derived from the prior ECO2 simulations, with the selection criterion being the median value.

Fourthly, to confirm the improvement in energy performance, constructions aimed at enhancing energy efficiency were carried out in selected older facilities serving the elderly and children in the Seoul metropolitan area. Blower door tests were conducted before and after the improvements to verify the effectiveness of the energy enhancement measures. Based on the results of the ECO2 simulation and blower door test conducted in the third and fourth stages, a method for selecting improvement factors was proposed.

Fifthly, employing a standard model as the framework, this study conducted a detailed analysis of the impact of specific elements on the energy performance improvement of senior centers. This analysis aims to delineate a prioritization process for selecting improvement elements and to evaluate their economic implications.

Finally, this study analyzed the energy performance and economic effect of each improvement element and suggested a method for selecting senior center energy performance improvement elements.

4. Senior Center Energy Performance Diagnosis

To assess the effects of the energy performance improvement measures implemented in the senior centers, an architectural survey of 20 buildings was conducted. Additionally, to evaluate the effectiveness of each energy performance improvement measure, the annual primary energy consumption per unit area was determined by conducting ECO2 simulations.

4.1. Senior Center Status

To examine the conditions of the existing senior centers, the architectural and equipment statuses of 20 buildings were surveyed via their drawings. For the building factors that could not be assessed solely through a visual inspection, such as walls and roofs, the applicable materials were estimated by referring to the relevant insulation application standards based on the year of completion.

4.1.1. Architectural Status

The architectural status of 20 senior centers in the metropolitan area was surveyed based on the year of building completion, building standards at the time of completion, and the visual inspection results. First, in terms of building age, as shown in Table 1, the A-senior center, completed in 1963, is the oldest, while the T-senior center, completed in 1998, is the newest. Most of the buildings were completed between 1980 and 1989.

Table 1. Result of survey on completion year of 20 senior centers in Seoul metropolitan area.

Construction Completion Year	Senior Center
Before 1970	A, B
1970–1979	C, D, E, F, G
1980–1989	H, I, J, K, L, M, N, O
1990–2000	Q, R, S, T

In addition, the external walls and roofs were estimated by analyzing design books, performing a visual inspection, and evaluating the building standards for the year of completion. Through a visual inspection, the shape, area, and space of the buildings were identified, and the materials used for the windows and doors, as well as the energy sources and capacities of the boilers, were analyzed. For the buildings for which it was not possible to obtain the architectural drawings because a long time had elapsed since their completion date, the insulation measures applied to the outer walls and roof were estimated by referring to the building information display and the year of completion [7,8].

The insulation composition and heat transfer coefficient of the outer walls were investigated. As shown in Table 2, the measures to prevent heat loss were initially implemented in 1979 under Article 25 of the Enforcement Rules of the Building Act [8,9]. In addition, in 1992, the measures to prevent heat loss were implemented under the "Rules on Equipment Standards for Buildings" and in 2008, under the "Criteria for Energy Saving Design of Buildings" [10,11]

Table 2. External wall insulation standards in Seoul metropolitan area by revision year.

Revision Year	External Wall Insulation Standard (Heat Transfer Coefficient U: W/m^{2}K)	Building Code Standard
1979	2.093 or 50 mm-thick insulation	Article 25 of Building Act Enforcement Rules
1980	0.581	Article 25 of Building Act Enforcement Rules
1984	0.581 or 50 mm-thick insulation	Article 19 of Building Act Enforcement Rules
1987	0.581 or 50 mm-thick insulation	Article 19 of Building Act Enforcement Rules—Central region
1992	0.581 or 50 mm-thick insulation	Article 21 of Regulations on Equipment Standards for Buildings—Central region

Regarding total floor area, as outlined in Table 3, the highest number of senior centers was observed within the range of 50 m^2 to 150 m^2.

Table 3. Total floor area survey results of 20 senior centers in Seoul metropolitan area.

Total Floor Area (m^2)	Senior Center
Less than 50	P
50–100 or lower	H, I, K, L, M, N, R
100–150 or lower	B, D, E, F, G, J, O, S
More than 150	A, C, Q, T

Therefore, the heat transfer coefficients of the outer walls were estimated by checking the age since the completion year of the senior centers and the implementation date of the standard for the outer wall insulations. According to the results, the estimated heat transfer coefficients of the outer walls before the improvements were 2.093 W/m^{2}K and 0.581 W/m^{2}K, as listed in Table 4. The external finishing materials were verified through a site survey, and the insulation for the exterior walls was assumed to be PF boards, with the insulation thickness calculated in reverse to meet the required overall heat transfer coefficient.

Table 4. Estimated heat transfer coefficients of the exterior walls of 20 senior centers in Seoul metropolitan area.

External Wall Insulation Standard (Heat Transfer Coefficient U: W/m^{2}K)	Senior Center
0.581	H, I, J, K, L, M, N, O, P, Q, R, S, T
2.093	A, B, C, D, E, F, G

The roof heat transfer coefficients were estimated by checking the roof insulation standards enforced in the Seoul metropolitan area by revision year and building age since the year of completion, as shown in Table 5.

Table 5. Roof insulation standards in Seoul metropolitan area by revision year.

Revision Year	Roof Insulation Standard (Heat Transfer Coefficient U: W/m^{2}K)	Building Code Standard
1979	1.047 or 50 mm-thick insulation	Article 25 of Building Act Enforcement Rules
1980	0.581	Article 25 of Building Act Enforcement Rules
1984	0.581 or 50 mm-thick insulation	Article 19 of Building Act Enforcement Rules
1987	0.407 or 50 mm-thick insulation	Article 19 of Building Act Enforcement Rules—Central region
1992	0.407 or 80 mm-thick insulation	Article 21 of Regulations on Equipment Standards for Buildings—Central region

The roof's heat transfer coefficient was estimated based on the completion time of the senior center and the building standards code at that time.

According to the estimates, the roof heat transfer coefficients before the improvements were 0.407 W/m^{2}K, 0.581 W/m^{2}K, and 1.047 W/m^{2}K, as shown in Table 6.

Table 6. Estimated roof heat transfer coefficient values of 20 senior centers in Seoul metropolitan area.

Roof Insulation Standard (Heat Transfer Coefficient U: W/m^{2}K)	Senior Center
0.407	N, O, P, Q, R, S, T
0.581	H, I, J, K, L, M
1.047	A, B, C, D, E, F, G

Regarding windows, this study identified the distinguishable materials, assessed the number of glass layers, and measured the glass thickness. In addition, to understand the heat transfer coefficients and compositions of the windows built in each year, we referred to the Korea Land and Housing Corporation's report "Research on Sound Insulation Design of External Windows", the literature, and the energy-saving design standards [8,10–13].

Table 7 summarizes the compositions and heat transfer coefficients of the windows estimated based on the visual inspection results and the information contained in the reference materials.

Table 7. Results of a survey of major window compositions and heat transfer coefficients of 20 senior centers in Seoul metropolitan area.

Window Insulation Standard (Heat Transfer Coefficient U: W/m^{2}K)	Window Composition	Senior Center
1.8	22 mm (5 mm + 12 mm (Air) + 5 mm), PVC	E, F, R
2.1	16 mm (5 mm + 6 mm (Low-E) + 5 mm), PVC	C, H
2.4	16 mm (5 mm + 6 mm (Air) + 5 mm), PVC	A, B, G, I, J, M, O, P
	10 mm (5 mm + 5 mm), PVC	D, K, L
4	16 mm (5 mm + 6 mm (Air) + 5 mm), AL	N, Q, S
6.6	5 mm, AL	T

A visual inspection confirmed that among the 20 senior centers, with the exception of the D, J, K, L, N, Q, S, and T senior centers, the other buildings had undergone window improvement work at least once in their lifetimes. It was confirmed that the improved polyvinyl chloride (PVC)-framed windows were superior to the aluminum or wooden windows that were typically installed at the time of building completion.

The heat transfer coefficients and compositions of the doors in these buildings were checked via a visual inspection and by referring to the "Energy saving plan design criteria", as listed in Table 8.

Hence, the architectural factors related to the energy improvement of the senior center were identified through field surveys, the literature review, and the legal history.

Table 8. Results of an investigation of major door compositions and heat transfer coefficients of 20 senior centers in Seoul metropolitan area.

Door Insulation Standard (Heat Transfer Coefficient *U*: W/m²K)	Door Composition	Senior Center
2.4	Wood	J
2.7	Steel	H, K, L, M, N, O, P, Q, R, S
5.5	Single glaze	A, B, C, D, E, F, G, I, T

4.1.2. Building Equipment Status

The equipment installed in the senior centers was investigated. The boilers which affect heating energy consumption, were analyzed, and it was found that new and renewable equipment was not installed in all 20 senior centers.

Table 9 lists the main boiler specifications determined via the visual inspections and the attached boiler construction signs. In addition, the visual inspections confirmed that electric boilers were installed as auxiliary equipment in addition to the main boilers.

Table 9. Major boilers installed in 20 senior centers in Seoul metropolitan area.

Boiler Type	Efficiency (%)	Senior Center
Gas	80–83	A, B, C, D, E, F, H, I, J, K, M, N, O, Q, R, S, T
Condensing	86–87	G, L, P

A thorough examination was conducted on the equipment installed within the senior centers. This investigation specifically focused on boilers, which are known to significantly impact heating energy consumption. The findings revealed that none of the 20 senior centers had incorporated new and renewable energy systems. Table 9 lists the boiler specifications determined via the visual inspections and the attached boiler construction signs. In addition, the visual inspections confirmed that electric boilers were installed as auxiliary equipment in addition to the main boilers.

4.2. Energy Performance Diagnosis—Method

To confirm the changes in building energy performance after the improvements to the architectural and facility factors of the 20 senior centers, the differences in the primary energy consumption per unit area of these buildings were analyzed before and after the implementation of the improvements to each building. To assess the variations in primary energy consumption, ECO2s, a building energy simulation program extensively utilized in Korea, were conducted.

ECO2 is a window-based program that calculates building energy requirements and consumption using monthly average weather data. When inputs such as building orientation, architectural features, mechanical systems, renewable energy, building usage, operating hours, and management methods are provided, the program can estimate the building's monthly energy requirements and predict its energy consumption. Energy consumption is categorized into heating, cooling, hot water, lighting, and ventilation. These calculations can then be used to project the building's primary energy consumption and carbon dioxide emissions [14].

4.2.1. Simulation Methods

ECO2 is a building energy consumption assessment program based on ISO 13790 [15] and DIN 18599 [14,16–20]. The ECO2 program is mainly used to perform building energy and zero-energy building assessments [21]. It has also been used for comparative analyses of actual energy consumption and other program results in several studies [22–25]. Thus, this study examined the pre- and post-renovation energy demand, energy consumption, and primary energy consumption of the exterior walls, roofs, windows, and boilers in 20 senior centers that were the recipients of domestic support projects.

4.2.2. Selection of Improvement Factors—Methods

The selection of improvement factors, namely exterior wall insulation, roof insulation, windows, doors, and boilers for the senior centers, was based on prior identifications within this study. The selection of improvement factors was conducted via a literature review to confirm the effects before and after energy performance improvement, as shown in Table 10.

Table 10. Comparison of the literature review—selection of improvement factors for energy performance improvement.

Reference	Contents
[26–28] [29] Major topic	Energy improvement effect of PF (phenolic foam) board exterior wall insulation A study on materials for improved energy conservation in buildings—PF (phenolic foam) board Improvement of exterior wall insulation
[30,31] [29] Major topic	Energy improvement effect of PF (phenolic foam) board roof insulation A study on materials for improved energy conservation in buildings—PF (phenolic foam) board Improvement of roof insulation
[24] [32–34] Major topic	Thermal transmittance rate by window type A study on windows for improved energy conservation in buildings—low-e, double-glazed glass Improvement of window
[24] Major topic	Thermal transmittance rate by door type Improvement of door
[35–37] Major topic	A study on boilers for improved energy conservation in buildings—condensing boiler Improvement of boiler
This work	Selection of improvement factors for energy performance improvement

To improve energy performance, options such as reducing the window area, installing shading devices, adding windproof structures, installing nighttime insulation devices for the windows, implementing inverter control for the rotating equipment, applying LED lighting, and using standby power cut-off outlets could be considered. However, in this study, these measures were not considered due to the difficulty of applying them to small, aging buildings.

4.3. Energy Consumption by Improvement Factor

The improvement factors selected through the literature review were applied to the ECO2 simulations to compare the energy demand, energy consumption, and primary energy consumption before and after the improvement of the exterior walls, roofs, windows, and boilers.

4.3.1. Wall Insulation

The changes in building energy performance after the application of the exterior wall improvements to the 20 senior centers were analyzed. The exterior wall configurations utilized in the ECO2 simulation input were based on the exterior wall conditions outlined in Table 2. The conditions for post-improvement of the exterior walls involved the application of widely used PF boards, ensuring fire resistance performance as required by the relevant regulations [38–40]. Table 11 lists the analysis results of the energy demand, energy consumption, and primary energy consumption before and after the application of PF boards for the improving the external walls. The average decrease in primary energy consumption after the application of the external wall improvements was found to be 25.9%. The senior center with the least improvement effect was analyzed as O. The median value of the primary energy consumption reduction was 14.6% and was with the K-senior center.

Table 11. ECO2 simulation results for energy demand, energy consumption, and primary energy consumption before and after the application of external wall improvements.

Improvement Effect (Analysis of 20 Models)		Demand or Consumption per Unit Area per Year (Unit: kWh/m^{2}y)						
		Energy Demand		Energy Consumption		Primary Energy Consumption		
		Before	After	Before	After	Before	After	Improvement Rate (%)
Lowest	O	114.4	104.7	143.5	129.0	201.9	186.4	8.7
Median	K	127.3	110.4	174.3	148.5	222.5	194.5	14.6
Highest	D	207.8	126.4	296.9	176.7	357.6	225.9	59.3
Average		150.8	116.7	205.3	155.8	260.5	205.6	25.9

Figure 3 shows the ECO2 simulation input screen for the K-senior center, the improvement rate of which was found to be the median value. The heat transfer coefficients of the exterior walls of the K-senior center before and after the application of the exterior wall improvements were 0.581 W/m^{2}K and 0.340 W/m^{2}K, respectively. Therefore, based on the simulation results, the predicted primary energy consumption per unit area per year would decrease by 28 kWh/m^{2}y after the application of PF boards for the exterior wall improvements (Table 11).

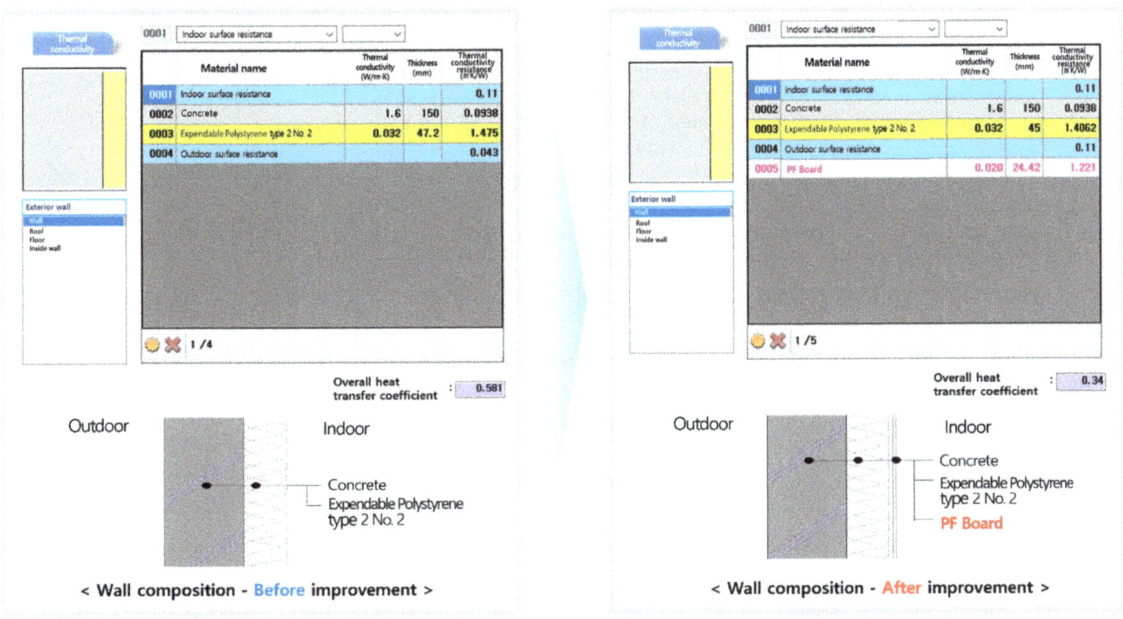

Figure 3. ECO2 simulation input screen before and after wall improvement at K-senior center.

4.3.2. Roof Insulation

Table 12 shows the results of analyzing the changes in energy performance due to roof improvements for the 20 senior centers using ECO2 simulations, with reference to the values in Table 6. Regarding the material for roof improvement, a PF board, known for its practical effect of enhancement, was applied [38,39]. The average improvement rate of the roofs after the application of a PF board was 6.4%, and the N-senior center exhibited the lowest improvement effect. In addition, the I-senior center exhibited the median improvement effect with an improvement rate of 4.8%. For I senior center, the heat transfer coefficient values before and after the application of the roof improvements were 0.581 W/m^{2}K and 0.150 W/m^{2}K, respectively.

Table 12. ECO2 simulation results of energy demand, energy consumption, and primary energy consumption before and after the application of roof improvements.

Improvement Effect (Analysis of 20 Models)		Demand or Consumption per Unit Area per Year (Unit: kWh/m^{2}y)						
		Energy Demand		Energy Consumption		Primary Energy Consumption		
		Before	After	Before	After	Before	After	Improvement Rate (%)
Lowest	N	142.4	139.1	201.3	196.7	247.8	237.7	4.2
Median	I	125.4	118.8	176.2	166.7	230.9	220.4	4.8
Highest	G	234.5	207.8	290.9	255	371.2	330.8	12.2
Average		150.8	140.4	205.3	190.9	260.5	243.4	6.4

Hence, according to the simulation outcomes, our projections indicate a reduction of 10.5 kWh/m^{2}y in the annual primary energy consumption per unit area subsequent to the implementation of PF boards for roof enhancement (Table 12).

4.3.3. Window

For windows, an ECO2 simulation analysis was conducted for senior centers D, J, K, L, N, Q, S, and T, excluding the 12 centers where window improvement projects had already been carried out. For the post-window improvement condition, low-e double-glazed glass and PVC frames, currently the most widely utilized in energy-saving design standards, were implemented [11,41–43]. According to the results of the ECO2 simulations, the average window improvement rate was 8.0%, and the J-senior center exhibited the least improvement. The K-senior center exhibited the median improvement effect with an improvement rate of 4.4% (Table 13).

Table 13. ECO2 simulation results of energy demand, energy consumption, and primary energy consumption before and after the application of window improvements.

Improvement Effect (D, J, K, L, N, Q, S, T)		Demand or Consumption per Unit Area per Year (Unit: kWh/m^{2}y)						
		Energy Demand		Energy Consumption		Primary Energy Consumption		
		Before	After	Before	After	Before	After	Improvement Rate (%)
Lowest	J	120.3	119.6	171.1	170.9	224.8	224.3	0.1
Median	K	127.3	121.1	174.3	167.0	222.5	213.7	4.4
Highest	S	137.0	110.5	170.9	139.7	222.5	185.6	22.3
Average		141.5	130.6	195.0	181.6	247.2	231.4	8.0

The windows of the K-senior center were composed of 10 mm-thick (5 mm + 5 mm) PVC, and the heat transfer coefficient was 3.1 W/m^2 K. After the improvement with low-e, double-glazed glass, the heat transfer coefficient became 2.1 W/m^2 K. Therefore, based on the simulation results, we anticipated that the primary energy consumption per unit area per year would diminish by 8.8 kWh/m^{2}y with the window enhancement measures.

4.3.4. Door

ECO2 simulations were performed on the door systems of 17 senior centers, with reference to the values in Table 8. The senior centers designated as C, F, and N, where structural alterations to the doors were unfeasible, were excluded from the analysis. As for the conditions after the door improvements, the heat transfer coefficients were applied by referring to the energy-saving design standard. For the door materials, the same materials as before the improvements were applied, considering the structural characteristics. [11].

The ECO2 simulation results indicated that the average door improvement rate was 1.8%, and the R-senior center exhibited the lowest improvement. Among the 17 senior centers considered in this analysis, the S-senior center exhibited the median value, and the corresponding improvement rate was 1.2% (Table 14).

Table 14. ECO2 simulation results of energy demand, energy consumption, and primary energy consumption before and after door improvements.

Improvement Effect (A, B, D, E, G, H, I, J, K, L, M, N, O, P, Q, R, S, T)		Demand or Consumption per Unit Area per Year (Unit: kWh/m^{2}y)						
		Energy Demand		Energy Consumption		Primary Energy Consumption		
		Before	After	Before	After	Before	After	Improvement Rate (%)
Lowest	D	207.8	206.5	296.9	295.4	357.6	355.6	0.6
Median	S	137.0	135.4	170.9	168.3	222.5	219.8	1.2
Highest	R	115.0	110.5	160.7	154.1	210.7	203.5	3.5
Average		146.3	143.8	199.8	196.7	255.2	250.5	1.8

The doors used in the S-senior center were made of steel, and the overall heat transfer coefficient was 2.7 W/m^2 K. After the door improvements, the applied overall heat transfer coefficient of the doors was 1.7 W/m^2 K. The simulation predicts that the annual primary energy consumption per unit area will decrease by 2.7 kWh/㎡y when applying door renovations.

4.3.5. Boiler

An ECO2 simulation analysis was conducted on 17 senior centers, excluding G, L, and P, where condensing boilers was already applied, as referenced in Table 9. The conditions for boiler improvement involved the application of condensing boilers characterized by low nitrogen oxide emissions and a high thermal efficiency [44,45].

The ECO2 simulation results indicated that the average improvement rate was 6.5%, as listed in Table 15, and the C-senior center exhibited the lowest improvement effect. The N-senior center showed the median improvement effect among the 17 senior centers, with a corresponding improvement rate of 8.2% (Table 15).

Table 15. ECO2 simulation analysis results of boilers before and after improvements: energy demand, energy consumption, and primary energy consumption.

Improvement Effect (A, B, C, D, E, F, H, I, J, K, M, N, O, Q, R, S, T)		Demand or Consumption per Unit Area per Year (Unit: kWh/m^{2}y)						
		Energy Demand		Energy Consumption		Primary Energy Consumption		
		Before	After	Before	After	Before	After	Improvement Rate (%)
Lowest	C	209.1	209.1	269.4	262.5	337.8	331.3	2.0
Median	N	142.4	142.4	201.3	187.8	247.8	229.0	8.2
Highest	I	125.4	125.4	176.2	158.3	230.9	212.0	8.9
Average		149.0	149.0	202.8	188.9	257.0	241.6	6.5

Figure 4 shows the ECO2 simulation input screens of the N-senior center, which exhibited the median improvement effect. In the N-senior center, 18.6 kW gas boilers were installed on both the first and second floors before the improvement. After the improvement, they were replaced with condensing boilers of the same capacities. Therefore, the simulation predicts that the annual primary energy consumption per unit area will decrease by 18.8 kWh/㎡·y with the boiler renovation.

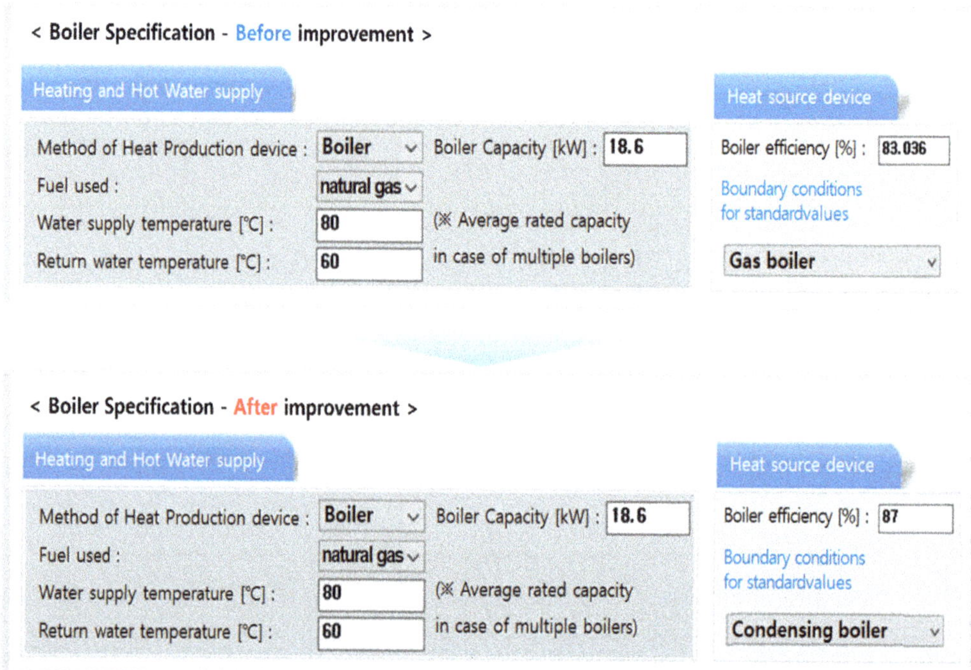

Figure 4. ECO2 simulation input screens of N-senior center before and after boiler improvement.

5. Standard Models

Subsequently, a standard model for the senior centers was selected to validate the earlier ECO2 simulation results and analyze the economic impact of the primary energy improvement measures.

5.1. Selection of Standard Models

Among the 20 senior centers, the standard models were determined based on the ECO2 simulation results. Following an analysis of the average characteristics related to floor area, energy demand, energy consumption, and primary energy consumption, which had the most significant impact on annual primary energy consumption, the standard models were selected, as outlined in Table 16. Table 16 specifies the average characteristics, designated as J, K, and S, for each condition, representing the standard models for the senior centers.

Table 16. Selected standard models based on median values of each category of 20 senior centers in the metropolitan area.

Division	Standard Model
Total floor area	S
Annual energy requirement per unit area	
Annual energy consumption per unit area	K
Annual primary energy consumption per unit area	J

Table 17 shows the results of the ECO2 analysis of the primary energy consumption before and after the energy performance improvements were applied to the standard models J, K, and S. According to the results of a standard model energy performance improvement analysis, as shown in Figure 5, the improvements to the exterior walls had the strongest effects on models J and K, and the window improvements had the strongest effect on model S as shown by dotted boxes. Overall, the element that showed the greatest improvement was determined to be the windows.

Table 17. ECO2 simulation results showing primary energy consumption after application of improvements to outer walls, roofs, windows, doors, and boilers in standard models J, K, and S.

Standard Model	Annual Primary Energy Consumption per Unit Area (Unit: kWh/m^{2}y)									
	Wall		Roof		Window		Door		Boiler	
	Before	After	Before	After	Before	After	Before	After	Before	After
J	224.8	202.5	224.80	214.20	224.80	224.30	224.80	220.00	224.80	210.50
K	222.5	194.5	222.50	212.10	222.50	213.70	222.50	215.80	222.50	204.50
S	222.5	203.2	222.50	217.10	222.50	185.60	222.50	219.80	222.50	211.60
Average	223.3	200.1	223.3	214.5	223.3	207.9	223.3	218.5	223.3	208.9

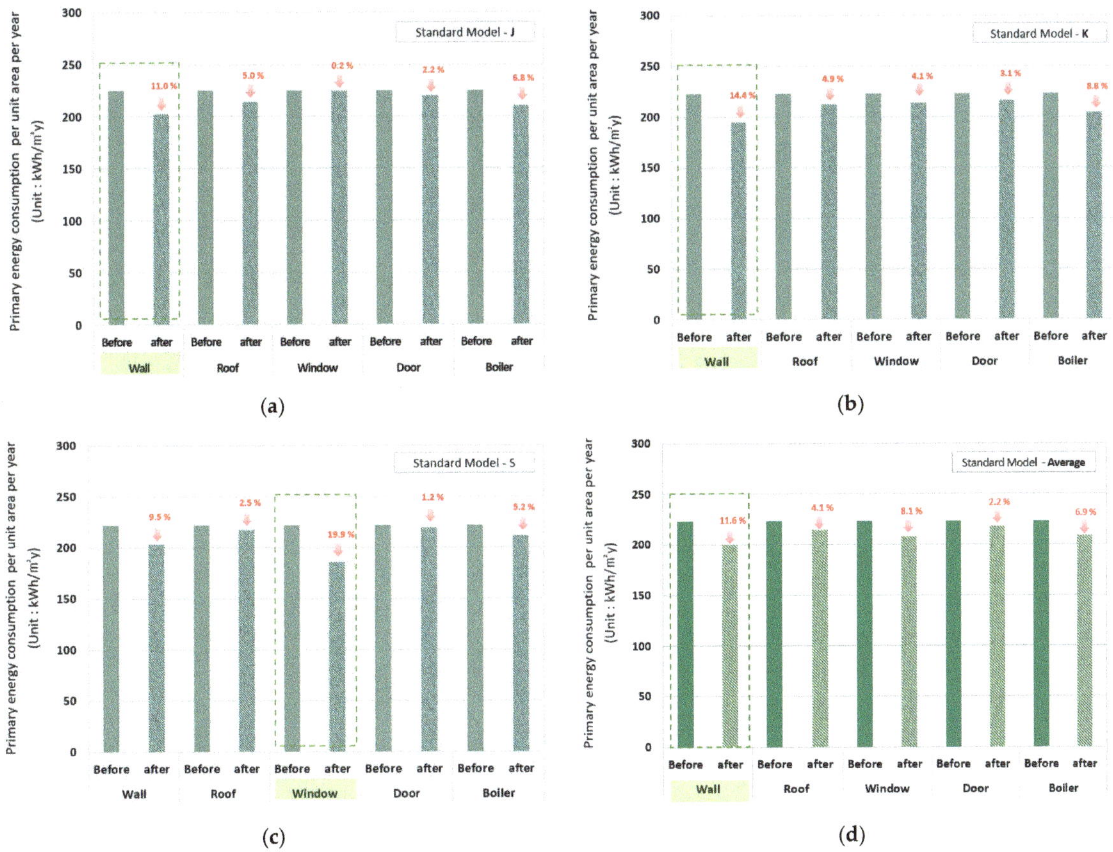

Figure 5. ECO2 primary energy consumption analysis results by applying exterior wall, roof, window, door, and boiler improvements to standard models J, K, and S: (**a**) ECO2 primary energy consumption analysis results of standard models J, (**b**) K, and (**c**) S. (**d**) ECO2 simulation results of average primary energy consumption of standard models J, K, and S.

Next, the energy performance improvement effect of the window improvements was verified.

5.2. Verification of Energy Performance Improvement

To validate the improvements observed in the windows with the most significant impact, this study selected aged senior centers within the metropolitan area and conducted simulations and verification experiments. Figure 6 presents diagrams of the α- and β-senior centers, where the verification experiment was conducted. To determine the composition of the windows in the α- and β-senior centers, the window thickness was measured using an EDTM Glass-Chek Pro GC3000 instrument. As shown in Table 18, the windows of the α-senior center were 6 mm thick and composed of PVC frames, while the windows of the β-senior center were 6 mm thick and composed of aluminum frames. For the window and door improvements, a heat transfer coefficient of 1.8 W/m^{2}K, an overall thickness of 22 mm (5 mm + 12 mm (Air) + 5 mm), and PVC frames were applied.

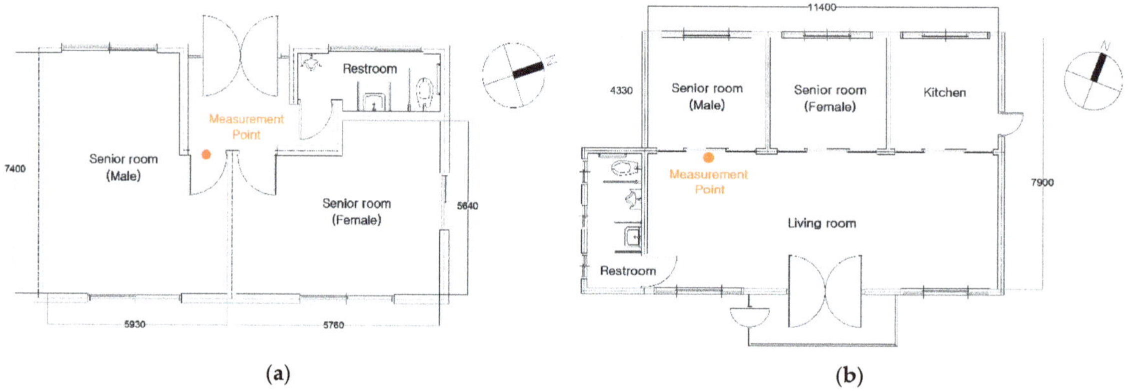

Figure 6. Drawing of blower door test site: (**a**) α-senior center with an area of 114 m^2 and (**b**) β-senior center with an area of 115 m^2.

Table 18. Construction completion years of α- and β-senior centers, and compositions of the applied windows.

Senior Center	Construction Completion Year	Window Insulation Standard (Thermal Transmittance *U*: W/m^{2}K)	Window Composition
α	1996	5.3	6 mm, PVC
β	2000	6.6	6 mm, AL

5.2.1. Simulation Analysis

To verify the improvement effect, an ECO2 simulation analysis was conducted. According to the ECO2 simulation, in the case of the α-senior center, the heating energy improvement rate of primary energy consumption was 28.5%, as illustrated in Figure 7. In the case of the β-senior center, the heating energy improvement rate of primary energy consumption was 17.1%, as illustrated in Figure 8. Therefore, the total energy demand and consumption improved when the windows were improved. The energy improvement had a greater impact on heating energy consumption than on cooling energy consumption [46,47].

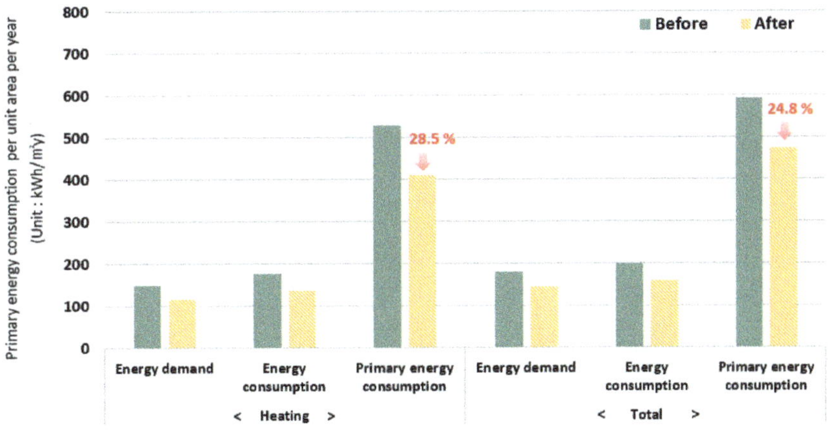

Figure 7. ECO2 simulation results of α-senior center after window improvements.

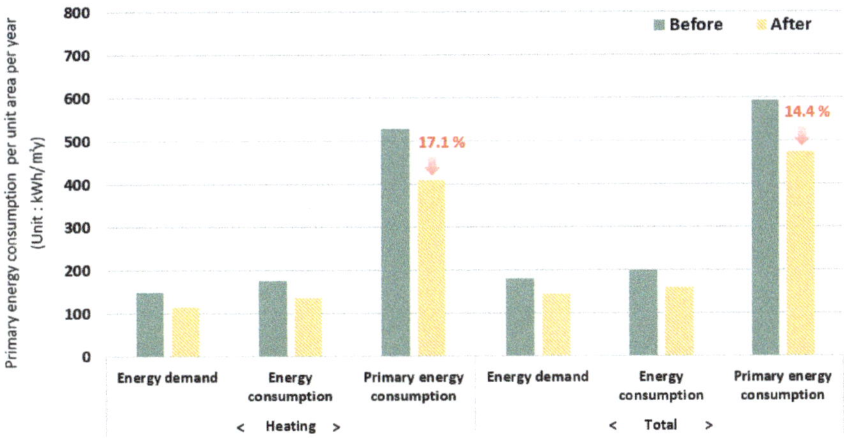

Figure 8. ECO2 simulation results of β -senior center after window improvements.

In this study, the results of the heating energy performance improvements by simulation were verified by measuring the airtightness before and after the window improvements [26,27,48].

5.2.2. Blower Door Test

The improvement in airtightness due to window improvements affects both the heating energy demand and consumption. The strong correlation between airtightness and energy consumption was confirmed by Jeon [48] and Choi [26]. Therefore, airtightness measurement experiments were conducted to verify the energy performance improvements of the windows and doors. The measurement was taken in the male room (39.3 m^2) of the α-senior center and the male room (16.9 m^2) of the β-senior center, as shown in Figure 6. The airtightness test was conducted after installing a blower fan (TEC Minneapolis Blower Door™ System, Minneapolis, MN, USA) at the door, as shown in (a) and (c) of Figure 9. The blower door test was conducted under indoor and outdoor pressure difference conditions ranging from 10 to 65 Pa, following the positive/negative pressurization method (ASTM E779) [28–30]. The airtightness was measured more than eight times using the pressurization and depressurization methods for the pressure range of 10–65 Pa.

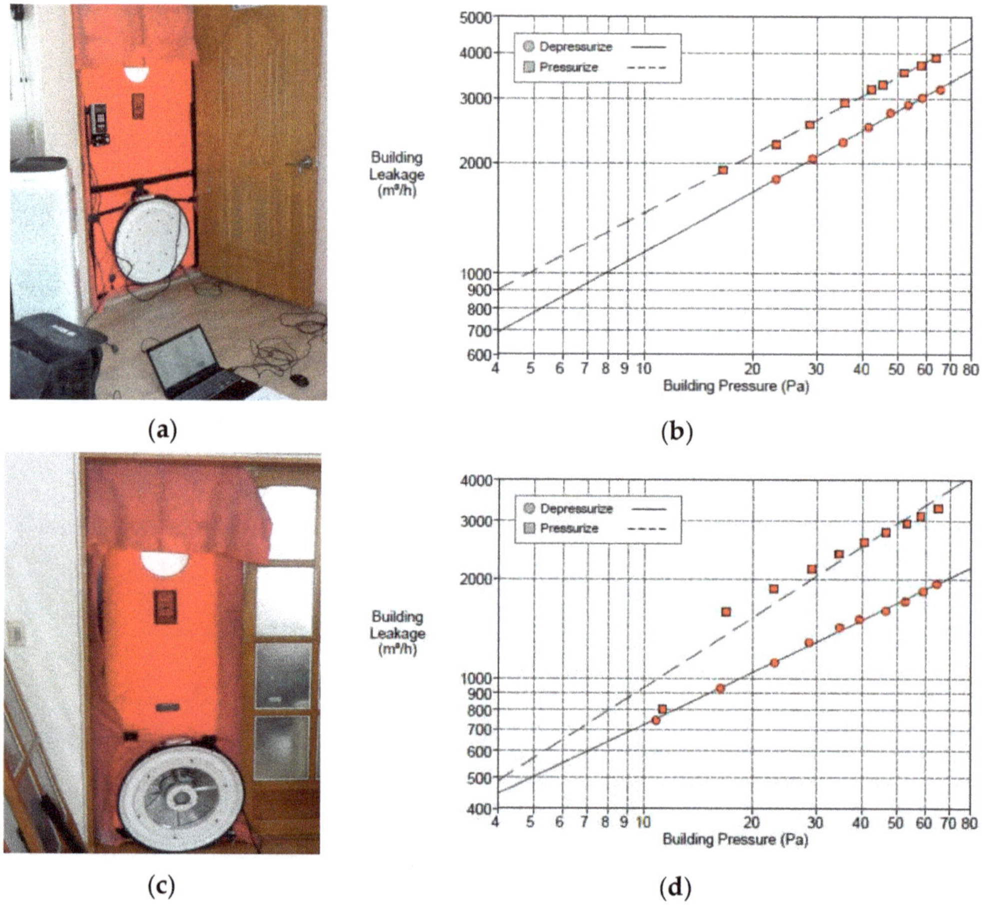

Figure 9. Experimental setup and results of blower door test before window construction: (**a**) blower door test conducted at α-senior center, and (**b**) test results at α-senior center. (**c**) Blower door test conducted at β-senior center, and (**d**) test results at β-senior center.

According to the results of the blower door test conducted before the window improvements, the building infiltration in the α-senior center was 2271 m^3/h with a pressure difference of 50 Pa during depressurization, and 3435 m^3/h with a pressure difference of 50 Pa during pressurization. In the β-senior center, the building infiltration was 1693 m^3/h with a pressure difference of 50 Pa during depressurization and 2945 m^3/h with a pressure difference of 50 Pa during pressurization.

The results of the blower door test conducted after the window improvements are depicted in Figure 10. In the α-senior center, the building infiltration was 1310 m^3/h with a pressure difference of 50 Pa during depressurization, and 1636 m^3/h with a pressure difference of 50 Pa during pressurization. In the β-senior center, the building infiltration was 1461 m^3/h with a pressure difference of 50 Pa during depressurization, and 2724 m^3/h with a pressure difference of 50 Pa during pressurization.

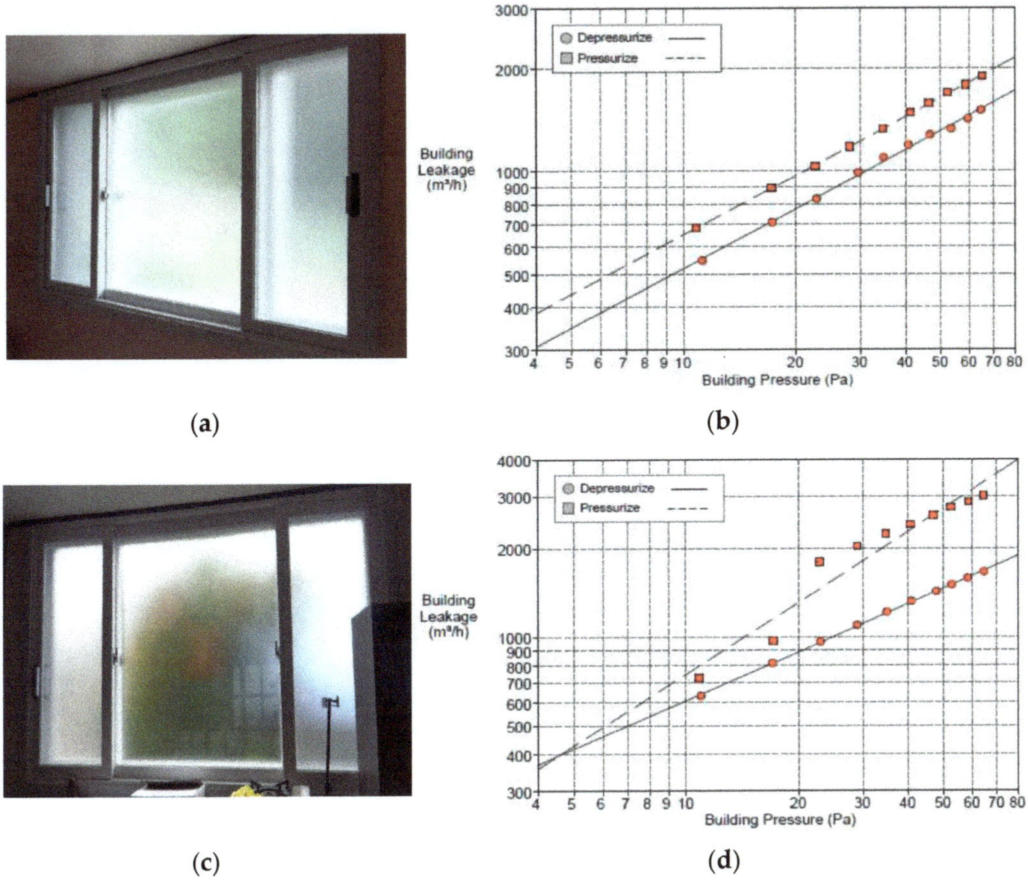

Figure 10. Blower door test after window improvement: (**a**) photograph of window constructed in α-senior center, and (**b**) blower door test results under depressurization and pressurization at α-senior center. (**c**) Photograph of window constructed in β-senior center, and (**d**) blower door test results under depressurization and pressurization at β-senior center.

The results of the blower door test for the α- and β-senior centers were analyzed and converted into air change per hour (ACH) as shown in Table 19.

Table 19. Air change per hour (ACH) of α- and β-senior centers according to the blower door test results. Pressure difference = 50 Pa.

Category		Depressurization ACH (1/h)	Pressurization ACH (1/h)	Average ACH (1/h)
α-senior center	Before	25.16	31.18	28.17
	After	11.90	14.86	13.38
	Improvement rate	52.70%	52.34%	52.50%
β-senior center	Before	39.21	68.19	53.70
	After	33.83	63.07	48.45
	Improvement rate	13.72%	7.51%	9.78%

The measured air change per hour (ACH) showed an improvement of 52.34% for the α-senior center, which was approximately 35.24% higher than the simulation result. For the β-senior center, the measured improvement was approximately 4.62% lower than the simulated improvement effect of 9.78%. Therefore, we concluded that the ACH could be improved by at least 9.78% by replacing the existing single-pane windows.

Therefore, when prioritizing energy performance improvement factors in senior centers, an energy consumption analysis through a detailed diagnosis of each performance improvement factor is the most reliable method, but if a detailed diagnosis is difficult, it is suggested to carry out improvements based on the process presented in this study.

The increase in a heating load due to an infiltration is calculated using the following equation.

$$\dot{Q} = \rho c_p V (t_i - t_o) \quad (1)$$

Here, ρ represents the density, c_p denotes the specific heat of air at a constant pressure, n is the air changes per hour, V is the volume of the space, t_i is the indoor temperature, and t_o is the outdoor temperature. The indoor temperature is assumed to be 26 °C.

The reduction in the heating load due to the infiltration improvement is 33.1 kWh/(m^2 y) from the ECO2 simulation and 32.1 kWh/(m^2 y) from the blower door test. As shown in Table 20, the two results show an error of 3.0%, indicating that the ECO2 simulation accurately predicts the actual heating load.

Table 20. Conditions of α center for heating load calculations due to infiltrations.

		Average Outdoor (°C)	Air Density (kg/m^3)	Operation Time per Day (h)	Day of Use per Month (d)	ACH 50 Pa (1/h)	ACH 10 Pa (1/h)	Infiltration (cmh)	Heat Loss (kWh)	Annual Heat Loss per Area (kWh/m^2 y)
December	Before	0.9	1.287	8	21	28.17	5.634	509.26	770.78	19.61
	After					13.38	2.676	241.88	366.10	9.32
January	Before	−2.1	1.301	8	22	28.17	5.634	509.26	913.83	23.25
	After					13.38	2.676	241.88	434.04	11.04
February	Before	0.2	1.290	8	19	28.17	5.634	509.26	718.49	18.28
	After					13.38	2.676	241.88	341.26	8.68

As described in Figure 11, the first step is to check whether the window has been remodeled through a visual inspection. Next, if the windows have not been remodeled, priority is given to improving the windows, and if the windows have already been remodeled, improvement of the exterior wall insulation, replacement with a high-efficiency condensing boiler, and replacement of the roof insulation are suggested.

This study also included an economic analysis of the standard model to analyze not only the effect of improving energy performance, but also the economic burden caused by the construction.

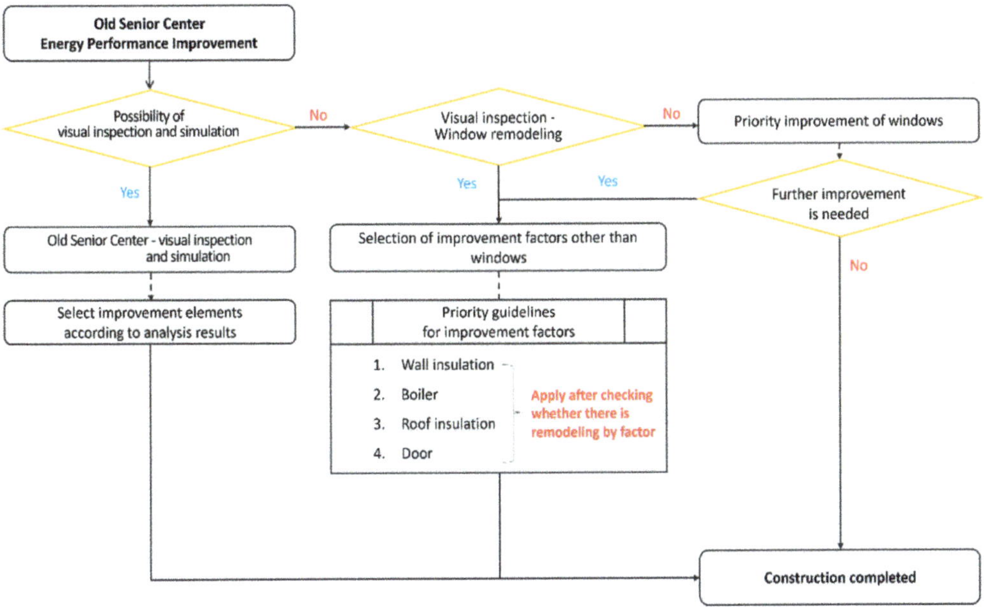

Figure 11. Energy performance improvement factor selection step considering the energy improvement effect.

5.3. Economic Analysis of the Standard Model

To analyze the economic effect of improvements, the annual heating energy cost and life cycle cost of the standard model were conducted [31–34].

In the economic analysis, product price, labor cost, demolition cost, additional costs (asbestos removal costs, structural maintenance costs, electric capacity expansion costs, etc.), and waste disposal cost were considered [34,35].

The unit cost analysis results for the J-senior center are shown in Table 21, for the S-senior center in Table 22, and for the K-senior center in Table 23. The results show that the highest improvement construction cost is for the exterior wall, followed by the window, roof, and boiler, respectively.

Table 21. Unit cost analysis of energy improvement construction in J-senior center.

	Wall (Area: 173.28 m^2)	**Window (Area: 15.17 m^2)**	**Roof (Area: 43.68 m^2)**	**Boiler (18.6, 34.9 kW)**
Product price	20,152 (153,315 KRW/m^2)	3795 (329,791 KRW/m^2)	2323 (70,121 KRW/m^2)	1115
Labor cost	2592 (37,429 KRW/m^2)	628 (54,572 KRW/m^2)	207 (6256 KRW/m^2)	358
Demolition cost	2328	158	587	215
Additional cost	6823	1327	759	-
Waste disposal cost	682	133	76	44
Total cost	32,577	6041	3953	1732

Costs as of 18 April 2023 (unit: USD 1 = KRW 1318.3).

Table 22. Unit price analysis of energy improvement construction in S-senior center.

	Wall (Area: 132.35 m²)	**Window (Area: 48.67 m²)**	**Roof (Area: 29.9 m²)**	**Boiler (18.6, 18.6 kW)**
Product price	15,392 (153,315 KRW/m²)	12,175 (329,791 KRW/m²)	1590 (70,121 KRW/m²)	956
Labor cost	1979 (37,429 KRW/m²)	2015 (54,572.3 KRW/m²)	142 (6256 KRW/m²)	314
Demolition cost	1778	278	402	188
Additional cost	5211	4257	520	-
Waste disposal cost	521	426	52	188
Total cost	24,882	19,151	2706	1646

Costs as of 18 April 2023 (unit: USD 1 = KRW 1318.3).

Table 23. Energy improvement construction unit price analysis results of K-senior center.

	Wall (Area: 132.35 m²)	**Window (Area: 16.87 m²)**	**Roof (Area: 51.72 m²)**	**Boiler (22.8 kW)**
Product price	13,921 (153,315 KRW/m²)	4220 (329,791 KRW/m²)	2751 (70,121 KRW/m²)	546
Labor cost	1790 (37,429 KRW/m²)	698 (54,572.3 KRW/m²)	245 (6256 KRW/m²)	169
Demolition cost	1608	118	695	102
Additional cost	4713	1476	899	-
Waste disposal cost	471	148	90	21
Total cost	22,504	6660	4680	839

Costs as of 18 April 2023 (unit: USD 1 = 1318.3 KRW).

As shown in the table, the construction costs for improving elderly facility elements are substantial. However, when undertaking energy performance improvement projects for elderly care facilities, the government provides partial financial support for the construction costs [36,37]. Currently, in order to reduce greenhouse gas emissions, the government is implementing green remodeling policies and energy improvement projects in facilities for the elderly and children [49–51]. Therefore, participation in various government support projects is expected to reduce the construction costs of such improvements.

The lifecycle analysis of this study excluded the asset value appreciation rate, focusing solely on heating energy consumption, which accounts for the highest usage [31,33,34,52]. The heating energy, maintenance, and replacement costs were analyzed using the present value analysis method, taking into account the inflation rate [1].

To determine the annual heating consumption rate, the primary heating energy consumption obtained from the ECO2 simulation was computed [31,34,53]. Heating energy costs were calculated by analyzing the consumption of primary fuel, LNG, and electricity.

Equations (2) and (3) were used for the annual heating energy consumption per unit area to derive the gas heating rates and power usage rates.

The formula for the gas heating cost is given by Equation (2).

$$C_f = C_{af}\, E_{hec,f}\, A_s \tag{2}$$

The variables in Equation (2) are defined as follows:

C_f: Heating energy cost (USD).
C_{af}: Average gas heating unit cost (USD/Mcal).
$E_{hec,f}$: Annual heating energy consumption per unit area (kWh/m^2).
A_s: Total floor area of the senior center (m^2).

The gas unit price was determined based on the seasonal gas cost in the metropolitan area, 0.0620 USD/Mcal during the winter season and 0.0613 USD/Mcal for the other seasons [54].

The formula for the electric power cost is expressed as Equation (3).

$$C_p = C_{ap}\, E_{hec,p}\, A_s + C_B \tag{3}$$

The variables in Equation (3) are defined as follows:

C_p: Electric power energy cost (USD).
C_{ap}: Average electric power unit cost (USD/kWh).
$E_{hec,p}$: Annual heating energy consumption per unit area (kWh/m^2).
A_s: Total floor area of the senior center (m^2).
C_B: Additional factor over basic cost (USD).

In Equation (3), an electric unit cost of 0.074 USD/kWh was applied by referring to seasonal electricity rate data, and for a base rate of C_B, USD 0.690 was applied. By using Equations (2) and (3), the energy costs according to the annual heating energy consumption per unit area before and after the improvements were calculated [31,34,53].

In this study, a life cycle cost analysis was conducted for a 40-year period. In accordance with the building's age, remodeling or reconstruction of the exterior walls, windows, and roof was undertaken [54–57]. For the life cycle cost analysis, the present value coefficient and the present value coefficient of annuity were computed using Equations (4) and (5).

The present value coefficient formula (3) was applied to the equipment maintenance and replacement costs. The assumption is made that boiler replacements occur every ten years. Therefore, it was assumed that boiler replacements occurred four times during a 40-year lifespan [58].

The discount rate of 1.89%, representing the average annual increase in household gas and electricity tariffs from 2012 to 2022, was applied to calculate the present value coefficient and annuity present value coefficient [1].

The present value coefficient formula is expressed as Equation (4).

$$F_C = \frac{1}{(1+r)^n} \tag{4}$$

The variables in Equation (4) are defined as follows:

F_C: Present value coefficient (-).
r: Average consumer price inflation rate between 2012 and 2022 (-).

The formula for the annuity present value coefficient is given in Equation (5).

$$F_{EC} = \frac{(1+r)^n - 1}{r(1+r)^n} \tag{5}$$

The variables in Equation (5) are defined as follows:

F_{EC}: Annuity present value coefficient (-).
r: Average consumer price inflation rate between 2012 and 2022 (-).

Furthermore, as mentioned in previous research, a supplementary life cycle analysis was conducted to reflect the sharp inflation in 2022, incorporating a 5.5% inflation rate for that year [1].

Based on the results of the life cycle cost analysis of the heating costs over 40 years for the enhancement of the exterior walls, windows, and roofs, as summarized in Table 24, it was confirmed that the implementation of the enhancements provided no economic benefit. However, when receiving construction subsidies, the improvement in heating costs was most significantly observed in the window enhancement of the S model. In addition, when accounting for the construction cost subsidies, the wall factor yielded the greatest average improvement benefit across the J, K, and S standard models.

Table 24. Results of life cycle cost analysis conducted over 40 years for improvement of each factor in the standard models J, K, and S.

Improvement Factor	Senior Center	Annuity Present Value Coefficient	Energy Cost		* Profit (USD)
		** 1.89%	Before (USD/y)	After (USD/y)	** 1.89%
Wall	J	27.89	762	642	−29,230
	K		482	393	−20,022
	S		833	721	−21,758
Window	J		762	762	−6041
	K		482	460	−6046
	S		833	653	−14,214
Roof	J		762	707	−2419
	K		482	450	−3788
	S		833	803	−1869

Costs as of 18 April 2023 (unit: USD 1 = 1318.3 KRW). * Amount may change depending on the building condition. Installation cost (including mechanical, electrical, and labor). ** 1.89%: average consumer price inflation rate between 2012 and 2022.

Table 25 shows the results of the present value analysis of the life cycle costs of boiler maintenance and replacement over a 40-year period. The cost of replacing an aging boiler was analyzed based on the model from Company B.

Table 25. Results of life cycle analysis of boiler replacement over 40 years following boiler improvement of standard models J, K, and S.

Inflation Rate	Year	Present Value Coefficient	Construction Cost by Year When Considering the Present Value Factor					
			J		K		S	
			Before	After	Before	After	Before	After
1.89%	* Construction cost		1019	1732	1019	839	1019	1496
	10	0.829	845	1436	845	696	845	1241
	20	0.688	700	1191	700	577	700	1029
	30	0.570	581	987	581	478	581	853
	40	0.473	482	819	482	397	482	707
	Sum		3626	6164	3626	2986	3626	5326
Construction Profit			−2538		640		−1700	

Costs as of 18 April 2023 (unit: USD 1 = 1318.3 KRW). * Costs of condensing boilers are subsidized by the government.

As shown in Table 24, only the K model was analyzed to be economically viable. In the case of the K model, the building was small, so a boiler with a small capacity was sufficient, making it economically feasible.

Table 26 presents the results of the 40-year life cycle cost analysis of the heating costs based on boiler operations.

Table 26. Life cycle analysis results considering heating costs over 40 years after boiler improvements in standard models J, K, and S.

Improvement Factor	Senior Center	Present Value Coefficient of Annuity	Energy Cost		Energy Profit (USD)	* Total Profit (USD)
		** 1.89%	Before	After	** 1.89%	** 1.89%
Boiler	J	27.89	762	717	1255	−1.238
	K		482	442	1116	1756
	S		833	793	1116	−584

Costs as of 18 April 2023 (unit: USD 1 = 1318.3 KRW). * Amount may change depending on the building condition. Installation cost (including mechanical, electrical, and labor). ** 1.89%: average consumer price inflation rate between 2012 and 2022.

Therefore, when construction costs are not supported, the economic benefit is negative in all the cases. However, replacing a standard boiler with a condensing boiler resulted in the smallest deficit.

As a result of the life cycle cost analysis, as shown in Figure 12, only the boiler for the K model was found to generate benefits in total costs over 40 years.

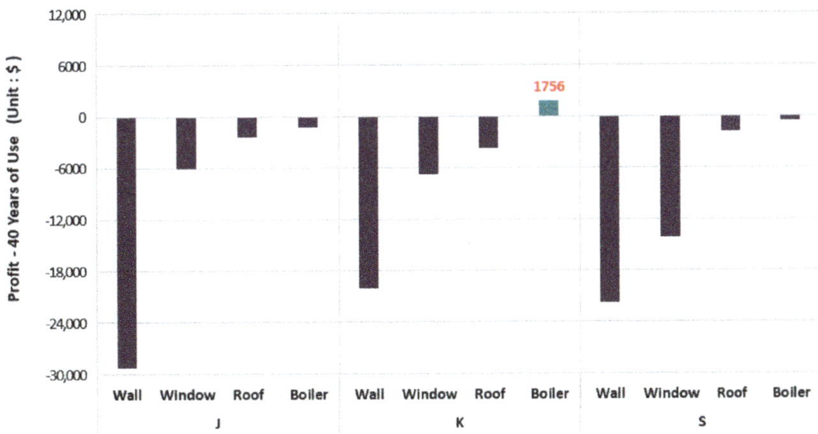

Figure 12. Result of analysis of economic feasibility of applying each improvement factor in the case of non-subsidized construction costs relating to welfare centers for the elderly.

In the case of improvement works, as shown in Figure 13, when only the net energy improvement cost is analyzed, such as when government support or remodeling work is mandatory, the S model's windows have the highest energy profit amount, followed by the J model's external wall insulation improvement work, which shows the next highest profit amount.

Therefore, when prioritizing the energy performance improvement factors for senior centers while considering the economic effects [59,60], it is suggested to make improvements according to the process shown in Figure 14. As described in Figure 14, the first step is to determine whether the construction is eligible for support and to identify the intended effects of the improvements. This process suggests that boiler improvements should be prioritized when considering the economic effects, if no support is available.

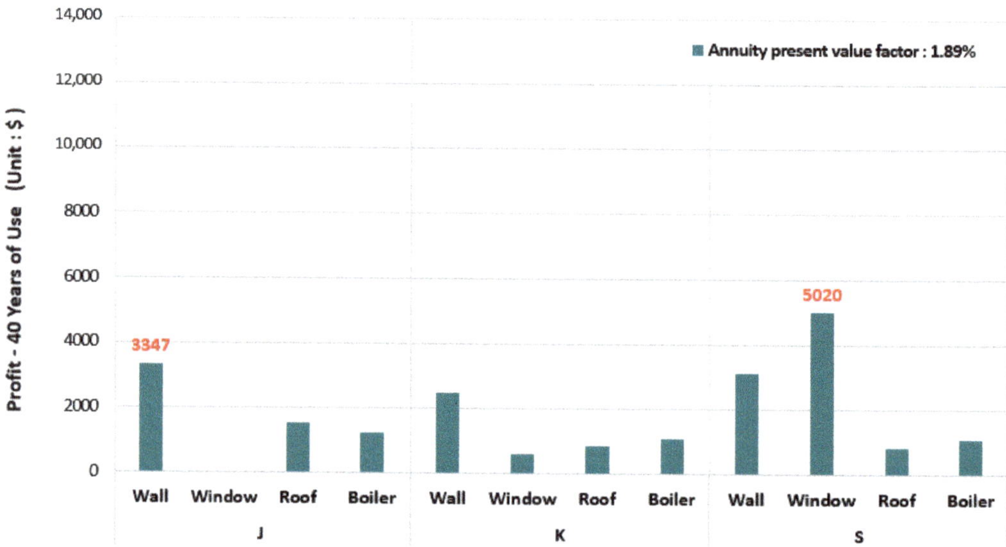

Figure 13. Results of economic analysis of the application of each improvement factor in senior centers when construction costs are subsidized.

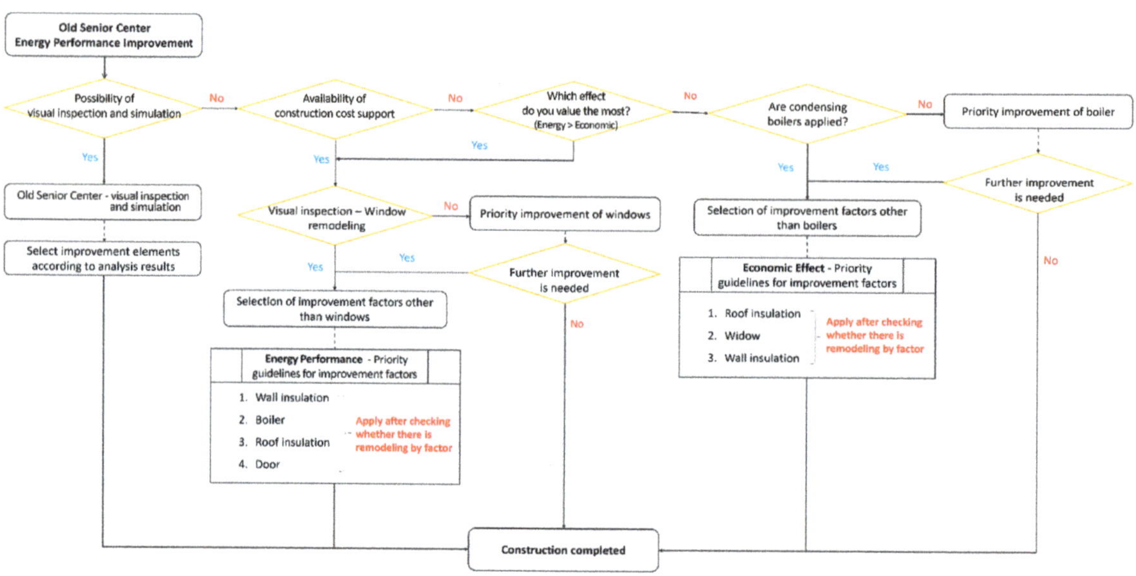

Figure 14. Energy performance improvement factor selection step considering energy improvement effects and economic improvement effects.

6. Discussion

This study evaluated energy performance improvements in aging senior centers by analyzing the impact of various building retrofits, with each element's energy-saving effects assessed through the ECO2 simulation tool. The results indicated the following average improvement rates in primary energy consumption: walls (25.9%), windows (8.0%), boilers (6.5%), roofs (6.0%), and doors (1.8%). These findings demonstrate the importance of enhancing building envelope components, particularly walls and windows, to achieve significant energy conservation.

After interpreting these findings, it becomes evident that retrofitting the walls and windows was the most effective measure in reducing energy consumption by minimizing heat loss, air infiltration, and maintaining stable indoor temperatures. Windows, in particular, proved crucial due to their potential for heat loss reduction, thus enhancing the overall energy efficiency. Boiler replacements also demonstrated considerable improvement in heating efficiency, reducing heating energy demands significantly. These results align with the previous research, which highlights the importance of building envelope retrofits in enhancing energy efficiency in facilities that serve elderly populations.

In terms of practical applications, this study provides a prioritized framework for energy performance improvements in senior centers and other aging public facilities. By indicating that windows and walls should be prioritized for maximum energy savings, this framework offers a practical guideline for facility retrofits within budget constraints. Economically, the 40-year life cycle analysis identified boiler replacements as the most cost-effective retrofit. When financial support is limited, focusing on boilers, followed by roof, window, and wall upgrades, could optimize returns based on operational cost reductions.

With government subsidies, the recommended prioritization shifts to windows, exterior walls, boilers, and roofs, leveraging the financial support to align with greenhouse gas reduction initiatives and maximize energy performance. Current government initiatives aimed at reducing carbon emissions present an ideal opportunity to implement comprehensive retrofits, particularly in building envelope improvements, to achieve substantial primary energy savings in senior centers.

There are some limitations to this study, particularly regarding data accuracy and scope. Due to a lack of preserved architectural records for many of the senior centers, the baseline insulation conditions were estimated based on the Building Act standards relevant to each building's construction year, which may limit precision. Additionally, variations in building configurations and material integrity were observed across the surveyed sites. Although the ECO2 simulations and blower door tests provided valuable insights, the lack of direct energy consumption data limited the accuracy of the projected energy savings. Future studies should integrate long-term energy monitoring systems to capture actual energy use before and after retrofits, thereby enhancing accuracy.

Further research could also consider the health-related outcomes of energy-efficient retrofits, such as indoor air quality improvements. Given that elderly residents spend extended time indoors, integrating systems like total heat exchangers could simultaneously improve air quality and energy efficiency, thereby enhancing resident health and comfort. Extending this study to various types of public facilities would also allow for a more comprehensive comparison of the energy performance improvements under different conditions.

In conclusion, this study provides a prioritized approach to retrofitting senior centers and other aging buildings, which can effectively achieve energy cost savings and carbon reduction goals. The findings highlight windows and walls as key components for energy efficiency and suggest that limited-budget retrofits should focus on maximizing the energy impacts by prioritizing these elements. This study serves as a practical reference for researchers and practitioners aiming to optimize energy improvements in aged facilities and emphasizes that renewing aged structures is essential, given both the environmental and economic imperatives. As such, this research contributes a foundational model that future studies can build upon to develop sustainable, energy-efficient strategies for aging buildings.

7. Conclusions

In this study, a current status survey was conducted, targeting 20 senior centers in the metropolitan area, and the energy performance improvement factors were selected via a literature review. An ECO2 simulation was performed to analyze the changes in energy consumption, and an economic analysis was performed for the J, K, and S standard models. Accordingly, a method for selecting the energy performance improvement factors was proposed.

Based on these findings, the main results of this study are as follows:

(1) As a result of the ECO2 simulation analysis, the following measures had the greatest effect on primary energy consumption, in descending order: outer walls, windows, boilers, roofs, and doors. The greatest effect for exterior wall improvements was observed in senior center D, and the average energy consumption improvement resulting from the application of PF boards was 25.9%.
(2) As a result of the ECO2 simulation analysis, the senior center with the greatest improvement effect relating to windows and doors was S, and the average improvement rate was 8% when low-e, double-glazed windows were installed.
(3) The standard models J, K, and S senior centers were selected based on the average type of floor area, energy demand, energy consumption, and primary energy consumption. For the energy consumption improvement effect of the performance improvement measures on the standard models, the effect of the outer wall improvements on model J was 11.0%, and the effect of the window and door improvements on model S was the greatest at 19.9%. For model K, the effect of outer wall improvement was the highest at 14.4%.
(4) A blower door test was conducted to verify the airtight performance of the buildings after the window improvements. According to the results of a calculation of the number of ventilations per hour, the α-senior center exhibited an improvement effect of 52.5%, which was about 35.2% higher than the simulation result. For the β-senior center, the improvement effect was about 4.6% lower than the 9.8% effect obtained in the simulation, but the effect of pressurization was similar to that obtained in the simulation.
(5) According to the results of the energy performance improvement effect analysis in this study, the improvement factors should be selected in the following order: windows, outer walls, boilers, roofs, and doors.
(6) As a result of the life cycle cost analysis of the heating costs over a 40-year period, the energy cost improvements of the J, K, and S standard models were the greatest after window improvements. However, in the life cycle cost analysis conducted by including the construction costs, only the boiler improvement of the K model generated a profit of USD 1756.
(7) According to the results of the economic feasibility analysis in this study, when carrying out improvement work without government support, priority should be given to improving the boiler, as it resulted in the lowest deficit.
(8) In recent years, construction costs have risen excessively, and it has been confirmed that some energy performance improvements are not economically feasible in terms of life cycle costs. However, due to the aging of buildings, renewal work has become a necessity rather than an option. When carrying out renewal work within a limited budget, it is expected that this study will serve as a reference for prioritizing areas for improvement by considering energy performance enhancements first.

In future research, we plan to measure energy consumption before and after the application of various improvement factors during the conversion of the existing senior centers into green buildings, with the aim of enhancing their energy performance. In addition, we plan to analyze the effects of installing a total heat exchanger, which is effective at improving indoor air quality and energy consumption in consideration of the health of the elderly, whose activities are restricted to indoor settings if the outdoor air is contaminated. Finally, the data were verified using confidential measurement data that affect heating energy consumption, rather than direct data on energy consumption. Therefore, a comparison with actual usage is limited.

Author Contributions: Conceptualization, A.N.; methodology, A.N.; experiment, A.N.; software, A.N., Y.I.K. and A.N.; verification, Y.I.K.; formal analysis, A.N.; investigation, A.N.; resources, A.N.; data Curation, A.N. and A.N.; writing—original draft preparation, A.N. and A.N.; writing—review and editing, Y.I.K.; visualization, A.N.; director, Y.I.K.; project management, Y.I.K.; funding, A.N. All authors have read and agreed to the published version of the manuscript.

Funding: This study was supported by the Research Program funded by Seoul National University of Science and Technology.

Data Availability Statement: The original contributions presented in the study are included in the article, further inquiries can be directed to the corresponding author.

Conflicts of Interest: The authors declare no conflicts of interest.

References

1. Consumer Price Inflation. Available online: https://www.index.go.kr/unify/idx-info.do?idxCd=4226 (accessed on 30 May 2023).
2. Kerosene Price Increase Rate. Available online: https://www.agrinet.co.kr/news/articleView.html?idxno=315642 (accessed on 30 May 2023).
3. Winter Heating Cost Analysis. Available online: https://idp.theminjoo.kr/board/view/briefing/1946 (accessed on 30 May 2023).
4. 2022–2023 Cold Disease Emergency Room Monitoring System Operation Results. Available online: https://www.kdca.go.kr/board/board.es?mid=a20304020300&bid=0004 (accessed on 30 May 2023).
5. *MOLIT, the 2nd Green Building Master Plan in Korea*; Ministry of Land, Infrastructure and Transport: Sejong-si, Republic of Korea, 2019.
6. Building Life History Management System. Available online: https://blcm.go.kr/stat/customizedStatic/CustomizedStaticSttst.do (accessed on 30 May 2023).
7. Building License Information. Available online: https://www.eais.go.kr/moct/awp/aec01/AWPAEC01L01 (accessed on 30 May 2023).
8. Ministry of Land, Infrastructure and Transport. *Building Act Enforcement Rules*; Ordinance No. 1107; Ministry of Land, Infrastructure and Transport: Sejong, Republic of Korea, 2022.
9. *Building Rules*; Presidential Decree No. 19045; Ministry of Land, Infrastructure and Transport, Korea: Sejong, Republic of Korea, 2022.
10. *Rules on Building Facility Standards, Etc.*; Ministry of Construction Ordinance No. 506; Ministry of Land, Infrastructure and Transport: Sejong, Republic of Korea, 1992.
11. Building Energy Conservation Design Standards; Notification of the Ministry of Construction and Transportation No. *2008-5*; Ministry of Land, Infrastructure and Transport: Sejong, Republic of Korea, 2008.
12. Korea Land & Housing Corporation. *A Study on Sound Insulation Design of Exterior Windows*; Korea Land & Housing Corporation: Jinju-si, Republic of Korea, 1992.
13. Choi, D.S.; Lee, M.E.; Chun, H.C. Energy Consumption Characteristics and Evaluation of Thermal Insulation Performance in Accordance with Built Year of Apartment Complex. *KIEAE J.* **2014**, *14*, 79–86. [CrossRef]
14. Korea Energy Agency. ECO2-Manual. Korea Energy Agency. Building Energy Evaluation Program. Available online: https://www.energy.or.kr/web/kem_home_new/info/data/open/kem_view.asp?q=22116 (accessed on 30 May 2023).
15. *ISO 13790*; Energy Performance of Buildings—Calculation of Energy Use for Space Heating. International Organization for Standardization (ISO): Geneva, Switzerland, 2008.
16. Jeon, S.-M.; Choi, G.-S.; Kang, J.-S. A Method for the Classification of Calculation Method for Energy Consumption of School Buildings on ECO2 Software. In Proceedings of the 2019 SAREK Summer Annual Conference, Yongpyong, Republic of Korea, 25–28 June 2019; pp. 509–512.
17. Kim, T.Y.; Kim, Y.H.; Hwang, J.H. Comparison of Primary Energy Requirement of Neighborhood Facilities according to Heat Source Types and Window Wall Ratio by Usage Purpose. *J. KIAEBS* **2017**, *11*, 425–439.
18. Kim, M.K.; Park, H.S.; Song, K.D. Evaluation and Analysis of Building Energy Rating System Accroding to Insulation Performance of Building Envelope in Regional and Building Form of Apartment House. *J. Air-Cond. Refrig. Eng.* **2013**, *25*, 49–54.
19. Energy Demand, Consumption, Consumption, Primary Energy Consumption. Available online: https://www.phiko.kr/bbs/board.php?bo_table=z3_01&wr_id=5 (accessed on 30 May 2023).
20. *DIN 18599*; Energy Efficiency Of Buildings—Calculation of the Energy Needs for Heating, Cooling, Ventilation, Domestic Hot Water, Lighting and Room Lighting Appliances. German Institute for Standardization (DIN): Berlin, Germany, 2020.
21. Ministry of Land, Infrastructure and Transport. *Regulations on Building Energy Efficiency Rating Certification and Zero Energy Building Certification*; Ministry of Land, Infrastructure and Transport Ordinance No. 878; Ministry of Land, Infrastructure and Transport: Sejong, Republic of Korea, 2020.
22. An, Y.S.; Kim, J.; Joo, H.J.; Han, G.W.; Kim, H.; Lee, W.; Kim, M.H. Retrofit of renewable energy systems in existing community for positive energy community. *Energy Rep.* **2023**, *9*, 3733–3744. [CrossRef]
23. Kim, Y.; Yu, K.H. Study on the Certification Policy of Zero-Energy Buildings in Korea. *Sustainability* **2020**, *12*, 5172. [CrossRef]

24. Choi, B.E.; Shin, J.H.; Lee, J.H.; Kim, S.S.; Cho, Y.H. Establishment of Passive Energy Conservation Measure and Economic Evaluation of Fenestration System in Nonresidential Building of Korea. *Int. J. Polym. Sci.* **2017**, *2017*, 8681737. [CrossRef]
25. No, S.T.; Kim, J.Y. A Comparison of Detailed and Simple Building Energy Analysis Tools for Eco-Friendly Office Building Design. *Appl. Mech. Mater.* **2012**, *178–181*, 3–6. [CrossRef]
26. Kim, J.G.; Ahn, B.L.; Lee, N.E.; Han, H.S.; Jang, C.Y.; Jang, S.H.; Choi, S.C.; Haan, C.H. The Evaluation of Energy Efficiency According to Windows Thermal Performance in Deteriorated Housing. *J. Korean Sol. Energy Soc.* **2012**, *32*, 446–454.
27. Kwak, H.J.; Jang, H.I.; Lee, H.S.; Eom, J.Y.; Suh, S.J. Correlation Analysis Between Fenestration Energy Consumption Efficiency Rating System and Building Energy Consumption. *Korean J. Air-Cond. Refrig. Eng.* **2013**, *25*, 338–345.
28. Mathur, U.; Damle, R. Impact of air infiltration rate on the thermal transmittance value of building envelope. *J. Build. Eng.* **2021**, *40*, 102302. [CrossRef]
29. Undram, M.; Song, D.S.; Kim, J.W. A Review of the Test Methods for Air-tightness Performance of Building Components Using the Blower Door System. *Korean J. Air-Cond. Refrig. Eng.* **2019**, *31*, 216–226.
30. Mun, J.H.; Lee, J.I.; Kim, M.S. Estimation of Infiltration Rate (ACH Natural) using Blower Door Test and Simulation. *Energies* **2021**, *14*, 912. [CrossRef]
31. Jeong, M.G.; Kim, G.H. A Comparative Analysis of Life Cycle Cost on the Window Glass and the Insulation Film Coated Glass for Window. *J. Korea Inst. Build. Constr.* **2014**, *14*, 583–588. [CrossRef]
32. Lee, S.H.; Lee, J.S. An Economic Evaluation Study of Office Remodeling and Green-remodeling Projects: A Simulation Approach to a Rental Office in GBD, Seoul. *J. Archit. Inst. Korea Plan. Des.* **2018**, *34*, 23–38.
33. Choi, S.W.; Kim, J.Y.; Park, H.S.; Kim, J.T. The Energy Saving Effect and Economic Assessment of Office Building according to the Building Envelope Remodeling. *J. Korea Inst. Ecol. Archit. Environ.* **2012**, *12*, 85–92.
34. Lee, K.W.; Kim, Y.I. Selection of Energy Improvement Factors and Economic Analysis of Standard MDU Complexes in Korean Metropolitan Regions. *Energies* **2022**, *15*, 4042. [CrossRef]
35. Korea On-Line e-Procurement System. Available online: https://www.g2b.go.kr/index.jsp (accessed on 30 May 2023).
36. Green Remodeling Public Project Performance. Available online: https://www.greenremodeling.or.kr/business/bus3000.asp (accessed on 30 May 2023).
37. Korea Housing Finance Corporation. *BiWeekly Housing Finance Insight*; No. 2020-12; Korea Housing Finance Corporation: Busan, Republic of Korea, 2020.
38. Kim, S.E.; Song, Y.W.; Park, J.C. Comparison of Energy Performance Changes through Green Remodeling Cases of Daycare Centers. *J. Korean Soc. Air-Cond. Refrig. Eng.* **2022**, *51*, 26–33.
39. Choi, M.J.; Jeon, J.W.; Lee, D.H. The Analysis of Energy Performance for Certification Case of Zero Energy Building through Green Remodeling in Korea. *J. KIAEBS* **2021**, *15*, 213–223.
40. Cho, H.Y.; Kang, J.S.; Ahn, H.S. Case Study of Building Energy Efficiency Performance Evaluation for Green Remodeling of Public Buildings. In Proceedings of the Architectural Institute of Korea Conference, Jeju, Republic of Korea, 1–2 December 2022; Volume 42, pp. 938–939.
41. Kang, J.A.; Lee, S.H.; Kim, D.U.; Son, H.J.; Seo, D.H.; Kim, T.H.; Jeong, Y.S.; Kwak, Y.H. Study on the Definition of Reference Building Based on Analysis of Design of Retail Buildings. *IOP Conf. Ser. Earth Environ. Sci.* **2019**, *238*, 012043. [CrossRef]
42. Jeong, Y.S.; Jung, H.K.; Jang, H.K.; Yu, K.H. A Study on the Reference Building based on the Building Design Trends for Non-residential Buildings. *J. Korean Sol. Energy Soc.* **2014**, *34*, 1–10. [CrossRef]
43. Kim, C.S. The Evaluation of Energy Efficiency of Apartment Units after Conversion of Balconies into an Integrated Part of Interior Living Space by Computing with ECO2 Software. *KIEAE J.* **2016**, *16*, 11–16. [CrossRef]
44. Park, C.I.; Kim, L.H. A Development of Test Method on the Energy Consumption Efficiency of Domestic Gas Boiler below 70 kW. *J. Energy Eng.* **2016**, *25*, 73–82. [CrossRef]
45. Yu, B.H.; Seo, B.M.; Moon, J.W.; Lee, K.H. Analysis of the Part Load Ratio Characteristics and Gas Energy Consumption of a Hot Water Boiler in a Residential Building under Korean Climatic Conditions. *J. Air-Cond. Refrig. Eng.* **2015**, *27*, 455–462.
46. Hwang, S.H. A Study on the Alternative Selection Method for Energy Efficiency Improvement of Old Detached House. *J. Korean Sol. Energy Soc.* **2019**, *39*, 45–55. [CrossRef]
47. Ju, Y.H.; Park, J.I.; Dong, W.H.; Lee, J.H. Analysis of Heating Energy Performance Improvement according to the Survey on the Actual Condition of Old Residential Buildings and the Improvement Plan for Remodeling—Focusing on aging houses in Jeonju. *J. Korean Soc. Living Environ. Syst.* **2021**, *28*, 576–585. [CrossRef]
48. Jeon, J.S.; Kim, M.J.; Kim, S.J.; Park, J.I.; Choi, J.M. Development of Building Airtightness Prediction Model Using Machine Learning. *J. Korean Soc. Living Environ. Syst.* **2022**, *29*, 573–586. [CrossRef]
49. Seoul News. Available online: https://news.seoul.go.kr/env/archives/521602 (accessed on 30 May 2023).
50. Proliferation of Low-Carbon Buildings in Seoul. Available online: http://www.kharn.kr/news/article.html?no=18840 (accessed on 30 May 2023).
51. Joint Response with Autonomous Districts for Public Welfare Safety of the Energy Vulnerable Class. Available online: https://news.seoul.go.kr/gov/archives/544975 (accessed on 30 May 2023).
52. Kim, Y.D.; Cha, H.S.; Kim, K.R.; Shin, D.W. Evaluation Method of Green Construction Technologies Using Integrated LCC and LCA Analysis. *J. Korea Inst. Constr. Eng. Manag.* **2011**, *12*, 91–102.

53. Kim, J.M.; Lee, J.H.; Lee, D.H. The Economic Comparison through LCC Analysis on each Graded Alternatives for Green Remodeling of Public Building. *J. Korea Inst. Constr. Eng. Manag.* **2018**, *19*, 38–49.
54. Gas Rate Table by Region. Available online: https://www.seoulgas.co.kr/front/payment/gasPayTable.do (accessed on 30 May 2023).
55. Choi, M.C.; Cho, G.H.; Kim, G.T.; Han, S.H.; Kim, J.Y. A Study on the Improvement of Energy Efficiency of Existing Apartment House through Cost Benefit Analysis. *J. KIAEBS* **2017**, *11*, 376–391.
56. Choi, D.S.; Jeon, H.C.; Cho, K.H.; Yoo, J.S. A Study on Barracks Remodeling in Energy Reduction and Economic Feasibility. *J. Korean Soc. Living Environ. Syst.* **2016**, *23*, 283–291. [CrossRef]
57. Kim, H.Y.; Kim, S.W. A Basic Study on the Character of the Replacement and Repair Cycle of Infill Constituent Appeared in the Repair Process of the Apartment House. *J. Archit. Inst. Korea Plan. Des.* **2006**, *22*, 21–30.
58. Korea Gas Safety Corporation. *Reinforcement of Gas Water Heater Efficiency Rating Standard and Resetting of Household Gas Boiler Efficiency Rating Standard*; Korea Energy Management Corporation: Ulsan, Republic of Korea, 2012.
59. Olman, A.M.; Cristina, M.; Agustin, G.B.; Maria, C.; Daniel, G. Energy Savings after Comprehensive Renovations of the Building: A Case Study in the United Kingdom and Italy. *Energies* **2021**, *14*, 6460. [CrossRef]
60. Raj, B.P.; Meena, C.S.; Agarwal, N.; Saini, L.; Hussain Khahro, S.; Subramaniam, U.; Ghosh, A. A review on numerical approach to achieve building energy efficiency for energy, economy and environment (3E) benefit. *Energies* **2021**, *14*, 4487. [CrossRef]

energies

MDPI

Article

Assessing the Influence of Occupancy Factors on Energy Performance in US Small Office Buildings

Seddigheh Norouziasl, Sorena Vosoughkhosravi, Amirhosein Jafari * and Zhihong Pang

Bert S. Turner Department of Construction Management, Louisiana State University, Baton Rouge, LA 70803, USA; snorou1@lsu.edu (S.N.); svosou1@lsu.edu (S.V.); zpang@lsu.edu (Z.P.)
* Correspondence: ajafari1@lsu.edu

Abstract: Office buildings are responsible for about 35% of the total electricity in the US and over 70% of building energy consumption occurs during occupancy periods. Therefore, understanding occupancy behavior is crucial for reducing building energy consumption. However, given the stochastic nature of occupant behavior, identifying which occupancy parameters have the most impact on energy consumption poses a considerable challenge. This study aims to investigate and quantify the impact of various occupancy parameters on the energy performance of a US small-sized office building using an EnergyPlus-based nationwide energy simulation. First, dynamic occupancy schedules are created based on different occupancy parameters using an agent-based model. Next, the generated dynamic occupancy schedules are integrated into a small office building model from the Department of Energy's prototypes. This creates a dataset of occupancy parameters and building energy performance across various climate zones. Finally, various feature selection and statistical analysis methods are applied to the generated dataset. This helps identify significant occupancy parameters and quantify their impact on building energy performance across different climate zones. According to the results of the study, buildings located in cool marine, mixed marine, and warm marine climate zones had lower total energy consumption compared to other zones. Additionally, feature selection methods identified "Occupant Density" as the primary significant variable impacting energy consumption, across all climate zones. These findings offer valuable insights into the influential occupancy parameters across various climate zones, highlighting the importance of tailoring occupancy schedules to enhance energy efficiency. They provide practical guidance that can be directly applied to optimize energy consumption and achieve significant energy savings in small office settings with different weather conditions.

Keywords: building simulation; occupant behavior; energy consumption; dynamic occupancy schedules

1. Introduction

Office buildings account for about 35% of overall electricity consumption in the US [1]. Crucially, over 70% of building energy consumption occurs during occupancy periods, underlining the importance of understanding and optimizing occupancy behavior for improving building energy efficiency [2]. Analyzing the various occupancy parameters and their impacts on building energy performance is essential for moving toward smart and energy-efficient buildings [3]. Therefore, it is necessary to identify which occupancy parameters have a significant impact on building energy performance and quantify their impact on building energy consumption [4].

Some studies have focused on the role of occupant behavior in building energy performance simulation and presented a review of the existing literature. For instance, Yan et al. [5] reviewed the literature in terms of monitoring occupant behavior and collecting the occupancy data, modeling occupant behavior, evaluating the occupancy models, and integrating the models into building performance simulation tools. They emphasized the importance of developing a standardized framework for describing and modeling occupant behavior in buildings. Yang et al. [6] explored occupancy sensing technologies and methods for modeling occupant behavior in institutional buildings. They highlighted the challenges associated with implementing occupancy sensing and monitoring in institutional buildings, primarily because of the substantial number of occupants, their considerable variability, and the diverse functions of these buildings for certain techniques and methodologies. Norouziasl et al. [7] conducted a systematic review of the literature regarding modeling and simulation tools for human-building energy-related interaction. They established a framework for inputs and outputs in modeling occupant behavior and outlined the most effective techniques for simulating occupant behavior in building energy performance. Bäcklund et al. [8] focused on campus buildings, highlighting the evolving behaviors of occupants influenced by smart building systems. Their semisystematic literature review emphasized the significant impact of such systems on energy use, promoting a shift towards more energy-aware behaviors. This research underscores the potential for integrating smart systems into building management to optimize energy consumption and enhance educational environments. Additionally, Vosoughkhosravi et al. [9] provided a systematic review of the use of the American Time Use Survey (ATUS) in modeling occupant behavior. In this review, the authors investigated occupant behavior models and approaches developed based on ATUS. They offered a comprehensive analysis of modeling methods, required inputs and outputs, as well as the most practical occupant behavior methods.

Among the studies aiming to simulate occupant energy-related behavior, Chen et al. [10] developed an agent-based occupancy simulator to simulate the stochastic behavior of occupants, including occupants' presence and movement. They employed a homogeneous Markov chain model to simulate the stochastic occupancy schedules for each office room and the whole building. Then, the generated occupancy schedules were used in the EnergyPlus and obFMU simulation to evaluate the impacts of occupant behavior on building energy performance [11]. Putra et al. [12] also developed an agent-based model (ABM) to study building occupant behavior during load shedding, simulating occupants' adaptive actions and their impact on building energy consumption. In a similar study, Jia et al. [13] investigated the impact of actual and modeled occupant behavior information on building performance simulation. They used an agent-based modeling approach to simulate occupant behavior and conducted a cosimulation with a building energy model. Their study highlighted the significant influence of different occupant behavior inputs on building energy performance. Another study by Parys et al. [14] focused on integrating stochastic models of occupant behavior with dynamic building simulations. The authors reviewed various methods for this integration, emphasizing the importance of accurately modeling occupant behavior to improve the precision of energy performance predictions in office buildings. By coupling dynamic building simulations with stochastic occupant behavior models, the study aimed to address the variability and unpredictability of human actions, which significantly impact energy consumption. This integrated methodology helped in creating more realistic and reliable simulations, ultimately leading to better-informed decisions for energy-efficient building design and operation. Almeida et al. [15] also studied the uncertainty in occupant behavior in building energy models. They found that energy consumption could vary significantly based on different occupancy schedules and environmental preferences, highlighting the importance of accurate occupant behavior modeling. In their 2019 study, Gunay et al. [16] developed an occupancy learning algorithm for terminal heating units. They investigated how occupant behavior impacts the energy performance of HVAC systems by utilizing both field data and simulation models. Their findings highlighted the importance of accurately modeling occupant behavior to optimize HVAC system performance and improve energy efficiency in buildings. In a similar study by Li et al. [17], they explored the use of radio frequency identification (RFID) technology to measure and monitor occupancy in buildings. The authors developed an RFID-based system designed to provide real-time occupancy data, which can be used to optimize HVAC operations. By accurately tracking the presence and movement of occupants, the system allows for demand-driven HVAC control, which adjusts heating and cooling based on actual occupancy rather than predefined schedules. This approach can significantly improve energy efficiency by reducing unnecessary heating or cooling of unoccupied spaces, leading to potential energy savings and enhanced comfort for building occupants. In addition, Chen et al. [18] proposed two stochastic Markov chain models using real data to simulate the occupancy schedule in commercial buildings. These models simplified transition probability calculations and offered occupancy-based energy models for single-zone and multi-zone offices. Page et al. [19] developed an algorithm for simulating occupant presence in buildings using an inhomogeneous Markov chain model. The model was then integrated with building energy simulation as an input to account for future occupant behavior. By applying this model to occupancy data from private offices, the study demonstrated the key aspects of occupant presence, including arrival and departure times, and intermediate periods of absence in energy consumption patterns. A number of studies have concentrated on enhancing energy efficiency in buildings through improved occupancy modeling and predictive analysis. For instance, Oldewurtel et al. [20] investigated the potential of using occupancy information to realize a more energy efficient building climate control. In their research, a model predictive control (MPC) framework was employed to assess the energy savings potential of office buildings with different occupancy types. This comparative analysis considered different building types, HVAC systems, seasonal variations, and occupancy patterns to evaluate their respective effects on energy-saving potential. In another study, Rafsanjani et al. [21] conducted research on the influence of occupants' energy-consuming behaviors, such as arrival, departure and electricity-use patterns, in commercial buildings and quantified their potential for energy savings. The proposed study combined occupancy sensing with building energy data to assess the feasibility of the developed approach in identifying occupant-specific energy consumption information. Erickson et al. [22] addressed the inefficiencies of existing climate control systems that rely on maximum occupancy numbers, often resulting in unnecessary heating or cooling of infrequently used rooms. They utilized the occupancy data to develop multivariate Gaussian and agent-based models for predicting occupancy patterns and then implemented optimal control strategies to reduce the energy consumption of the HVAC system. Recent developments in urban-building energy modeling (UBEM) underscore the significant impact of occupant behavior on energy consumption within urban environments. Banfi et al. [23], in their comprehensive review, emphasized the limitations of static occupant profiles often utilized in current modeling practices. Advocating for dynamic and stochastic models, the study examines the integration challenges and the need for more sophisticated occupant behavior models to enhance the accuracy and relevance of urban energy simulations.

Existing research has primarily focused on the general effects of occupancy schedules on building energy performance. However, these studies often do not fully explore the influence of specific occupancy-related parameters, such as arrival and departure times, lunch breaks, and the frequency and duration of meetings on energy consumption, particularly within office settings. This gap limits the applicability of such studies for creating accurate, actionable energy management strategies tailored to routine human behaviors. Moreover, the body of research considering occupancy schedules rarely extends its analysis to compare these effects across diverse climatic conditions. The United States presents a unique landscape with a wide range of climate zones, each presenting distinct challenges and opportunities for energy management in office buildings. Comparative analysis across different climate zones is crucial but has been rarely covered in the research literature. Gaining insight into these differences is important. Understanding how occupancy schedules influence energy consumption in different climates can significantly enhance the development of localized, climate-specific energy conservation measures. Such detailed and comparative research is critical not only for advancing theoretical knowledge but also for informing policymakers and building managers. Tailored strategies could subsequently be developed to optimize energy use in office buildings nationwide, potentially leading to substantial reductions in energy costs and environmental impacts. This research gap presents a significant opportunity for a pioneering study that could set new directions for future energy efficiency initiatives and policies.

To address the mentioned gaps, this study aims to analyze the impacts of different occupancy parameters (e.g., the time occupants arrive at or leave the workplace, the time and number of meetings, and the time and duration of lunch breaks, among others) on energy consumption in office buildings across various climate zones in the US. In this study, an agent-based model (ABM) [24] is used to generate dynamic occupancy schedules from various sets of occupancy parameters to reflect stochastic occupancy behavior. In addition, the small-sized office building in the Department of Energy prototype Commercial Building Prototype Model (CBPMs) [25] is used for energy simulation. The generated stochastic occupancy schedules, as well as the office model, are integrated into the EnergyPlus simulation model to create a dataset of occupancy parameters and building end-use energy performance in different climate zones in the US. This dataset is used to select the most significant occupancy variables impacting building energy consumption using feature selection techniques. Furthermore, this research provides key insights that are invaluable for building designers, facility managers, and policymakers by delineating the critical occupancy parameters that substantially affect energy consumption in office buildings. This knowledge authorizes stakeholders to formulate specialized, climate-responsive strategies that not only optimize energy efficiency but also promote broader sustainability goals. Additionally, a thorough comprehension of these occupancy influences is crucial for crafting effective policies aimed at diminishing energy consumption and enhancing the energy efficiency of design and operational practices in American office buildings. By identifying these key parameters, the study equips stakeholders with the necessary tools to implement strategic interventions that can lead to significant energy savings and operational efficiencies, particularly in diverse climatic conditions across the United States.

2. Research Method

This study adopts a four-step methodology to analyze the impacts of different occupancy parameters on energy consumption, as illustrated in Figure 1. First, several dynamic occupancy schedules were created based on specific occupancy parameters using an agent-based model. Then, the generated dynamic occupancy schedules are integrated into the DOE prototype small-sized office building model. This prototype was developed by the Pacific Northwest National Laboratory (PNNL) [26] and the US Department of Energy's Building Energy Codes Program (BECP) to estimate how changes in energy codes and standards can lead to energy savings [27]. This prototype provides EnergyPlus IDF models for different office buildings designed based on the energy codes (i.e., The American Society of Heating, Refrigerating and Air-Conditioning Engineers (ASHRAE) Standard 90.1-2019 Energy Efficiency Standard for d Buildings Except Low-Rise Residential Buildings).

A nationwide energy simulation was conducted to generate a database of occupancy parameters and building end-use energy performance in different US climate zones. Then, the impact of different occupancy parameters on building energy performance was analyzed using a sensitivity analysis and four different feature selection methods. In the subsequent sections, this study investigates the occupancy parameters utilized to generate the occupancy schedule. Next, the integration process of the occupancy schedule, considering various climate zones, into the EnergyPlus simulation model is elaborated. Finally, this study presents the employed feature selection methods to analyze the outcomes of the energy performance simulation and identify the most influential variables.

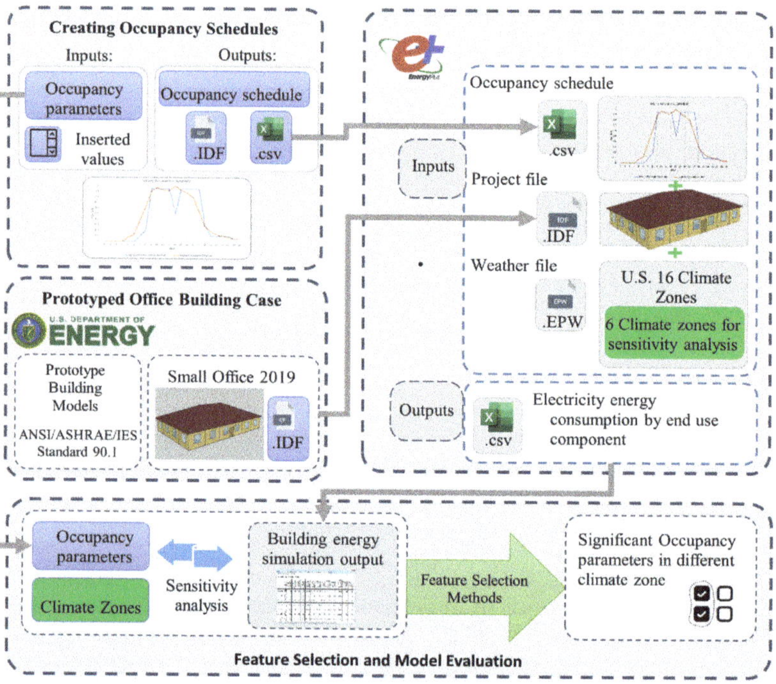

Figure 1. Dynamic occupancy schedule and building energy performance simulation framework.

2.1. Creating Occupancy Schedules

Occupancy in an office building can be defined with various parameters, such as the time occupants arrive at or leave the workplace, the time and number of meetings, and the time and duration of lunch breaks throughout the working hours for a specific day [28]. In the first step, different occupancy parameters are used to create various occupancy schedules using an agent-based model. This simulation approach offers a significant advantage over traditional case studies by allowing for the exploration of multiple scenarios, thereby providing a more comprehensive understanding of how different occupancy dynamics can influence building energy consumption. For this purpose, a web-based occupancy simulator tool is used, which was developed by Lawrence Berkeley National Laboratory (LBNL) [10]. The web-based occupancy simulator tool utilizes a Markov-chain model to simulate occupant movements and generate stochastic schedules. This innovative tool has been validated in real-world scenarios. The validation studies, as outlined in research by Luo et al. [29], demonstrate the tool's efficacy in accurately reflecting occupant behavior in commercial buildings, thereby providing reliable data for building simulation to optimize energy usage and operational efficiency. To closely mirror real-world occupancy patterns, the occupancy schedule is crafted by meticulously integrating as many relevant parameters as possible. The occupancy schedule can be generated by integrating the occupancy parameters and office layout parameters in the web-based occupancy simulator application. This simulator is a user-friendly application that uses a Markov Chain (MC) model to simulate occupancy in buildings. It takes in high-level inputs on occupant density, space area, event arrangements, etc., and simulates occupancy movements inside the building to generate stochastic and dynamic occupancy schedules for each space. This detailed simulation ensures that the generated occupancy schedules closely approximate actual daily activities and interactions within office environments. The generated occupancy schedule may vary daily, reflecting the inherent nature of occupants and their energy-related behavior in office buildings. The output results of these schedules can be downloaded in CSV format to facilitate integration with energy simulation software, such as EnergyPlus 22.1.0 [30].

Table 1 presents the occupancy parameters utilized in this study to generate the occupancy schedules (the parameters are selected based on the advancement of the current LBNL occupancy simulator), including variable names, definitions, units, and values. As illustrated in this table, three types of occupants are considered in the simulation: regular staff, administrators, and managers. Each type of occupant has different arrival and departure times, and the duration that each occupant type spends in various spaces varies.

Table 1. Considered occupancy parameters for the considered office building.

No.	Variable Name	Variable Definition	Variable Unit	Values *	
1	Occupant_Density	Number of people per area	person/m^2	1 = 0.05 * **2 = 0.06** 3 = 0.10	4 = 0.14 5 = 0.20 6 = 0.30
2	Occupant_Percent	Percentage of each occupant type: Regular staff/Manager/Administrator	Percentage	* **1 = 40/30/30** 2 = 20/40/40 3 = 30/35/35 4 = 50/25/25 5 = 60/20/20 6 = 50/10/40 7 = 45/20/35	8 = 35/40/25 9 = 30/50/20 10 = 50/40/10 11 = 45/35/20 12 = 35/25/40 13 = 30/20/50
3	Meeting_Count	Number of meetings per day (Min–Max)	Count	1 = 1–3 2 = 2–4 * **3 = 3–5**	4 = 4–6 5 = 5–7
4	Meeting_Attend	Number of people per meeting (Min–Max)	People	1 = 2–4 2 = 3–5 * **3 = 4–6**	4 = 5–7 5 = 6–8
5	Meeting_Duration	Probability of duration of meeting for the following numbers (30, 60, 90, 120)	Percentage	1 = 5, 60, 20, 15 2 = 10,65, 15,10 * **3 = 15, 70, 10, 5**	4 = 20, 65, 10, 5 5 = 25, 60, 10, 5 6 = 30, 55, 10, 5
6	Staff_Arriv_Depar	Regular Staff: Arrival time/Departure time	Time	1 = 6:30/15:30 2 = 7:00/16:00 * **3 = 7:30/16:30**	4 = 8:00/17:00 5 = 8:30/17:30
7	Admin_Arriv_Depar	Administrator: Arrival time/Departure time	Time	1 = 7:00/16:00 2 = 7:30/16:30 * **3 = 8:00/17:00**	4 = 8:30/17:30 5 = 9:00/18:00
8	Manag_Arriv_Depar	Manager: Arrival time/Departure time	Time	1 = 8:00/16:30 2 = 8:30/17:00 * **3 = 9:00/17:30**	4 = 9:30/18:00 5 = 10:00/18:30
9	Arriv_Depar_Vari	Arrival/departure time variation	Minutes	1= 0 min 2 = 15 min * **3 = 30 min**	4 = 45 min 5 = 60 min
10	Lunch_Time	Lunch or short-term leaving start time	Time	1 = 11:00 2 = 11:30 * **3 = 12:00**	4 = 12:30 5 = 13:00
11	Lunch_Start_Vari	Lunch or short-term leaving Start time variation	Minutes	1= 0 min 2 = 15 min * **3 = 30 min**	4 = 45 min 5 = 60 min
12	Lunch_Duration	Lunch or short-term leaving duration	Minutes	1 = 30 min 2 = 45 min * **3 = 60 min**	4 = 75 min 5 = 90 min

Table 1. *Cont.*

No.	Variable Name	Variable Definition	Variable Unit	Values *	
13	Lunch_duration_Vari	Lunch or short-term leaving duration variation	Minutes	1 = 5 min 2 = 10 min *** 3 = 15 min**	4 = 20 min 5 = 25 min
14	Staff_Room_Stay	(Regular Staff) Percentage of time that occupants stay in each space	Percentage	1 = 50, 20, 10, 10, 10 2 = 55, 20, 10, 10, 5 3 = 60, 15, 10, 10, 5 4 = 65, 15, 10, 5, 5	*** 5 = 70, 10, 10, 5, 5** 6 = 75, 10, 10, 5, 0 7 = 80, 10, 5, 5, 0
15	Admin_Room_Stay	(Administrator) Percentage of time that occupants stay in each space	Percentage	1 = 35, 10, 35, 10, 10 2 = 40, 10, 30, 10, 10 3 = 45, 10, 30, 5, 10 *** 4 = 50, 10, 30, 5, 5**	5 = 55, 10, 25, 5, 5 6 = 60, 10, 25, 5, 0 7 = 65, 10, 20, 5, 0
16	Manag_Room_Stay	(Manager) Percentage of time that occupants stay in each space	Percentage	1 = 35, 10, 40, 5, 10 2 = 40, 10, 35, 5, 10 3 = 45, 5, 35, 5, 10 *** 4 = 50, 5, 35, 5, 5**	5 = 55, 5, 30, 5, 5 6 = 60, 5, 30, 5, 0 7 = 65, 5, 25, 5, 0
17	Own_Stay_Duration	Average stay time at Own office	Minutes	1 = 30 min 2 = 45 min *** 3 = 60 min**	4 = 75 min 5 = 90 min
18	Other_Stay_Duration	Average stay time at Other offices	Minutes	1 = 10 min 2 = 15 min *** 3 = 20 min**	4 = 25 min 5 = 30 min
19	Meeting_Stay_Duration	Average stay time at Meeting rooms	Minutes	1 = 30 min 2 = 45 min *** 3 = 60 min**	4 = 75 min 5 = 90 min
20	Auxiliary_Stay_Duration	Average stay time at Auxiliary room	Minutes	1 = 10 min 2 = 15 min *** 3 = 20 min**	4 = 25 min 5 = 30 min
21	Outdoor_Stay_Duration	Average stay time at Outdoor	Minutes	1 = 10 min 2 = 20 min *** 3 = 30 min**	4 = 40 min 5 = 50 min
22	Time_Step	Simulation time step	Minutes	1 = 5 min *** 2 = 10 min**	3 = 15 min 4 = 20 min

* Baseline scenario.

In this study, the baseline scenario was established by considering the default values for occupancy parameters according to the LBNL occupancy simulator tool [31] which is also illustrated in Table 1. As an example, the average square meters per person in the baseline scenario is 0.05 person/m^2, and 40% of occupants in the baseline are regular staff, 30% are administrators, and the other 30% of occupants are managers. To enhance the realism of the simulation, occupancy schedules are meticulously generated in the web-based occupancy simulator using the one-at-a-time (OAT) method [32] by varying individual occupancy parameters while keeping the remaining parameters at their baseline values. This process results in the creation of a total of 104 distinct annual occupancy schedules for each office space. These schedules are designed to closely replicate real-world occupant behavior, and will be used in EnergyPlus model to analyze the impact of each parameter on building energy consumption with a simulation time step of 10-min as illustrated in the baseline scenario.

2.2. Prototyped Office Building Case

This study used the US DOE Commercial Prototype. Building Models (CPBM) for small offices to analyze the impact of various occupancy parameters. The prototype includes the building model for different climate locations (16 US climate zones) based on

ASHRAE Standard 90.1-2019 [33]. The small office has a rectangular shape layout with a total area of 5500 ft2 (511 m^2) and consists of five thermal zones (four perimeter zones and one core zone), as illustrated in Figure 2. It has to be noted that one thermal zone can have one or more subspaces, though the thermal condition is maintained the same for the subspaces within the same thermal zone [34]. The office layout that is designed for this study is as follows: Zones 1, 2, and 3 contain the individual office spaces, Zone 4 contains the meeting room and lobby space, and the core zone contains the auxiliary and corridor area. This layout is shown in Figure 2. The generated stochastic occupancy schedules for spaces within each zone will be combined for use within that specific zone. For example, the occupancy schedules for Office 1, Office 2, and Office 3 will be aggregated to form the combined occupancy schedule for Zone 1.

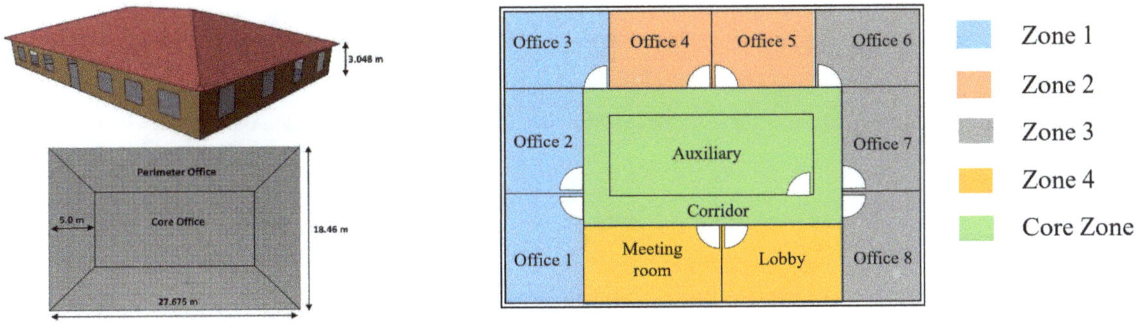

Figure 2. Modified DOE prototype model layout for a small office.

The small office model adheres to the energy efficiency standards established by ASHRAE 90.1, with the specific version (e.g., ASHRAE 90.1-2019) determining the envelope properties. For instance, the exterior wall U-value is approximately 0.057 Btu/h·ft^2·°F, reflecting an insulation level consistent with an R-value of about 17.5 ft^2·h·°F/Btu. This ensures that the building minimizes heat loss through the walls, contributing to overall thermal performance. The roof has an even lower U-value, around 0.027 Btu/h·ft^2·°F (equivalent to an R-value of approximately 37 ft^2·h·°F/Btu). This higher insulation level is essential in reducing heat transfer and limiting heat gain from the environment, especially in warmer climates where the roof is often exposed to direct sunlight. Windows in the small office model are another important component of the building envelope. The U-value for the windows is about 0.38 Btu/h·ft^2·°F. Additionally, the g-value, which indicates the amount of solar radiation admitted through the glazing, is 0.30. This balance allows beneficial daylight into the building while limiting unwanted solar heat gain, which can significantly impact cooling loads during the summer months.

2.3. Integrated Energy Simulation Model

To investigate the impact of each occupancy parameter on building energy performance, a total of 104 occupancy schedules were created to perform sensitivity analysis [35]. To incorporate occupancy schedules as inputs into the EnergyPlus model, a systematic approach is followed. First, a total of 104 annual occupancy schedule files are formatted as CSV files. These files are then organized within the EnergyPlus IDF project folder. Subsequently, a custom Python script is developed to automate the EnergyPlus simulation process. This script dynamically modifies the EnergyPlus IDF file, iterates through the schedule files one by one from the designated folder, and stores the results of each simulation in dedicated output folders, one for each unique occupancy schedule. This automatic workflow ensures the systematic integration of occupancy schedules into the EnergyPlus model, offering an organized approach for conducting simulations and managing their outputs.

To account for the effect of climate zones on energy performance simulation, 16 International Energy Conservation Code (IECC) climate zones in the US are integrated into EnergyPlus as well. These climate zones vary from "Very Hot Humid" to "Very Cold" and "Arctic" climate zone [36]. Table 2 presents the list of the 16 climate zones in the US, their climate zone ID, and the representative city. The energy performance of various building energy systems is then simulated for each climate zone, considering the various occupancy schedules. Finally, a dataset of 1664 data points (104 occupancy schedules and 16 climate zones) with 22 independent variables (occupancy parameter variables), and various systems' energy consumption variables consisting of heating, cooling, lighting, equipment, and fans are generated. The generated dataset is used as input for feature selection methods.

Table 2. US 16 climate zones [36].

No.	Climate Zone	Thermal Climate Zone Name	Weather Location
1	1A	Very Hot Humid	Honolulu, HI, USA
2	2A	Hot Humid	Tampa, FL, USA
3	2B	Hot Dry	Tucson, AZ, USA
4	3A	Warm Humid	Atlanta, GA, USA
5	3B	Warm Dry	El Paso, TX, USA
6	3C	Warm Marine	San Diego, CA, USA
7	4A	Mixed Humid	New York, NY, USA
8	4B	Mixed Dry	Albuquerque, NM, USA
9	4C	Mixed Marine	Seattle, WA, USA
10	5A	Cool Humid	Buffalo, NY, USA
11	5B	Cool Dry	Denver, CO, USA
12	5C	Cool Marine	Port Angeles, WA, USA
13	6A	Cold Humid	Rochester, MN, USA
14	6B	Cold Dry	Great Falls, MT, USA
15	7	Very Cold	International Falls, MN, USA
16	8	Subarctic/Arctic	Fairbanks, AK, USA

2.4. Feature Selection and Model Evaluation

Feature selection is an important task in identifying variables that can significantly impact the performance of the simulation model [37]. Previous studies on building energy performance simulation have used different methods for feature selection, including expert knowledge and judgment [38], correlation matrix [39], boosting tree algorithm to rank variables [40], and linear and monotonic correlation [41], among others. The generated dataset regarding different sets of occupancy schedules and climate zones is used as input for the feature selection models to determine the significant independent variables (occupancy schedules) to predict the dependent variables (energy consumption in different climate zones). In this study, four feature selection methods are utilized to identify the significant features regarding the dependent variables of building energy consumption:

- Multivariate linear regression (MVLR): A multivariate linear regression model expresses a d-dimensional continuous response vector as a linear combination of predictor terms plus a vector of error terms with a multivariate normal distribution. The "mvregress" function can be used to create a multivariate linear regression model [42]. While MVLR assumes linearity and can be influenced by multicollinearity and outliers, these challenges—multicollinearity and the influence of outliers—are prevalent in many types of statistical modeling, not just in MVLR. We have addressed this by careful variable selection and data preprocessing to minimize their impact.

- Least absolute shrinkage and selection operator (LASSO): LASSO constructs a dataset with redundant predictors and identifies those predictors. The "LASSO" function finds the coefficients of a regularized linear regression model using 10-fold cross-validation and the elastic net method [43]. LASSO may prioritize simpler models potentially at the cost of excluding some correlated predictors. However, this characteristic helps in enhancing model interpretability and reducing overfitting, which are crucial for the predictive robustness of the approach.
- Neighborhood component analysis (NCA) feature selection method: Neighborhood component analysis (NCA) is a supervised learning algorithm for choosing features with the goal of increasing the predictive power of regression and classification algorithms. The "fscnca" and "fsrnca" functions of the Statistics and Machine Learning Toolbox perform neighborhood component analysis feature selection with regularization to develop feature weights for the objective function that reduces the average leave-one-out classification or regression loss over the training data [44]. Despite NCA's computational demand, it is chosen for its effectiveness in smaller, well-defined datasets where feature interdependencies are critical, aligning well with our study's scope.
- Feature ranking method using the Relief algorithm: Relief is a feature selection technique that uses a filter-method approach to identify significant variables and is highly sensitive to feature interactions. Each feature in Relief is given a feature score, which can be used to rank and choose the highest scoring features for feature selection. These scores can also be used as feature weights to direct further modeling. The algorithm penalizes the predictors that result in different values to neighbors of the same class, and rewards predictors that provide different values to neighbors of different classes [45,46]. Although Relief's performance may be affected by noisy data, it is highly effective for datasets like ours where interaction among features is a significant factor. Proper parameter setting, based on extensive testing, ensures optimal feature selection.

The results of the four feature selection (also known as feature ranking) methods for various building energy systems (i.e., heating, cooling, lighting, equipment, and fans) are identified. In addition, the identified significant features, as results of feature selection methods, are integrated into a multiple linear regression model to compute the corresponding R-square values across different climate zones using the below equation:

$$R^2 = 1 - \frac{SSR}{SST} \tag{1}$$

$$SSR = \sum (\hat{y}_i - \overline{y})^2 \tag{2}$$

$$SSE = \sum (\hat{y}_i - y_i)^2 \tag{3}$$

where R^2, the coefficient of determination, is the proportion of the variation in the dependent variable that is predictable from the independent variables. SSR is the sum of squares of residuals, and SST is the sum of the distance the data are away from the mean all squared [47,48]. This value is utilized to compare the performance of feature selection methods, where a higher R^2 value represents a better fitting of the algorithm, and, on the other hand, a lower R^2 value represents a larger discrepancy between the actual and predicted results.

3. Results and Discussion

3.1. Building Energy Performance on Baseline Occupancy

In this study, the baseline scenario was established by considering the average occupancy parameters. Figure 3 compares the baseline stochastic occupancy schedule with

ASHRAE Standard 90.1-2019 [49], and simulated occupancy. According to the results (for the baseline occupancy schedule), this small office building accommodates up to 56 (considering the minimum required space for each person in an office area, every office room can occupy up to 7 occupants) occupants who typically arrive at the office around 8:00 a.m. and depart around 5:00 p.m. As demonstrated in Figure 3, the occupancy schedule aligns with the office schedule recommended by ASHRAE Standard 90.1-2019, demonstrating a remarkable similarity to real-world occupancy patterns; however, the stochastic occupancy schedule provides more realistic occupancy patterns in office buildings [28,49]. This close alignment not only validates the simulation's accuracy but also highlights its potential to predict actual building usage with high fidelity. The occupancy levels fluctuate throughout the day, with specific hours experiencing higher occupancy rates. Notably, between 9:00 a.m. and 11:00 a.m. and again from 2:00 p.m. to 4:00 p.m., the occupancy schedule reaches its peak.

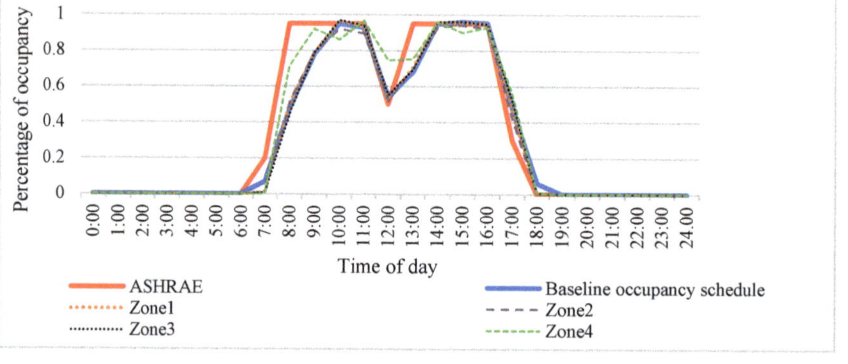

Figure 3. Baseline occupancy schedule vs. ASHRAE schedule.

Figure 4 illustrates the baseline occupancy schedule of each zone. The core zone and Zone 4, which include auxiliary, corridor, meeting room, and lobby, generally maintain lower occupancy levels compared to other zones. In addition, as shown in Figure 4, Zones 1, 2, and 3, which contain the office spaces, have more occupants since the average time spent by occupants in office spaces is longer than in other areas. It can be seen that the number of occupants decreases at 12:00 p.m. due to the lunch break.

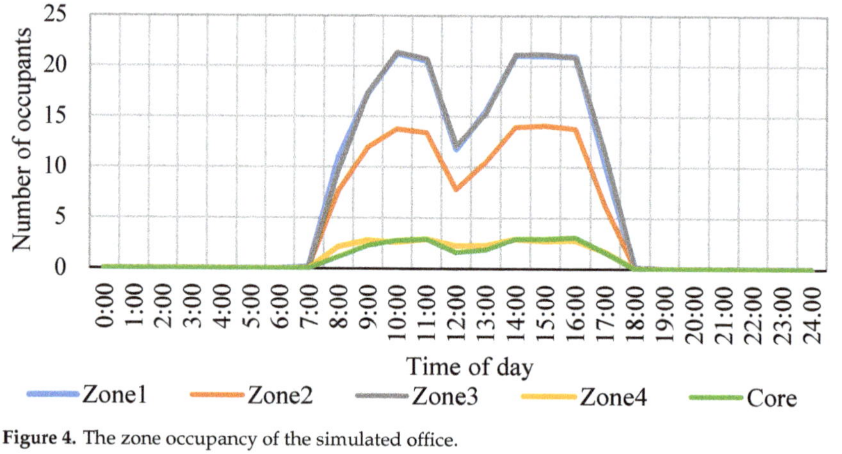

Figure 4. The zone occupancy of the simulated office.

To illustrate the average energy consumption during different seasons, we selected Tampa, FL (CZ 2A), as an example case due to the high building energy consumption in this climate zone. Figure 5 displays the average daily total electricity consumption of the simulated office building (including the electricity consumption for lighting, heating, cooling, equipment, and fans, in accordance with the baseline occupancy schedule) for the hot–humid climate zone (2A), for spring (1 March to 31 May), summer (1 June to 31 August), fall (1 September to 30 November), and winter (1 December to 28 February) seasons.

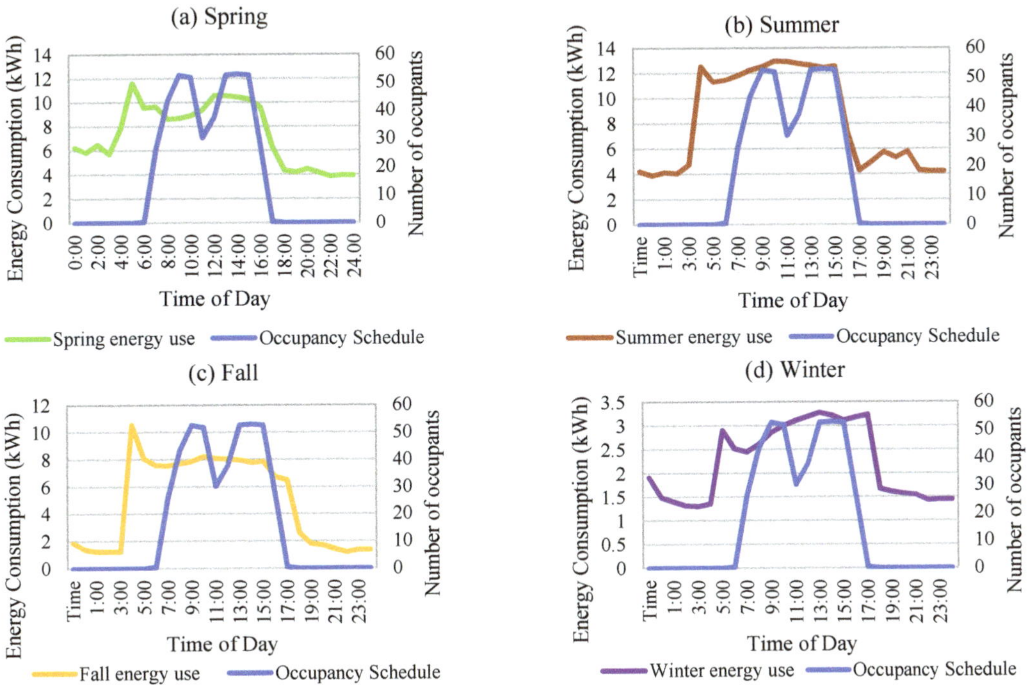

Figure 5. Simulated energy consumption and baseline occupancy schedule in (**a**) spring, (**b**) summer, (**c**) fall, and (**d**) winter seasons.

The graph reveals variations in the building's energy consumption across seasons, ranging from 1.5 to 13.5 kWh with respect to the baseline occupancy schedule. As shown, the building exhibits lower energy consumption in winter and fall seasons compared to summer and spring, primarily due to the need for higher cooling and air conditioning in the hot–humid climate zone.

Figure 6 illustrates the energy performance of various building energy systems (i.e., heating, cooling, lighting, equipment, and fans) in different climate zones. It can be seen that the total building energy can range from 35,000 to 47,000 kWh. According to this figure, buildings in 5C (cool marine), 4C (mixed marine), and 3C (warm marine) climate zones have the lowest total energy consumption among the other climate zones. One reason is the moderate and relatively stable temperatures characteristic of marine climates. The absence of extreme temperature fluctuations reduces the demand for heating or cooling, resulting in lower energy consumption. On the other hand, the energy consumption of buildings in hot–humid climate zones is higher than in other climate zones. This is due to the high demand for cooling and the necessity of using air conditioning, dehumidifying, and circulating the air in these climate zones.

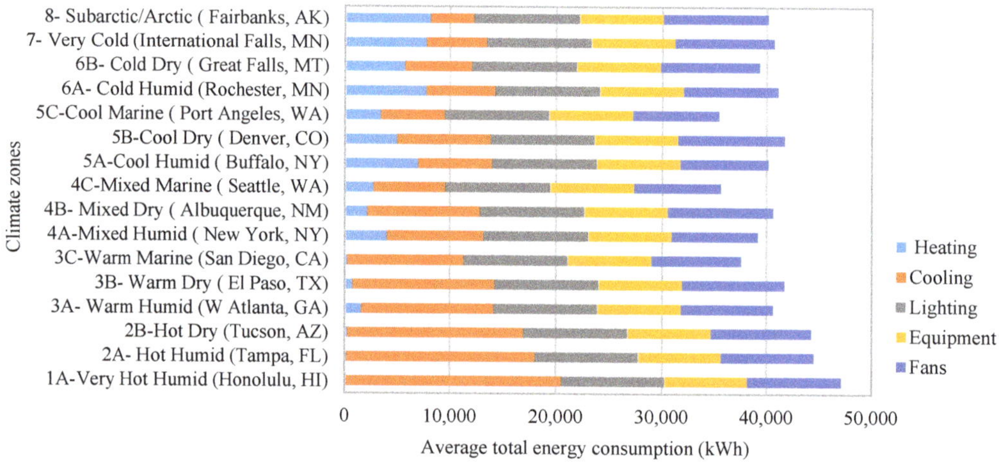

Figure 6. Energy consumption of the simulated office buildings in various climate zones.

As shown in Figure 6, the energy consumption of heating and cooling is very sensitive to climate conditions, resulting in significant variations. The average heating energy consumption in Fairbanks, AK (CZ 8), exceeds 6000 kWh, while in hot climate zones (e.g., 1A and 2A), heating energy consumption is negligible. In contrast, cooling energy consumption accounts for a large share of energy consumption, roughly 17,600 kWh, in very hot and humid climate zones and around 4400 kWh in cold climate zones. In addition, the energy consumption of lighting systems is mainly determined by indoor lighting needs, which are relatively consistent across various climate zones. Regardless of the climate, buildings require lighting for adequate illumination, resulting in comparable energy consumption for lighting systems. Similarly, the energy consumption of equipment, such as office equipment and appliances, is influenced more by occupant behavior and usage patterns rather than climate. Therefore, equipment energy consumption remains consistent across all climate zones. Additionally, fan energy consumption is about 9000 kWh with minimal variations in different climate zones. However, the demand for fans is slightly higher in dry climates to circulate the air and create a perceived cooling sensation for occupants, attributable to the low moisture content in the air in these dry climates.

The performance of the various end-use energy consumption in 16 climate zones considering the baseline occupancy parameters is illustrated in Figure 7. According to Figure 7a, heating electricity consumption is significantly high in humid climate zones (4A, 5A, and 6A). As shown in Figure 7b, cooling accounts for the largest share of electricity consumption in office buildings in many climate zones, ranging from very hot (e.g., 1A) to cool climate zones (e.g., 5B) compared to other end uses. Figure 7c,d demonstrate the electricity consumption for lighting and equipment in various climate zones. Although there are some slight differences regarding the comparison of lighting and equipment between climate zones, the results for these categories suggest that climate zones do not significantly impact lighting and equipment usage. This consistency across different climates can be attributed to the standardized nature of lighting and equipment operation in office settings. Unlike heating or cooling systems, which are directly influenced by external temperature variations and climate-specific requirements, lighting and equipment demands are predominantly driven by fixed office hours and internal activities that do not vary substantially with climate. Lighting consumption is largely determined by the daily work schedules that remain constant regardless of external weather conditions. Similarly, the usage of office equipment such as computers, printers, and other peripherals is aligned more closely with personnel presence and operational requirements rather than environmental factors. This explains the relative uniformity in energy usage for these categories across various climate zones observed in the study. Lastly, Figure 7e displays the electricity consumption for fans, revealing that dry climate zones (3B, 4B, and 5B) exhibit the highest electricity consumption for fan usage.

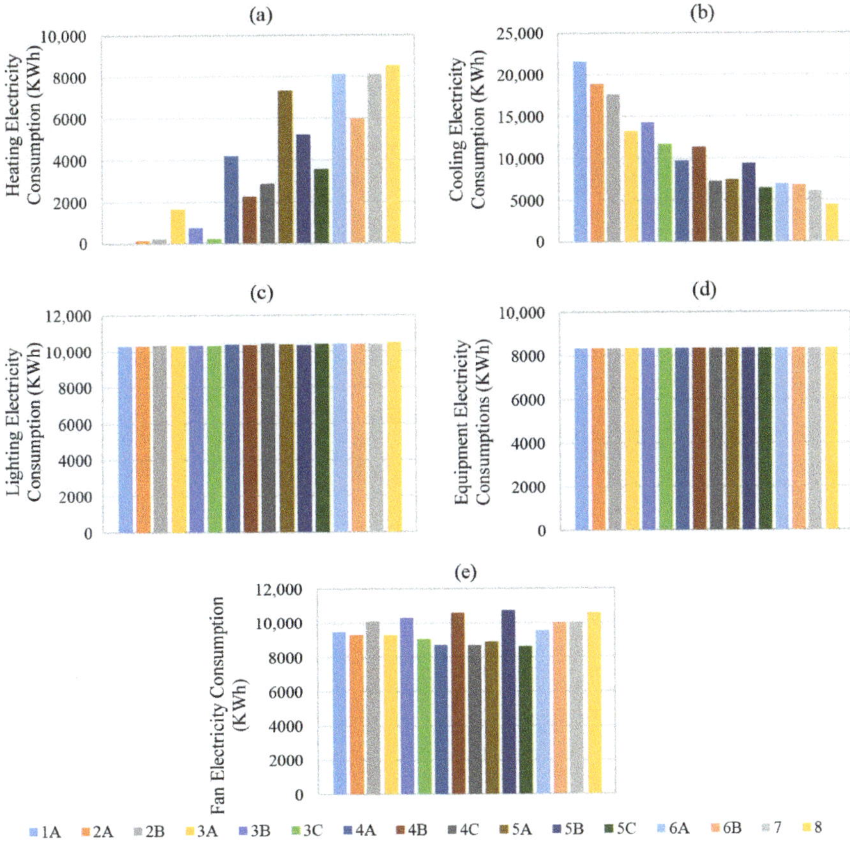

Figure 7. Average annual energy consumption of end-use systems for various climate zones. (**a**) Heating electricity consumption; (**b**) Cooling electricity consumption; (**c**) Lighting electricity consumption; (**d**) Equipment electricity consumption; (**e**) Fan electricity consumption.

Our findings indicate significant variations in building energy consumption across different climate zones, primarily driven by occupancy patterns and climatic conditions. This observation aligns with Chen et al. [50], who investigated the impact of future climate changes on office buildings across diverse climate zones in China, highlighting a similar trend of varying energy demands due to climatic differences. However, our study extends these findings by incorporating a wider range of climate zones in the US and employing dynamic occupancy schedules, which offer a more detailed understanding of energy consumption patterns. Unlike the static models used in this study, our dynamic approach reflects real-world variability and provides a new perspective on energy optimization strategies tailored to specific climate conditions.

Additionally, Meng et al. [51] investigated heating energy consumption in office buildings across various climate zones. Their findings confirm our results, which suggest that heating demands vary considerably, emphasizing the need for region-specific energy management strategies. Our study contributes further by examining both heating and cooling demands in a dynamic context, thereby improving the understanding of total energy consumption dynamics under changing climatic conditions.

3.2. Sensitivity Analysis of Occupancy Parameters

In this study, given the constraints on time and resources, six climate zones ranging from hot to cold were chosen to analyze occupancy parameters and assess the effectiveness of feature selection methods. The selected climate zones are presented in Table 3.

Table 3. Selected climate zones for sensitivity analysis.

No.	Climate Zone	Thermal Climate Zone Name	Weather Location
1	2A	Hot Humid	Tampa, FL, USA
2	3B	Warm Dry	El Paso, TX, USA
3	3C	Warm Marine	San Diego, CA, USA
4	4A	Mixed Humid	New York, NY, USA
5	5A	Cool Humid	Buffalo, NY, USA
6	6A	Cold Humid	Rochester, MN, USA

A sensitivity analysis was performed to analyze the impact of changing each occupancy schedule parameters on building energy performance. These parameters are analyzed by changing them while keeping all the other variables constant. The results are presented here.

Occupant Density: One of the occupancy parameters that may greatly impact building energy consumption is occupant density. Occupant density can be defined as the number of people per area (person/m^2). In this study, six different scenarios were considered to analyze the impact of occupant density on building energy performance:

- Scenario 1: 0.05 person/m^2, equal to 8 occupants using the office.
- Scenario 2: 0.06 person/m^2, equal to 16 occupants using the office.
- Scenario 3: 0.1 person/m^2, equal to 24 occupants using the office.
- Scenario 4: 0.14 person/m^2, equal to 40 occupants using the office.
- Scenario 5: 0.2 person/m^2, equal to 56 occupants using the office.

Figure 8 illustrates the annual building energy consumption across various climate zones, considering different occupant densities. Energy consumption for cooling, fans, and lighting tends to increase with higher occupant densities, while equipment and heating consumption show relatively smaller fluctuations. In warmer climate zones such as 2A, 3B, and 3C, the cooling and fan energy consumption exhibit high sensitivity to occupant density. Increasing the occupant density or number of occupants in these zones results in higher energy consumption for cooling and fans. This is primarily because a larger occupant density generates more heat and increases the cooling demand to maintain thermal comfort.

In addition, increasing the occupant density contributes to increased fan energy consumption to enhance air circulation in more crowded spaces. On the other hand, colder climate zones like 6A and 5A show higher heating energy consumption, with a negative correlation to occupant density. This can be attributed to the heat generation from occupants in more crowded spaces, resulting in reduced reliance on heating systems for maintaining comfortable temperatures.

This observed trend aligns with the findings of Zhao Dong et al. [52], who explored the impact of occupant behavior, particularly density, on energy consumption in office buildings. Their study provides a basis for understanding how increased occupancy contributes to higher energy demands. By referencing this broader scope of research, we can highlight that our findings are consistent with established trends in the field, suggesting that occupant density is a crucial factor in building energy dynamics. This also emphasizes the importance of incorporating occupant density considerations in the design and operation of HVAC systems, potentially through strategies such as demand-controlled ventilation and occupancy-based HVAC operation to optimize energy efficiency in office spaces.

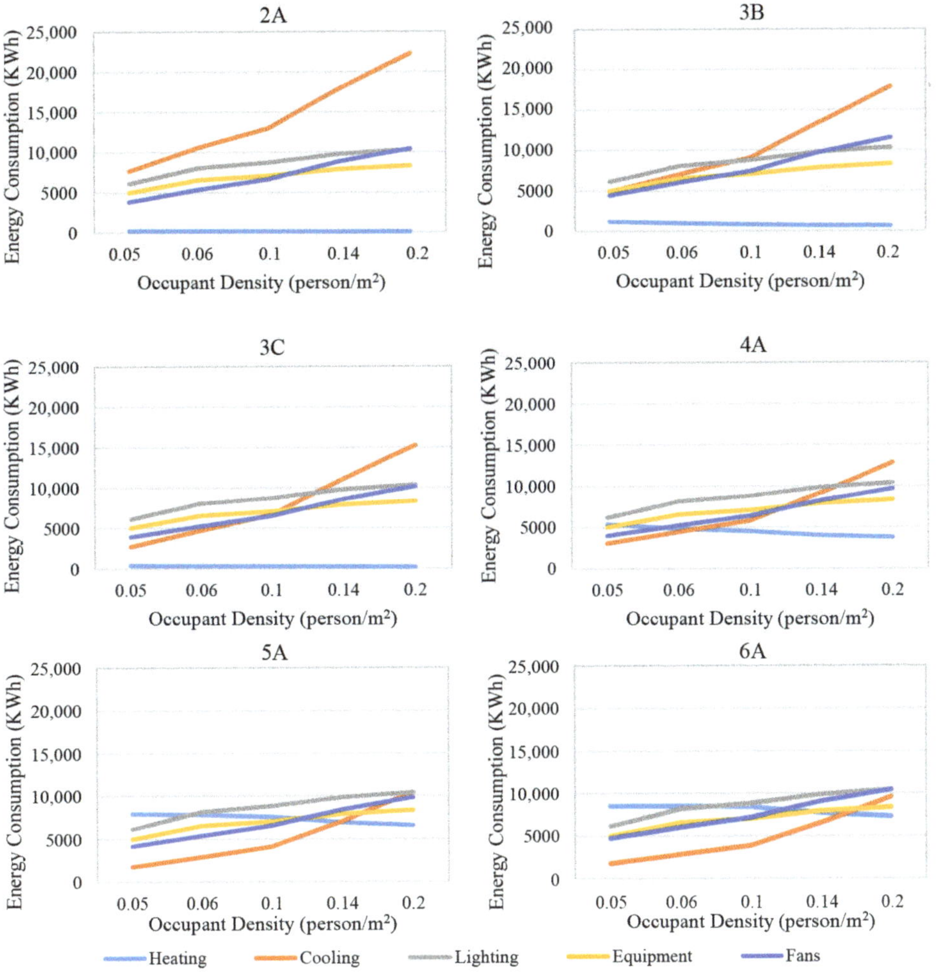

Figure 8. The impact of occupant density on building energy performance.

Arrival and departure: To analyze the impact of the arrival and departure time of occupants on building energy consumption, five distinct scenarios were considered:

- Scenario 1: Regular staff arrive and depart at 6:30 and 15:30, respectively.
- Scenario 2: Regular staff arrive and depart at 7:00 and 16:00, respectively.
- Scenario 3: Regular staff arrive and depart at 7:30 and 16:30, respectively.
- Scenario 4: Regular staff arrive and depart at 8:00 and 17:00, respectively.
- Scenario 5: Regular staff arrive and depart at 8:30 and 17:30, respectively.

Considering the average working hours of 8 h per day with a 1 h lunch break, regular staff work 9 h in the office building. Figure 9 presents the annual energy consumption of building systems in six climate zones considering different arrival and departure times for regular staff. Although the total working hours of regular staff are the same in all scenarios, the annual energy consumption is different. For example, in scenarios 1 and 2, the regular staff arrive at the office between 6:30 a.m. and 7:00 a.m. and leave the building between 3:30 p.m. and 4:00 p.m. It results in lower energy consumption due to taking advantage of cooler morning temperatures. In most of the climate zones, the early morning hours tend to be relatively cooler compared to the later part of the day. By starting work earlier, occupants can benefit from the cooler temperature and natural ventilation, reducing the need for extensive cooling and fan usage during the day. Consequently, in all climate zones, the cooling and fan energy consumption increases from scenario 1 to scenario 5, when shifting the arrival and departure times from 6:30 a.m. to 8:30 a.m. and 3:30 p.m. to 5:30 p.m., respectively. In contrast, there is a slight reduction in lighting and equipment energy consumption in all climate zones from scenario 1 to 5.

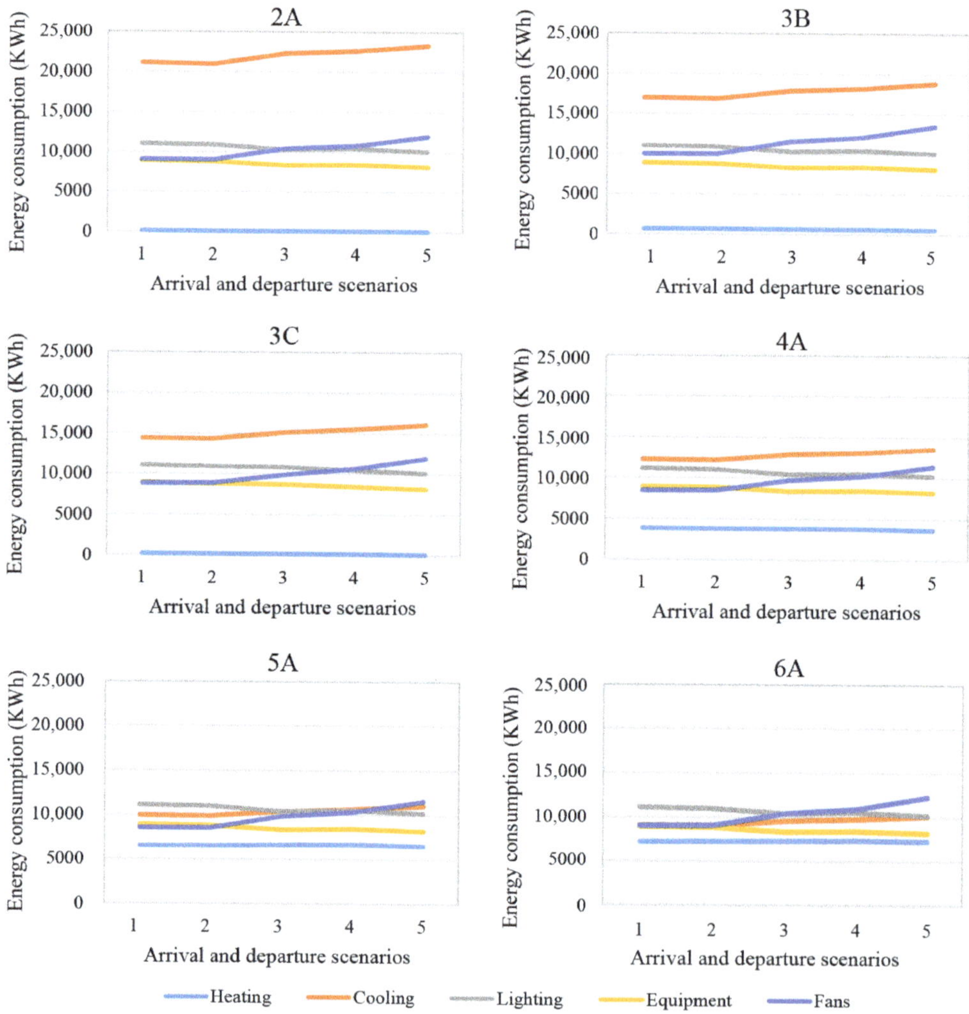

Figure 9. The impact of arrival and departure times of regular staff on building energy performance.

This observation aligns with findings from a study conducted by Gu et al. [53], which demonstrated that energy consumption varies significantly with different occupancy levels due to the presence and movement of occupants within the building. Specifically, the study highlighted how adjustments in the arrival and departure times can influence the operational schedules of building systems, thereby affecting the energy consumption for lighting, cooling, and heating. While the study of Gu et al. provided crucial insights, our study advances this by incorporating a broader analysis across multiple climate zones and more varied work schedule scenarios. This comprehensive approach not only allows for a detailed comparison across regions but also enhances the applicability of our findings to diverse environmental conditions and work patterns. These results highlight the potential for strategic scheduling to optimize energy efficiency in office buildings, corroborating the importance of considering occupant behavior patterns in energy management strategies.

Occupants' stay-time: Occupants' stay time in their own office is selected for analysis, where five different scenarios were considered to investigate the impact of occupants' stay-time in six climate zones:

- Scenario 1: Occupants stay in their own office for 30 min, on average.
- Scenario 2: Occupants stay in their own office for 45 min, on average.
- Scenario 3: Occupants stay in their own office for 60 min, on average.
- Scenario 4: Occupants stay in their own office for 75 min, on average.
- Scenario 5: Occupants stay in their own office for 90 min, on average.

Figure 10 presents the energy performance of various energy systems across six different climate zones, considering different durations of occupants' stay-time in their offices. As shown in the figure, for all six climate zones, there are minor changes in the energy consumption of the cooling system and fans when the duration of occupants' stay-time increases from 30 min to 90 min. As occupants stay in a space for a longer duration, their body heat gradually increases the room's temperature. Therefore, the cooling system and fans may need to operate for a slightly extended period to maintain air quality and temperature.

Conversely, the remaining energy systems, including lighting, equipment, and heating systems, remain unchanged, indicating that an increase in the duration of occupants' stay-time did not have a significant impact on these systems. This is because lighting and equipment energy consumption is related to occupancy presence and specific activities rather than the duration alone. Within the considered range of 30 to 90 min, the occupancy duration does not significantly affect the usage patterns or energy consumption of lighting and equipment. Similarly, heating systems respond to the set-point temperature and occupancy needs in colder climates. Therefore, increasing the occupants' stay-time does not demand an adjustment in these systems, resulting in unchanged energy consumption. This observation indicates that for the range of stay times considered, the direct impact on energy consumption is limited. This stability offers potential for energy management strategies that focus more on occupancy-based controls rather than adjustments based on duration of stay alone. Future studies might explore more granular time increments or different types of activities to further refine our understanding of occupancy impact on energy use. Such insights could inform more targeted energy efficiency measures that align with actual usage patterns and contribute to broader energy sustainability goals.

Time-step: To analyze the impact of simulation time-step on the energy performance of various energy systems in different climate zones, four different time-steps are considered:

- Scenario 1: Time-step size of 5 min.
- Scenario 2: Time-step size of 10 min.
- Scenario 3: Time-step size of 15 min.
- Scenario 4: Time-step size of 20 min.

Figure 11 presents the simulation performance of heating, cooling, lighting, and other energy consumptions using different time-step sizes. As shown, when employing a 5 min time-step, the energy performance of cooling systems and fans exhibits a significant increase across all six climate zones compared to larger step sizes. The utilization of a smaller time-step allows for a more detailed representation of the dynamic behavior of cooling systems and fans. With a smaller time-step, the simulation captures shorter fluctuations in cooling demand and the need for increased fan operation to maintain thermal comfort. Consequently, the energy consumption of cooling systems and fans appears higher when using a 5 min time-step, reflecting their more responsive nature to varying conditions. This responsiveness underscores the importance of selecting an appropriate simulation time-step in energy modeling to capture the nuanced effects of environmental changes and occupant interactions within a building.

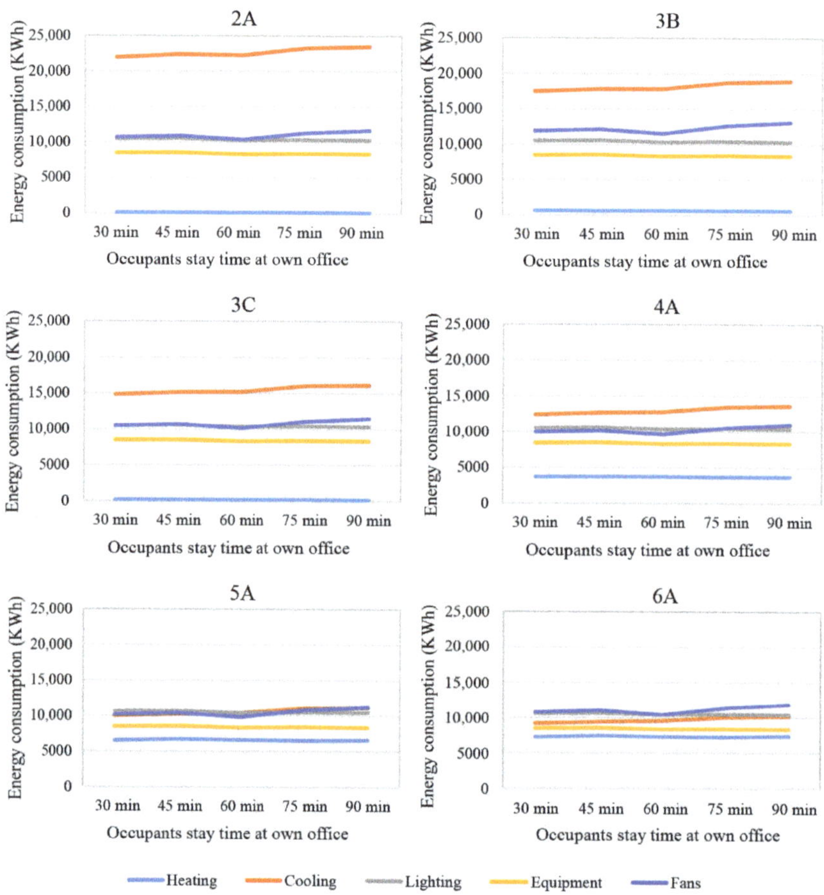

Figure 10. The impact of occupants' stay-time duration at their own office on building energy performance.

Additionally, the smaller time-step provides a clearer insight into the intermittent nature of heating requirements, thereby enhancing the ability to fine-tune heating operations to actual needs rather than sustained assumptions. This precision leads to energy conservation, particularly in climates where heating demands fluctuate significantly throughout the day. The energy consumption of lighting and equipment remains relatively constant regardless of the simulation time-step. This stability indicates that these systems' operational demands are less sensitive to shorter time fluctuations and more dependent on fixed schedules of occupant activity. Understanding these dynamics is crucial for building managers and designers aiming to optimize energy usage without compromising occupant comfort and productivity. Looking more closely at different time-steps could help us find new ways to save energy, especially in systems that automatically adjust based on real-time data.

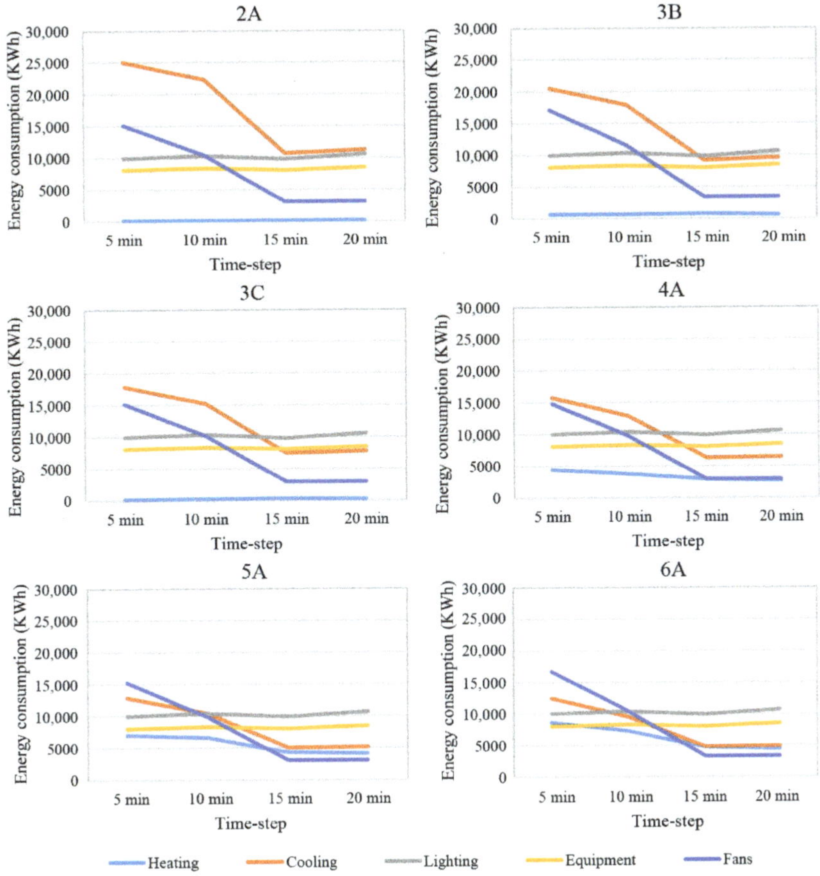

Figure 11. The impact of occupancy simulation time-step size on building energy performance.

The sensitivity analysis of four occupancy parameters was presented. Although a total of 22 parameters exist, it is important to note that the selection of these four parameters does not diminish the significance of the other parameters. The choice to analyze these particular parameters was driven by two reasons. Firstly, these four parameters were identified as having a high potential for influencing building energy consumption based on prior research and expert knowledge in the field. Secondly, considering the constraints of limited resources and time, focusing on a subset of parameters allowed for a more in-depth and targeted analysis.

3.3. Feature Selection Results

The generated dataset of building energy performance based on various climate zones and occupancy parameters served as input for feature selection methods to identify the most significant occupancy parameters affecting building energy consumption. In this regard, MVL, LASSO, NCA, and ReliefF feature selection methods were applied to the generated dataset. To assess the results of these feature selection methods, the selected features for each building energy system (i.e., heating, cooling, lighting, equipment, and fans) were incorporated into a multiple linear regression model to compute the corresponding R-square values. Subsequently, these calculated R-square values were evaluated against a predefined threshold of 0.05 as established in this study [54]. Features that demonstrated contributions greater than this threshold in the calculation of the R-square value were retained and considered significant variables. Conversely, features that exhibited contributions below the threshold in the calculation of the R-square value were eliminated from the list. This systematic process facilitated the identification of significant variables, allowing for a focused analysis of the features that exhibited a significant impact on the performance of the building energy systems.

Heating: This section focuses on the performance of heating energy consumption and presents the results of four feature selection methods to identify the significant occupancy parameters impacting heating energy consumption in six climate zones, as presented in Table 4. As shown in the table, all four methods consistently identified "Occupant-Density" as the primary significant variable across all six climate zones. Furthermore, "Time-Step" emerged as the second significant parameter in most cases, except for the "2A" climate zone where it did not exhibit the same significance.

Cooling: This section focuses on the performance of cooling energy consumption and presents the results of four feature selection methods to identify the significant occupancy parameters impacting cooling energy consumption in six climate zones, as presented in Table 5.

As shown in the table, all four methods consistently identified "Occupant-Density" and "Time-step" as the first two primary significant variables across all six climate zones. According to Table 5, it can be seen that the LASSO feature selection method presented a better performance across all six climate zones in identifying the significant features influencing cooling energy consumption. This method generated higher R-squared values compared to other methods. The selected features in the LASSO method include "Time-step", "Occupant_Density", "Staff_Arriv_Depar", "Lunch_Duration", and "Own_Stay_Duration". Conversely, the ReliefF method presented limited effectiveness in identifying significant features. This analysis focuses on influential occupancy parameters affecting cooling energy consumption in small office buildings across different climate zones and highlights the LASSO feature selection method as the preferred approach for identifying the features that have a significant impact on cooling energy consumption.

Lighting: This section focuses on the analysis of lighting energy performance and presents the results of four feature selection methods to identify the significant occupancy parameters impacting lighting energy consumption in six climate zones as presented in Table 6. As shown in the table, all four methods consistently identified "Occupant-Density" as the primary significant variable across all six climate zones. According to Table 6, it can be seen that both the NCA and LASSO feature selection methods presented a better performance across all six climate zones in identifying the significant features influencing lighting energy consumption. These methods generated higher R-squared values compared to other methods.

Conversely, the ReliefF method presented a low performance in identifying significant features. This analysis focuses on influential occupancy parameters affecting lighting energy consumption in small office buildings across different climate zones and highlights that both NCA and LASSO feature selection methods are preferable for identifying the features that have a significant impact on lighting energy consumption.

Equipment: This section focuses on the analysis of equipment energy performance and presents the results of four feature selection methods to identify the significant occupancy parameters impacting equipment energy consumption in six climate zones, as presented in Table 7. As shown in the table, all four methods consistently identified "Occupant-Density" as the primary significant variable across all six climate zones. According to Table 7, it can be seen that the LASSO feature selection method presented a better performance across all six climate zones in identifying the significant features influencing equipment energy consumption. This method generated higher R-squared values compared to other methods. The selected features in the LASSO method include "Occupant_Density", "Staff_Arriv_Depar", "Manag_Arriv_Depar", "Arriv_Depar_Vari", and "Time_Step". Conversely, the ReliefF method presented limited effectiveness in identifying significant features. This analysis focuses on influential occupancy parameters affecting equipment energy consumption in small office buildings across different climate zones and highlights the LASSO feature selection method as a preferred approach for identifying the features that have a significant impact on equipment energy consumption.

Table 4. Identified features for heating energy performance across different climate zones.

Climate Zone	MVLinear	LASSO	NCA	ReliefF
2A	• Occupant_Density • Staff_Arriv_Depar • Lunch_Time • Lunch_Duration	• Occupant_Density • Staff_Arriv_Depar • Admin_Arriv_Depar • Manag_Arriv_Depar • Own_Stay_Duration	• Occupant_Density • Admin_Arriv_Depar • Manag_Arriv_Depar	• Occupant_Density • Meeting_Stay_Duration • Manag_Room_Stay
R-squared	0.89596	0.90338	0.88389	0.87748
3B	• Occupant_Density • Time_Step • Staff_Arriv_Depar • Lunch_Duration • Lunch_Time	• Occupant_Density • Time_Step • Staff_Arriv_Depar • Admin_Arriv_Depar • Manag_Arriv_Depar	• Occupant_Density • Time_Step • Staff_Arriv_Depar	• Occupant_Density • Time_Step
R-squared	0.83587	0.83982	0.83773	0.82429
3C	• Occupant_Density • Time_Step • Staff_Arriv_Depar	• Occupant_Density • Time_Step • Staff_Arriv_Depar • Manag_Arriv_Depar • Own_Stay_Duration	• Occupant_Density • Time_Step • Manag_Arriv_Depar	• Occupant_Density • Time_Step
R-squared	0.89596	0.89345	0.88437	0.86468
4A	• Occupant_Density • Time_Step • Lunch_Duration	• Occupant_Density • Time_Step • Staff_Arriv_Depar • Manag_Arriv_Depar • Admin_Arriv_Depar	• Occupant_Density • Time_Step • Staff_Arriv_Depar • Manag_Arriv_Depar	• Occupant_Density • Time_Step
R-squared	0.88695	0.89287	0.89161	0.88692
5A	• Occupant_Density • Time_Step • Lunch_Duration	• Occupant_Density • Time_Step • Manag_Arriv_Depar • Staff_Arriv_Depar	• Occupant_Density • Time_Step • Manag_Arriv_Depar • Staff_Arriv_Depar	• Occupant_Density • Time_Step • Lunch_Duration_Vari
R-squared	0.87708	0.88016	0.87937	0.87729
6A	• Occupant_Density • Time_Step • Lunch_Duration • Outdoor_Stay_Duration	• Occupant_Density • Time_Step • Manag_Arriv_Depar • Auxiliary_Stay_Duration	• Occupant_Density • Time_Step • Manag_Arriv_Depar • Auxiliary_Stay_Duration	• Occupant_Density • Time_Step • Admin_Room_Stay • Own_Stay_Duration
R-squared	0.9034	0.90748	0.90694	0.90336

Fans: This section focuses on the performance of fan energy consumption and presents the results of four feature selection methods to identify the significant occupancy parameters impacting fan's energy consumption in six climate zones, as presented in Table 8. As shown in the table, three methods of MVLinear, NCA, and LASSO consistently identified "Time-Step" as the primary significant variable across all six climate zones. Furthermore, "Occupant-Density" emerged as the second significant parameter, except in the "ReliefF" method, which is the opposite. According to Table 8, both MVLinear and LASSO feature selection methods presented better performances across all six climate zones in identifying the significant features influencing fan energy consumption.

These methods generated higher R-squared values compared to other methods. The selected features in the MVLinear and LASSO methods include "Time_Step", "Occupant_Density", "Staff_Arriv_Depar", "Admin_Arriv_Depar", and "Own_Stay_Duration", and, conversely, the ReliefF method presented limited effectiveness in identifying significant features. The higher performance of MVLinear and LASSO in our study points to their robustness in handling diverse datasets and their capability in effectively isolating key factors that influence fan energy consumption. These methods prove particularly valuable in scenarios where predictive accuracy is important to developing energy management solutions that are both effective and scalable across different climate zones.

Table 5. Identified features for cooling energy performance across different climate zones.

Climate Zone	MVLinear	LASSO	NCA	ReliefF
2A	• Time_Step • Occupant_Density • Own_Stay_Duration	• Time_Step • Occupant_Density • Staff_Arriv_Depar • Lunch_Duration • Own_Stay_Duration	• Time_Step • Occupant_Density • Manag_Room_Stay • Lunch_Duration	• Occupant_Density • Time_Step • Own_Stay_Duration • Meeting_Stay_Duration
R-squared	0.85107	0.85221	0.85143	0.84634
3B	• Time_Step • Occupant_Density • Staff_Arriv_Depar • Own_Stay_Duration	• Time_Step • Occupant_Density • Staff_Arriv_Depar • Lunch_Duration • Own_Stay_Duration	• Occupant_Density • Time_Step • Staff_Arriv_Depar	• Occupant_Density • Time_Step
R-squared	0.83984	0.84153	0.84052	0.83586
3C	• Time_Step • Occupant_Density • Staff_Arriv_Depar	• Time_Step • Occupant_Density • Staff_Arriv_Depar • Lunch_Duration • Own_Stay_Duration	• Occupant_Density • Time_Step • Manag_Room_Stay • Staff_Arriv_Depar	• Occupant_Density • Time_Step • Own_Stay_Duration
R-squared	0.82795	0.83965	0.83164	0.82541
4A	• Time_Step • Occupant_Density • Staff_Arriv_Depar	• Time_Step • Occupant_Density • Staff_Arriv_Depar • Lunch_Duration	• Occupant_Density • Time_Step • Staff_Arriv_Depar • Lunch_Duration	• Occupant_Density • Time_Step • Own_Stay_Duration • Auxiliary_Stay_Duration
R-squared	0.81269	0.8148	0.81346	0.81127
5A	• Occupant_Density • Time_Step • Own_Stay_Duration	• Time_Step • Occupant_Density • Lunch_Duration • Own_Stay_Duration	• Occupant_Density • Time_Step • Staff_Arriv_Depar • Lunch_Duration • Own_Stay_Duration	• Occupant_Density • Time_Step • Own_Stay_Duration • Auxiliary_Stay_Duration
R-squared	0.78551	0.78816	0.78816	0.78456
6A	• Occupant_Density • Time_Step • Staff_Arriv_Depar • Own_Stay_Duration	• Time_Step • Occupant_Density • Lunch_Duration • Staff_Arriv_Depar • Own_Stay_Duration	• Time_Step • Occupant_Density • Staff_Arriv_Depar • Lunch_Duration • Own_Stay_Duration	• Occupant_Density • Time_Step • Own_Stay_Duration
R-squared	0.77699	0.77968	0.77968	0.77546

This analysis focuses on influential occupancy parameters affecting fan energy consumption in small office buildings across different climate zones and highlights the LASSO and MVLinear feature selection methods as preferred approaches for identifying the features that have a significant impact on fan energy consumption. By systematically identifying these key parameters, our approach aids facility managers and designers in implementing precision-driven energy conservation measures. Specifically, understanding the detailed impacts of occupancy parameters such as "Occupant-Density" and "Time-Step" allows for the optimization of HVAC operations and other energy systems to better match actual building use patterns, thus avoiding both overuse and underuse of energy resources.

Table 6. Identified features for lighting energy performance across different climate zones.

Climate Zone	MVLinear	LASSO	NCA	ReliefF
2A	• Occupant_Density • Manag_Arriv_Depar • Arriv_Depar_Vari • Time_Step	• Occupant_Density • Staff_Arriv_Depar • Manag_Arriv_Depar • Arriv_Depar_Vari • Time_Step	• Occupant_Density • Arriv_Depar_Vari • Staff_Arriv_Depar • Manag_Arriv_Depar • Time_Step	• Occupant_Density • Time_Step • Manag_Room_Stay
R-squared	0.83369	0.84654	0.84654	0.81792
3B	• Occupant_Density • Manag_Arriv_Depar • Arriv_Depar_Vari • Time_Step • Lunch_Start_Vari	• Occupant_Density • Staff_Arriv_Depar • Manag_Arriv_Depar • Arriv_Depar_Vari • Time_Step	• Occupant_Density • Arriv_Depar_Vari • Staff_Arriv_Depar • Manag_Arriv_Depar	• Occupant_Density • Time_Step • Manag_Room_Stay
R-squared	0.83252	0.84592	0.84592	0.81617
3C	• Occupant_Density • Manag_Arriv_Depar • Arriv_Depar_Vari • Time_Step • Lunch_Start_Vari	• Occupant_Density • Staff_Arriv_Depar • Manag_Arriv_Depar • Arriv_Depar_Vari	• Occupant_Density • Arriv_Depar_Vari • Staff_Arriv_Depar • Manag_Arriv_Depar • Admin_Arriv_Depar	• Occupant_Density • Time_Step • Manag_Room_Stay
R-squared	0.83225	0.846712	0.845331	0.81572
4A	• Occupant_Density • Manag_Arriv_Depar • Arriv_Depar_Vari • Time_Step • Lunch_Start_Vari	• Occupant_Density • Staff_Arriv_Depar • Manag_Arriv_Depar • Arriv_Depar_Vari	• Occupant_Density • Arriv_Depar_Vari • Staff_Arriv_Depar • Manag_Arriv_Depar	• Occupant_Density • Time_Step • Manag_Room_Stay • Auxiliary_Stay_Duration
R-squared	0.83203	0.84544	0.84544	0.81546
5A	• Occupant_Density • Manag_Arriv_Depar • Arriv_Depar_Vari • Time_Step • Lunch_Start_Vari	• Occupant_Density • Staff_Arriv_Depar • Manag_Arriv_Depar • Arriv_Depar_Vari	• Occupant_Density • Arriv_Depar_Vari • Staff_Arriv_Depar • Manag_Arriv_Depar • Admin_Arriv_Depar	• Occupant_Density • Time_Step • Manag_Room_Stay • Auxiliary_Stay_Duration
R-squared	0.83351	0.84682	0.84647	0.81456
6A	• Occupant_Density • Manag_Arriv_Depar • Arriv_Depar_Vari • Time_Step • Lunch_Start_Vari	• Occupant_Density • Staff_Arriv_Depar • Manag_Arriv_Depar • Arriv_Depar_Vari	• Occupant_Density • Arriv_Depar_Vari • Staff_Arriv_Depar • Manag_Arriv_Depar	• Occupant_Density • Time_Step • Manag_Room_Stay • Auxiliary_Stay_Duration • Outdoor_Stay_Duration
R-squared	0.83316	0.84647	0.84635	0.81632

The results of the feature selection analyses highlight critical occupancy parameters such as "Occupant-Density" and "Time-Step", which are consistently significant across multiple climate zones. This consistency across zones not only validates our feature selection methods but also reinforces the importance of these parameters in the energy efficiency profiling of office buildings. These parameters offer valuable insights into the design and operation of energy-efficient buildings, tailored to the unique characteristics of each climate zone. For instance, the consistent significance of occupant density suggests that strategies focused on optimizing space usage can lead to substantial energy savings. This is particularly relevant in denser office environments where the effective management of space and occupant schedules can reduce unnecessary energy expenditure during peak and off-peak hours. Moreover, the distinction in significant parameters across climate zones allows for the development of region-specific energy management strategies. In colder climates, during high heating demands, understanding the impact of occupancy timings can help in better scheduling of heating systems to align with actual office hours, thus avoiding wastage. Similarly, in hotter regions, cooling systems can be optimized based on detailed occupancy schedules to ensure that energy is not wasted cooling unoccupied spaces.

Table 7. Identified features for equipment energy performance across different climate zones.

Climate Zone	MVLinear	LASSO	NCA	ReliefF
2A	• Occupant_Density • Manag_Arriv_Depar • Arriv_Depar_Vari • Lunch_Start_Vari	• Occupant_Density • Staff_Arriv_Depar • Manag_Arriv_Depar • Arriv_Depar_Vari • Time_Step	• Occupant_Density • Arriv_Depar_Vari • Staff_Arriv_Depar • Manag_Arriv_Depar	• Occupant_Density • Time_Step • Manag_Room_Stay • Auxiliary_Stay_Duration • Outdoor_Stay_Duration
R-squared	0.83041	0.84465	0.84444	0.8087
3B	• Occupant_Density • Manag_Arriv_Depar • Lunch_Start_Vari	• Occupant_Density • Staff_Arriv_Depar • Manag_Arriv_Depar • Arriv_Depar_Vari • Time_Step	• Occupant_Density • Arriv_Depar_Vari • Staff_Arriv_Depar • Manag_Arriv_Depar	• Occupant_Density • Time_Step • Outdoor_Stay_Duration
R-squared	0.83041	0.84465	0.84444	0.8087
3C	• Occupant_Density • Manag_Arriv_Depar • Arriv_Depar_Vari • Lunch_Start_Vari	• Occupant_Density • Staff_Arriv_Depar • Manag_Arriv_Depar • Arriv_Depar_Vari • Time_Step	• Occupant_Density • Arriv_Depar_Vari • Staff_Arriv_Depar • Manag_Arriv_Depar	• Occupant_Density • Time_Step • Manag_Room_Stay • Auxiliary_Stay_Duration • Outdoor_Stay_Duration
R-squared	0.83041	0.84465	0.84444	0.8087
4A	• Occupant_Density • Manag_Arriv_Depar • Arriv_Depar_Vari • Lunch_Start_Vari	• Occupant_Density • Staff_Arriv_Depar • Manag_Arriv_Depar • Arriv_Depar_Vari • Time_Step	• Occupant_Density • Staff_Arriv_Depar • Manag_Arriv_Depar	• Occupant_Density • Time_Step • Manag_Room_Stay • Auxiliary_Stay_Duration • Outdoor_Stay_Duration
R-squared	0.83041	0.84465	0.84444	0.8087
5A	• Occupant_Density • Manag_Arriv_Depar • Arriv_Depar_Vari • Lunch_Start_Vari	• Occupant_Density • Staff_Arriv_Depar • Manag_Arriv_Depar • Arriv_Depar_Vari • Time_Step	• Occupant_Density • Arriv_Depar_Vari • Staff_Arriv_Depar • Manag_Arriv_Depar	• Occupant_Density • Time_Step • Manag_Room_Stay

Table 7. *Cont.*

Climate Zone	MVLinear	LASSO	NCA	ReliefF
R-squared	0.83041	0.84465	0.84444	0.8087
6A	• Occupant_Density • Manag_Arriv_Depar • Lunch_Start_Vari	• Occupant_Density • Staff_Arriv_Depar • Manag_Arriv_Depar • Arriv_Depar_Vari • Time_Step	• Occupant_Density • Arriv_Depar_Vari • Staff_Arriv_Depar • Manag_Arriv_Depar	• Occupant_Density • Time_Step • Manag_Room_Stay • Auxiliary_Stay_Duration • Outdoor_Stay_Duration
R-squared	0.83041	0.84465	0.84444	0.8087

In this study, we applied a variety of feature selection methods to enhance the accuracy and efficiency of predicting building energy consumption. This approach is supported by recent research, including the work by Henriques et al. [55], who utilized advanced feature-selection methods like recursive feature elimination and random forests to uncover atypical energy-consumption patterns in households. Their findings demonstrate the effectiveness of comprehensive feature selection in identifying significant predictors of energy usage, which aligns with our methodology for improving model performance. Additionally, the application of feature selection for support vector regression models, as discussed in a study by Zhao and Magoulès [56], highlights the reduction in model complexity and the improvement in predictive accuracy for building energy consumption. This reference further validates our choice of feature selection methods by illustrating their benefits in similar applications, thereby highlighting the relevance of our methodological choices in the context of current research trends.

Table 8. Identified features for fan energy performance across different climate zones.

Climate Zone	MVLinear	LASSO	NCA	ReliefF
2A	• Time_Step • Occupant_Density • Staff_Arriv_Depar • Admin_Arriv_Depar • Own_Stay_Duration	• Time_Step • Occupant_Density • Staff_Arriv_Depar • Admin_Arriv_Depar • Own_Stay_Duration	• Time_Step • Occupant_Density • Staff_Arriv_Depar • Auxiliary_Stay_Duration • Manag_Room_Stay	• Occupant_Density • Time_Step • Manag_Room_Stay • Own_Stay_Duration
R-squared	0.83267	0.83267	0.83092	0.81171
3B	• Time_Step • Occupant_Density • Staff_Arriv_Depar • Admin_Arriv_Depar • Own_Stay_Duration	• Time_Step • Occupant_Density • Staff_Arriv_Depar • Admin_Arriv_Depar • Own_Stay_Duration	• Time_Step • Occupant_Density • Staff_Arriv_Depar • Auxiliary_Stay_Duration • Manag_Room_Stay	• Occupant_Density • Time_Step • Manag_Room_Stay • Own_Stay_Duration
R-squared	0.82586	0.82586	0.82374	0.80272
3C	• Time_Step • Occupant_Density • Staff_Arriv_Depar • Admin_Arriv_Depar • Own_Stay_Duration	• Time_Step • Occupant_Density • Staff_Arriv_Depar • Admin_Arriv_Depar • Own_Stay_Duration	• Time_Step • Occupant_Density • Staff_Arriv_Depar • Auxiliary_Stay_Duration • Manag_Room_Stay	• Occupant_Density • Time_Step • Own_Stay_Duration

Table 8. *Cont.*

Climate Zone	MVLinear	LASSO	NCA	ReliefF
R-squared	0.82016	0.82016	0.81748	0.80914
4A	• Time_Step • Occupant_Density • Staff_Arriv_Depar • Admin_Arriv_Depar • Own_Stay_Duration	• Time_Step • Occupant_Density • Staff_Arriv_Depar • Admin_Arriv_Depar • Own_Stay_Duration	• Time_Step • Occupant_Density • Staff_Arriv_Depar • Auxiliary_Stay_Duration • Admin_Arriv_Depar	• Occupant_Density • Time_Step
R-squared	0.81928	0.81928	0.81827	0.79416
5A	• Time_Step • Occupant_Density • Staff_Arriv_Depar • Own_Stay_Duration	• Time_Step • Occupant_Density • Staff_Arriv_Depar • Admin_Arriv_Depar • Own_Stay_Duration	• Time_Step • Staff_Arriv_Depar • Occupant_Density • Auxiliary_Stay_Duration • Admin_Arriv_Depar	• Occupant_Density • Time_Step • Own_Stay_Duration
R-squared	0.81505	0.81505	0.81381	0.78929
6A	• Time_Step • Occupant_Density • Staff_Arriv_Depar • Admin_Arriv_Depar	• Time_Step • Occupant_Density • Staff_Arriv_Depar • Admin_Arriv_Depar • Own_Stay_Duration	• Time_Step • Staff_Arriv_Depar • Occupant_Density • Auxiliary_Stay_Duration	• Occupant_Density • Time_Step • Own_Stay_Duration
R-squared	0.81444	0.81444	0.81307	0.78684

These insights not only enhance the ability to create more adaptive and intelligent building management systems but also support the development of policies and standards that promote energy efficiency. By integrating these findings into the regulatory frameworks, it is possible to set more realistic and attainable energy usage benchmarks that reflect the real-world conditions of office buildings. This approach can lead to more sustainable energy practices and significant reductions in operational costs, contributing to broader environmental and economic benefits.

4. Conclusions

This study utilizes a nationwide energy simulation to analyze the impact of occupancy parameters on building energy performance across different US climate zones. Dynamic occupancy schedules are generated based on identified occupancy parameters using an ABM. The generated schedule, along with the DOE small office prototype model, are integrated into the BPS tool (i.e., EnergyPlus). The simulation results provide a dataset of occupancy parameters and building energy performance in various climate zones. This study employs sensitivity analysis and feature selection methods to assess the influence of occupancy parameters on the energy consumption of various building energy systems (i.e., heating, cooling, lighting, equipment, and fans).

The energy performance simulation across 16 climate zones revealed that buildings located in cool marine, mixed marine, and warm marine climate zones had lower total energy consumption compared to other zones. This lower energy consumption can be attributed to the moderate and stable temperatures characteristic of marine climates. On the other hand, buildings in hot–humid climate zones demonstrated significantly higher energy consumption, primarily due to the high demand for cooling and air conditioning systems. As a result of the sensitivity analysis, heating and cooling energy consumption were found to be sensitive to climate zones. However, lighting and equipment energy consumption remained relatively constant across climate zones, due to their dependence on occupancy presence and specific activities rather than climate zones. Fan energy consumption exhibited minimal variation across climate zones but was slightly higher in dry climates to create a perceived cooling sensation for occupants.

The sensitivity analysis of occupancy parameters revealed the impact of occupant density, arrival and departure times, and occupants' stay-time and simulation time-step on energy consumption. For instance, reducing occupancy density from 0.2 person/m^2 to 0.05 person/m^2 if possible (e.g., transition to a hybrid work environment) can result in annual energy savings ranging from 9000 to 12,000 kWh. In other words, an underoccupied office building can reduce energy consumption by 60 percent. Modifying arrival and departure times, particularly by starting work earlier at 6:30 a.m. instead of 8:00 a.m., resulted in lower cooling and fan energy consumption due to taking advantage of cooler temperatures. Also, in hot climate zones, by shifting the arrival and departure time to the early morning, an annual energy savings of up to 3000 kWh can be achieved. In addition, changing the duration of occupants' stay-time in their own office from 30 to 90 min had a minimal impact on building energy consumption across all climate zones, with only slight changes observed in cooling and fan energy consumption. Moreover, the simulation time-step size influenced the energy performance of cooling systems and fans, with smaller time-steps resulting in higher energy consumption for these systems. In other words, changing the simulation time-step from 20 min to 5 min can result in a 50% discrepancy, which is around 12,000 to 15,000 kWh in annual cooling and fan energy consumption.

Feature selection methods applied in this study effectively identified significant occupancy parameters impacting energy consumption, with "Occupant Density" consistently identified as the primary significant variable across all climate zones. Other variables such as "staff's arrival and departure time", "stay-time in own office", "lunch break time", and "simulation time-step" were also found to have a significant impact on the simulation results. These findings offer valuable insights into the influential occupancy parameters for different small offices across various climate zones, underscoring the importance of tailoring occupancy schedules to enhance energy efficiency. This study provides practical guidance that can be directly applied to optimize energy consumption and achieve significant energy savings in small office settings with different weather conditions.

The novel insights gained from this research highlight the critical role of detailed occupancy parameter analysis in understanding and optimizing building energy performance across diverse climate conditions. By employing advanced modeling techniques and comprehensive sensitivity analyses, this study improves our knowledge of how occupancy behaviors impact building energy consumption and provides a pathway to more effective energy management strategies tailored to specific climatic and operational contexts. Overall, this study not only clarifies the effects of various occupancy parameters on energy consumption but also underscores the importance of detailed modeling in crafting energy-efficient buildings. The findings emphasize the necessity of considering both the micro (occupancy behavior) and macro (climate influences) elements in building energy assessments to foster sustainability in the built environment. This research also provides actionable insights that could significantly refine current building energy standards and practices. Specifically, the findings suggest that ASHRAE standards could be updated to include more nuanced models of occupant behavior, particularly in relation to occupant density and dynamic occupancy schedules. We recommend that ASHRAE Standard 90.1 incorporate specific guidelines on adjusting HVAC operation based on real-time occupancy data, which our research shows can lead to substantial energy savings.

One limitation of this study is the focus solely on small office buildings, neglecting the analysis of other types of office buildings prevalent in the US. Additionally, the study uses a single building model for all 16 climate zones, which may limit the reflection of specific architectural and environmental adaptations typically required for optimal energy efficiency in diverse climates. This narrow scope restricts the generalizability of our findings to broader office building contexts. In addition, while we extensively investigated the influence of occupancy parameters on building energy consumption, our analysis did not encompass the impact of other factors such as heating, cooling, lighting, and energy control strategies. Moreover, the assumption that subspaces within each thermal zone maintain similar thermal conditions may not fully capture the variability in actual office environments, where different uses or solar exposures can affect zone-specific energy demands. Future research could explore these aspects comprehensively to provide a more holistic understanding of energy efficiency in diverse office building settings.

Author Contributions: Conceptualization, S.N. and A.J.; methodology, S.N. and AJ; software, S.N.; validation, S.N., A.J., S.V. and Z.P.; formal analysis, S.N.; investigation, S.N.; resources, S.N.; data curation, S.N.; writing—original draft preparation, S.N., S.V. and A.J.; writing—review and editing, S.N., S.V., A.J. and Z.P.; visualization, S.N. and S.V.; supervision, A.J. and Z.P.; project administration, A.J. All authors have read and agreed to the published version of the manuscript.

Funding: This research received no external funding.

Data Availability Statement: The data presented in this study are available on request from the corresponding author.

Conflicts of Interest: The authors declare no conflict of interest.

References

1. Alazazmeh, A.; Asif, M. Commercial building retrofitting: Assessment of improvements in energy performance and indoor air quality. *Case Stud. Therm. Eng.* **2021**, *26*, 100946. [CrossRef]
2. Li, J.; Yu, Z.J.; Haghighat, F.; Zhang, G. Development and improvement of occupant behavior models towards realistic building performance simulation: A review. *Sustain. Cities Soc.* **2019**, *50*, 101685. [CrossRef]
3. Vassiljeva, K.; Matson, M.; Ferrantelli, A.; Petlenkov, E.; Thalfeldt, M.; Belikov, J. Data-Driven Occupancy Profile Identification and Application to the Ventilation Schedule in a School Building. *Energies* **2024**, *17*, 3080. [CrossRef]
4. Ouf, M.M.; O'Brien, W.; Gunay, H.B. Improving occupant-related features in building performance simulation tools. *Build. Simul.* **2018**, *11*, 803–817. [CrossRef]
5. Yan, D.; O'Brien, W.; Hong, T.; Feng, X.; Gunay, H.B.; Tahmasebi, F.; Mahdavi, A. Occupant behavior modeling for building performance simulation: Current state and future challenges. *Energy Build.* **2015**, *107*, 264–278. [CrossRef]
6. Yang, J.; Santamouris, M.; Lee, S.E. Review of occupancy sensing systems and occupancy modeling methodologies for the application in institutional buildings. *Energy Build.* **2016**, *121*, 344–349. [CrossRef]
7. Norouziasl, S.; Jafari, A.; Zhu, Y. Modeling and simulation of energy-related human-building interaction: A systematic review. *J. Build. Eng.* **2021**, *44*, 102928. [CrossRef]
8. Bäcklund, K.; Molinari, M.; Lundqvist, P.; Palm, B.; Occupants, B. Their Behavior and the Resulting Impact on Energy Use in Campus Buildings: A Literature Review with Focus on Smart Building Systems. *Energies* **2023**, *16*, 6104. [CrossRef]
9. Vosoughkhosravi, S.; Jafari, A. American Time Use Survey in Modeling Occupant Behavior: A Systematic Review. In *Computing in Civil Engineering 2023*; American Society of Civil Engineers: Corvallis, OR, USA, 2024; pp. 77–84. [CrossRef]
10. Chen, Y.; Hong, T.; Luo, X. An agent-based stochastic Occupancy Simulator. *Build. Simul.* **2018**, *11*, 37–49. [CrossRef]
11. Chen, Y.; Luo, X.; Hong, T.; Chen, Y.; Luo, X.; Hong, T. *An Agent-Based Occupancy Simulator for Building Performance Simulation*; Lawrence Berkeley National Laboratory; LBL Publications: Berkeley, CA, USA, 2016; Available online: https://escholarship.org/uc/item/0047c6c3 (accessed on 1 July 2024).
12. Putra, H.C.; Andrews, C.J.; Senick, J.A. An agent-based model of building occupant behavior during load shedding. *Build. Simul.* **2017**, *10*, 845–859. [CrossRef]
13. Jia, M.; Srinivasan, R.; Ries, R.; Bharathy, G.; Weyer, N. Investigating the Impact of Actual and Modeled Occupant Behavior Information Input to Building Performance Simulation. *Buildings* **2021**, *11*, 32. [CrossRef]
14. Parys, W.; Saelens, D.; Hens, H. Coupling of dynamic building simulation with stochastic modelling of occupant behaviour in offices—A review-based integrated methodology. *J. Build. Perform. Simul.* **2011**, *4*, 339–358. [CrossRef]
15. Almeida, L.; Tam, V.W.Y.; Le, K.N.; She, Y. Effects of occupant behaviour on energy performance in buildings: A green and non-green building comparison. *ECAM* **2020**, *27*, 1939–1962. [CrossRef]
16. Gunay, H.B.; O'Brien, W.; Beausoleil-Morrison, I. Development of an occupancy learning algorithm for terminal heating and cooling units. *Build. Environ.* **2015**, *93*, 71–85. [CrossRef]
17. Li, N.; Calis, G.; Becerik-Gerber, B. Measuring and monitoring occupancy with an RFID based system for demand-driven HVAC operations. *Autom. Constr.* **2012**, *24*, 89–99. [CrossRef]
18. Chen, Z.; Xu, J.; Soh, Y.C. Modeling regular occupancy in commercial buildings using stochastic models. *Energy Build.* **2015**, *103*, 216–223. [CrossRef]
19. Page, J.; Robinson, D.; Morel, N.; Scartezzini, J.L. A generalised stochastic model for the simulation of occupant presence. *Energy Build.* **2008**, *40*, 83–98. [CrossRef]

20. Oldewurtel, F.; Sturzenegger, D.; Morari, M. Importance of occupancy information for building climate control. *Appl. Energy* **2013**, *101*, 521–532. [CrossRef]
21. Rafsanjani, H.N.; Ahn, C.R.; Chen, J. Linking building energy consumption with occupants' energy-consuming behaviors in commercial buildings: Non-intrusive occupant load monitoring (NIOLM). *Energy Build.* **2018**, *172*, 317–327. [CrossRef]
22. Erickson, V.L.; Lin, Y.; Kamthe, A.; Brahme, R.; Surana, A.; Cerpa, A.E.; Sohn, M.D.; Narayanan, S. Energy efficient building environment control strategies using real-time occupancy measurements. In *BuildSys '09: Proceedings of the First ACM Workshop on Embedded Sensing Systems for Energy-Efficiency in Buildings*; Association for Computing Machinery: New York, NY, USA, 2009; pp. 19–24. [CrossRef]
23. Banfi, A.; Ferrando, M.; Li, P.; Shi, X.; Causone, F. Integrating Occupant Behaviour into Urban-Building Energy Modelling: A Review of Current Practices and Challenges. *Energies* **2024**, *17*, 4400. [CrossRef]
24. Dziedzic, J.W.; Yan, D.; Sun, H.; Novakovic, V. Building occupant transient agent-based model–Movement module. *Appl. Energy* **2020**, *261*, 114417. [CrossRef]
25. Thornton, B.A.; Rosenberg, M.I.; Richman, E.E.; Wang, W.; Xie, Y.; Zhang, J.; Cho, H.; Mendon, V.V.; Athalye, R.A.; Liu, B. *Achieving the 30% Goal: Energy and Cost Savings Analysis of ASHRAE Standard 90.1-2010 (No. PNNL-20405)*; Pacific Northwest National Lab. (PNNL): Richland, WA, USA, 2011. [CrossRef]
26. Goel, S.; Rosenberg, M.; Athalye, R.; Xie, Y.; Wang, W.; Hart, R.; Zhang, J.; Mendon, V. *Enhancements to ASHRAE Standard 90.1 Prototype Building Models (No. PNNL-23269)*; Pacific Northwest National Lab. (PNNL): Richland, WA, USA, 2014.
27. Bae, Y.; Yoon, Y.; Malhotra, M.; Jung, S. *Prototype College Building Energy Model: Building Characteristics and Energy Simulation Results (No. ORNL/TM-2022/2532)*; Oak Ridge National Lab. (ORNL): Oak Ridge, TN, USA, 2022. Available online: https://www.osti.gov/biblio/1878673 (accessed on 1 July 2024).
28. Norouziasl, S.; Jafari, A.; Wang, C. An agent-based simulation of occupancy schedule in office buildings. *Build. Environ.* **2020**, *186*, 107352. [CrossRef]
29. Luo, X.; Lam, K.P.; Chen, Y.; Hong, T. Performance evaluation of an agent-based occupancy simulation model. *Build. Environ.* **2017**, *115*, 42–53. [CrossRef]
30. Alonso, M.J.; Dols, W.S.; Mathisen, H.M. Using Co-simulation between EnergyPlus and CONTAM to evaluate recirculation-based, demand-controlled ventilation strategies in an office building. *Build. Environ.* **2022**, *211*, 108737. [CrossRef]
31. Chen, Y.; Liang, X.; Hong, T.; Luo, X. Simulation and visualization of energy-related occupant behavior in office buildings. *Build. Simul.* **2017**, *10*, 785–798. [CrossRef]
32. Hernandez, H.; Ochoa, S. Adaptive Step-size One-at-a-time (OAT) Optimization. *ForsChem Res. Rep.* **2022**, *7*, 1–44. [CrossRef]
33. Standard 90.1. Available online: https://www.ashrae.org/technical-resources/bookstore/standard-90-1 (accessed on 1 July 2024).
34. EnergyPlusTM Input Output Reference. 2024. Available online: https://energyplus.net/assets/nrel_custom/pdfs/pdfs_v24.1.0/InputOutputReference.pdf (accessed on 1 July 2024).
35. Blight, T.S.; Coley, D.A. Sensitivity analysis of the effect of occupant behaviour on the energy consumption of passive house dwellings. *Energy Build.* **2013**, *66*, 183–192. [CrossRef]
36. Commercial Reference Buildings | Department of Energy. Available online: https://www.energy.gov/eere/buildings/commercial-reference-buildings (accessed on 1 July 2024).
37. Zhang, L.; Wen, J. A systematic feature selection procedure for short-term data-driven building energy forecasting model development. *Energy Build.* **2019**, *183*, 428–442. [CrossRef]
38. Leung, M.C.; Tse, N.C.F.; Lai, L.L.; Chow, T.T. The use of occupancy space electrical power demand in building cooling load prediction. *Energy Build.* **2012**, *55*, 151–163. [CrossRef]
39. Murrieum, M.; Jafari, A.; Akhavian, R. Building Energy Performance Prediction Using Machine Learning: A Data-Driven Decision-Making Framework for Energy Retrofits. In *Construction Research Congress 2020: Computer Applications—Selected Papers from the Construction Research Congress 2020*; American Society of Civil Engineers (ASCE): Reston, VA, USA, 2020; pp. 436–447. [CrossRef]
40. Kusiak, A.; Li, M.; Zhang, Z. A data-driven approach for steam load prediction in buildings. *Appl. Energy* **2010**, *87*, 925–933. [CrossRef]
41. Kapetanakis, D.S.; Mangina, E.; Finn, D.P. Input variable selection for thermal load predictive models of commercial buildings. *Energy Build.* **2017**, *137*, 13–26. [CrossRef]
42. Asadi, S.; Amiri, S.S.; Mottahedi, M. On the development of multi-linear regression analysis to assess energy consumption in the early stages of building design. *Energy Build.* **2014**, *85*, 246–255. [CrossRef]
43. Jain, R.K.; Damoulas, T.; Kontokosta, C.E. Towards Data-Driven Energy Consumption Forecasting of Multi-Family Residential Buildings: Feature Selection via The Lasso. In *Computing in Civil and Building Engineering—Proceedings of the 2014 International Conference on Computing in Civil and Building Engineering*; American Society of Civil Engineers: Reston, VA, USA, 2014; pp. 1675–1682. [CrossRef]
44. Yang, W.; Wang, K.; Zuo, W. Neighborhood Component Feature Selection for High-Dimensional Data Detection and Segmentation View project Human Recognition View project Neighborhood Component Feature Selection for High-Dimensional Data. *J. Comput.* **2012**, *7*, 161–168. [CrossRef]
45. Kowshalya, A.M.; Madhumathi, R.; Gopika, N. Correlation Based Feature Selection Algorithms for Varying Datasets of Different Dimensionality. *Wirel. Pers. Commun.* **2019**, *108*, 1977–1993. [CrossRef]

46. Zhang, Y.; Ding, C.; Li, T. Gene selection algorithm by combining reliefF and mRMR. *BMC Genomics* **2008**, *9*, S27. [CrossRef]
47. Chen, Y.; Guo, M.; Chen, Z.; Chen, Z.; Ji, Y. Physical energy and data-driven models in building energy prediction: A review. *Energy Rep.* **2022**, *8*, 2656–2671. [CrossRef]
48. Sun, Y.; Haghighat, F.; Fung, B.C.M. A review of the-state-of-the-art in data-driven approaches for building energy prediction. *Energy Build.* **2020**, *221*, 110022. [CrossRef]
49. Energy Standard for Buildings Except Low-Rise Residential Buildings. 2010. Available online: https://www.ashrae.org/technical-resources/standards-and-guidelines (accessed on 1 July 2024).
50. Chen, Y.; Li, M.; Xiong, M.; Cao, J.; Li, J. Future Climate Change on Energy Consumption of Office Buildings in Different Climate Zones of China. *Pol. J. Environ. Stud.* **2018**, *27*, 45–53. [CrossRef]
51. Meng, F.; Li, M.; Cao, J.; Li, J.; Xiong, M.; Feng, X.; Ren, G. The effects of climate change on heating energy consumption of office buildings in different climate zones in China. *Theor. Appl. Clim.* **2018**, *133*, 521–530. [CrossRef]
52. Dong, Z.; Zhao, K.; Hua, Y.; Xue, Y.; Ge, J. Impact of occupants' behaviour on energy consumption and corresponding strategies in office buildings. *IOP Conf. Ser. Earth Environ. Sci.* **2019**, *294*, 012076. [CrossRef]
53. Gu, J.; Xu, P.; Ji, Y. A Fast Method for Calculating the Impact of Occupancy on Commercial Building Energy Consumption. *Buildings* **2023**, *13*, 567. [CrossRef]
54. Prasetiyowati, M.I.; Maulidevi, N.U.; Surendro, K. Determining threshold value on information gain feature selection to increase speed and prediction accuracy of random forest. *J. Big Data* **2021**, *8*, 84. [CrossRef]
55. Henriques, L.; Lima, F.P.; Castro, C. Combining Advanced Feature-Selection Methods to Uncover Atypical Energy-Consumption Patterns. *Future Internet* **2024**, *16*, 229. [CrossRef]
56. Zhao, H.; Magoules, F. Feature selection for support vector regression in the application of building energy prediction. In Proceedings of the 2011 IEEE 9th International Symposium on Applied Machine Intelligence and Informatics (SAMI), Smolenice, Slovakia, 27–29 January 2011; pp. 219–223. [CrossRef]

MDPI AG
Grosspeteranlage 5
4052 Basel
Switzerland
Tel.: +41 61 683 77 34

Energies Editorial Office
E-mail: energies@mdpi.com
www.mdpi.com/journal/energies

www.ingramcontent.com/pod-product-compliance
Lightning Source LLC
Chambersburg PA
CBHW061833210926
51850CB00006B/876